ANTHROPOLOGICAL PAPERS

MUSEUM OF ANTHROPOLOGY, UNIVERSITY OF MICHIGAN

NO. 67

# THE NATURE AND STATUS OF ETHNOBOTANY

Edited by
RICHARD I. FORD

Assisting Editors:
Michael F. Brown
Mary Hodge
William L. Merrill

ANN ARBOR, MICHIGAN
1978

Printed in the
United States of America
ISBN: 0-932206-79-4

*Dedicated to*
Volney H. Jones

# CONTENTS

# ILLUSTRATIONS

Part III



Part IV



Part V

# TABLES

## Part I



## Part II



## Part III



## Part IV



## Part V

# PREFACE

Ten years ago Volney Jones formally relinquished his professorial obligations and curatorial duties at The University of Michigan. However, the leisure associated with retirement still eludes him. He continues an active professional life and an enthusiastic tutorial relationship with numerous students. The past decade has witnessed more students than ever before sharing his knowledge and enthusiasm for ethnobotany.

Although published references to pertinent works in his areas of interest—Hopi ethnohistory, North American Indian ethnography, the history of technology, the biographies of early anthropologists—are meticulously added to his voluminous files daily, his primary dedication is to ethnobotany. The importance of plants to people in prehistory and in the ethnographic present continues to occupy his attention and to absorb his thoughts. His conceptualization of modern ethnobotany and his contributions to methods of archaeobotany are legend and attract new adherents to ethnobotany each year. As a consequence of his reputation as the premier American ethnobotanist, this volume was organized to explicate the central issues in ethnobotany today.

The inspiration for this commemorative work came from William Merrill and Michael Brown, who wished to express their appreciation in a meaningful and lasting manner to Mr. Jones for his unselfish assistance to them. Despite the exigencies of fieldwork during the development of this book, they followed its progress with a paternal eye.

Our greatest regret is that many anthropologists who were influenced in their careers by Volney Jones were omitted from this volume because of its focus. However, we appreciate their enthusiastic acceptance of this project and the many kindnesses they extended to us during its execution.

The contributors to this festschrift are either former University of Michigan students or present colleagues in the Museum of Anthropology. James B. Griffin has been Volney's friend, confidant, and fellow graduate student and curator for more than 40 years. Professor Jones chaired Richard I. Ford's doctoral dissertation, and they have shared interests, space, and resources in the Ethnobotanical Laboratory for 12 years. The presence of Volney as a Michigan resource was an incentive for both Joyce Marcus and Kent

Flannery to join the curators in the Museum, and subsequently, they have shared his ideas about Mesoamerican agricultural origins and opinions about Michigan athletic fortunes.

Before relinquishing his regular teaching duties, Volney Jones was chairman of several dissertation committees, an active member of numerous graduate committees, and sponsor of the research of many M.A. students. He chaired the Ph.D. committee of Richard A. Yarnell and helped to supervise the graduate work of Peter Kunstadter, Robert Carneiro, Gertrude Dole, James E. Fitting, and M. Jean Black. In addition, he sponsored the research activities of Karen Cowan Ford whose published work, *Las Yerbas de la Gente*, he directed.

Since 1969, many University of Michigan ethnobotany students have benefited from Professor Jones' daily presence in the Museum and his unselfish willingness to share his vast fund of experience and his personal library with them. Deborah Pearsall completed her undergraduate education at Michigan and her doctoral work at the University of Illinois. Wilma Wetterstrom earned all of her degrees at Michigan. Ellen Messer received her doctorate at Michigan. Nancy E. Asch and David L. Asch are doctoral students currently associated with the Northwestern University Archaeological Program. Paul E. Minnis is also a Michigan doctoral student working with the Mimbres Foundation in New Mexico. C. Wesley Cowan is a graduate student in archaeology and ethnobotany at Michigan.

Mary Hodge has had the unusual privilege to work with Mr. Jones as administrative secretary and editor in the Museum of Anthropology and now as a graduate student in archaeology. Her dedication to the publication of this volume is a lasting tribute to their friendship.

It was with particular sadness that we learned about Alfred Whiting's unexpected death. He had worked closely with Volney as a graduate student in botany at Michigan in the 1930s and as a field associate in the Hopi Mesas. He was preparing a paper based upon his work at Navajo Mountain for this volume at the time of his passing.

Finally, this publication was made possible by the generous support of the present curators in the Museum of Anthropology. They allocated the funds for it even though they were well aware that, if he knew of their decision, Volney would condemn their action as frivolous. They join with the authors and editors in extending to Volney Jones their sincerest appreciation for what he has done for the Museum, for students of anthropology, and most of all, for the field of ethnobotany.

Richard I. Ford
Director
Museum of Anthropology
The University of Michigan

# VOLNEY HURT JONES, ETHNOBOTANIST: AN APPRECIATION

*James B. Griffin*
Museum of Anthropology, The University of Michigan

The professional career of Volney Hurt Jones is inextricably linked with the Ethnobotanical Laboratory in the Museum of Anthropology at the University of Michigan. The second of two research units established by Carl E. Guthe under the auspices of the National Research Council in the Museum of Anthropology, The Ethnobotanical Laboratory became operative in 1930, with Melvin R. Gilmore as the first director in March 1928. This was the major research program in the Division of Ethnology, with Gilmore as its curator. The laboratory was the subject of the first formal publication of the Museum of Anthropology (Gilmore 1932). Gilmore was the recognized authority on ethnobotany, and it was natural that Jones would come to Ann Arbor in 1931 to study under Gilmore and to augment his knowledge of anthropology.

Jones' collegiate career began at North Texas Agricultural College in Arlington, Texas where between 1923 and 1926 he completed a junior college course in agriculture and was active in campus affairs. He moved to the Agricultural and Mechanical College of Texas in 1927 and received a B.S. degree from that institution in 1929. He remained at that institution during 1929-1930 for part-time graduate work primarily under Professor Robert G. Reeves. He then transferred to the University of New Mexico where he received an M.A. in 1931 with a major in biology and a minor in English. He was a Graduate Fellow in the Department of Biology. His thesis on the "Ethnobotany of Isleta Pueblo" was under the direction of Professor E. F. Castetter with funds from the National Research Council granted to Castetter. He had thus moved from an emphasis on practical applications of botany to a research interest in the use and knowledge of plant material by American Indians.

## JONES AND THE UMMA ETHNOBOTANICAL LABORATORY

At Ann Arbor in 1931 he became the laboratory assistant to Gilmore, held the Homeopathic Hospital Guild Scholarship in Michigan Ethnology in the summer of 1933, doing fieldwork among Indians in Michigan and Ontario. In 1933-1934 he held a University Fellowship in Anthropology. Between 1931 and 1939 he took courses in anthropology and botany which were particularly pertinent to his research interests. His principal fieldwork was done in the Southwest between 1931 and 1939 and in Michigan in 1933 and 1934. He was a member of the Peabody Museum of Harvard expedition to Awatovi, Arizona where his study of the floral resources of the area and his strategy of the recovering and preservation of the plant materials from the excavations under the direction of J. O. Brew was outstanding.

I came to Ann Arbor in February, 1933 on a Fellowship in Aboriginal North American Ceramics, and Volney as an "oldtimer" helped me get settled and initiated into the Museum and campus operations. Our hours were 8 a.m. to 5 p.m. on weekdays, unless we were in class, and from 8 to noon on Saturdays. This was the norm for the University Museum staff, as very few participated in the teaching programs. In addition there were many hours spent in the evening working on our research projects. These hours soon became a habit and, as the correct thing to do, were almost impossible to break in later years!

While Gilmore was officially in charge of the operations in the Ethnobotanical Laboratory, his attendance was irregular because of his health, and Volney was the individual who knew the collections and did most of the research. We had a considerable amount of discussion on anthropological as well as operational and other problems over the years. What appreciation I had of prehistoric use of plants and other botanical materials came primarily from him, and he was almost always consulted about the ethnology of Eastern American Indian societies. In exchange, my opinions about Eastern United States prehistoric cultural relationships were a part of his fund of knowledge whether he accepted them or not. Our careers have been interlocked from the early 1930s until the present, when we are both retired but still active in our several ways.

For too many years our position on campus, and our livelihood, such as it was, was entirely dependent upon the support of Carl Guthe. We recognized this and when Guthe began to fly to Washington in the early days of commercial air travel, we suggested to him that we would be much happier if he made his trips by rail, which he did then do much of the time. Guthe was admirable in his loose hand over our research interests. While occasionally offering suggestions, he allowed us to pursue our work with a minimum of direction, assuming that we knew what we were doing.

Because of Jones' presence and his position as heir apparent to the emphasis on ethnobotany established by Gilmore, the library, notes, manuscripts, and photographs that the latter had collected were left to the Museum for the Ethnobotanical Laboratory when he died in 1940. Gilmore was not an easy man to work for, but Volney's respect for Gilmore and his own native tolerance and tact enabled him to tolerate and often enjoy the relationship. Gilmore was proud of his Welsh ancestry and he saw in Jones another Welshman who must, because of that heritage, inevitably be a worthwhile individual. While Jones was not a student of the Arikara, or other village Plains Indians, he did at least study Indian cultures, and not African or Asian people.

One of the results of World War II was the initiation of Jones into the formal course offerings of the Department of Anthropology. He had earlier guided the ethnobotanical research of registered students through a Museum Methods course which Carl E. Guthe had established in 1931 and a second course called Research and Special Work. This allowed Volney to guide the programs of Gretchen Beardsley, Evelyne Hope, and Robert L. Fonner among others, but none of them were able to continue ethnobotany as a career. In the summer of 1944, Mischa Titiev began work with the Office of Strategic Service and was assigned to work in India. This left the two-man Department in some difficulty because Titiev was scheduled to teach courses on the American Indian and the Peoples of Asia in summer school as well as those several courses he had scheduled for the 1944-1945 academic year. Leslie White as Acting Chairman finally acceded to the idea that Jones could teach the American Indian course both in the summer, and again as Middle American and South American Indians in the following year, while I was hesitantly permitted to teach the Peoples of Asia in the summer and North American Archaeology in the academic year. We did this without teaching titles or formal appointments in the College of Literature, Science and the Arts and with no salary stipend or addition from that source. Volney had a close association with White who for some years had adopted the attitude that Museum staff members were not qualified or able to teach formal courses, and in this he was supported by Titiev. When it became evident that students in the classes not only did not drop the courses but even prospered, White decided after consultation with Dean Keniston that the Department should be enlarged after the Dean had learned from outside sources that both of us were regarded nationally as respected anthropologists. This enlargement was opposed by Titiev but was accepted by the College so that in 1945 Jones was appointed Assistant Professor in the Department of Anthropology as well as Curator of Ethnology in the Museum of Anthropology. Earlier efforts had been made to add Jones and me formally to the Department teaching staff, but these efforts were not successful for several reasons, one of them being the intrusion of the war years.

In the fall of 1945 the first formal meetings of the enlarged Department were held, and Jones was requested to act as Acting Secretary to take notes of the meetings. He continued in this temporary activity until the first meeting of the Department in the fall of 1949 when J. P. Thieme was unanimously elected Secretary when *he* was not present. Jones continued to teach the North American Indian course until his retirement in 1968, so that this offering was under his jurisdiction for 24 years. It was at first a course which served both undergraduate and graduate students but it soon became apparent that it should be offered as a primarily undergraduate listing and a separate graduate course was proposed but was not accepted until 1957-1958. By 1948 Jones was offering a course in Primitive Technology as well as a course in Ethnobotany approved in 1945, which was jointly listed in the Botany Department. He also participated in a General Seminar in Anthropology for graduate students, which was offered for a number of years in the late 1940s and early 1950s, as well as continuing to participate in the specialized Museum courses. Jones also offered a popular American Indian course in the Extension Service beginning in Detroit in 1945 and continued to do so intermittently until 1968 in a number of Extension centers in southeastern Michigan.

The American Indian course was an excellent one with an emphasis on the several environments of the several ethnological culture areas. Some amount of time was spent on the prehistoric background of the historic Indian groups so that students could have some comprehension of the growth and development of Indian societies. This was an acclaimed course and a source of a large number of student enrollments which varied from 100 to 150 in Jones' last years of teaching. Most of those registering were not Anthropology majors so that a rational view of the American Indian was acquired by many individuals both on the campus and in Extension.

Jones was appointed Associate Professor in 1952 and continued in that capacity until 1968 when he retired. A number of attempts were made by me to a succession of Departmental Chairmen to have them propose Jones for promotion to Professor, but these efforts were not successful for reasons which I did not regard as acceptable. There was little appreciation of his research papers and reports, his course offerings, or his value to the students interested in North American research. His command of the North American

Fig. 1. Mischa Titiev, Joyce H. Jones, Sun Chief-Don Talayesva, Estelle Titiev, Volney H. Jones, and Alan Jones in Ann Arbor ca. 1940.

Fig. 2. Leslie A. White and Volney H. Jones in front of the University Museums Building, University of Michigan.

literature was encyclopedic; his bibliographic records and knowledge of their contents made him an extraordinary source for students and staff. His courses and emphasis on the interrelation of culture and environment served to attract graduate students to Michigan, and he trained some of the best ethnobotanical practitioners of the present day.

The Museum of Anthropology was able to have research assistants assigned to the several curators, and over the years a considerable number benefited from their association with Jones. Prominent among these are Alfred F. Whiting, Vorsila L. Bohrer, James Howard, Richard A. Yarnell, and Richard I. Ford, the last three receiving their Ph.D. degrees with Jones as chairman of their doctoral committees. In addition, Volney served on a great many Ph.D. committees in Anthropology and on a smaller number with the Departments of Botany and Geography. He was on most of the committees of students specializing in archaeology and on many of those specializing in ethnology. On these his contributions were valued by both staff and students.

Jones represented the Museum of Anthropology for three years on the General Committee of the Division of Biological Sciences. He had a two year appointment on the University Space Committee and was on the Summer Symposium Committee one year. He was on the Radiocarbon Curatorial Committee in addition to his post as the active curator of materials received. He was an active member of the Science Research Club, serving on various committees of that organization in addition to being its secretary-treasurer for two years. Much of his time was devoted to the Michigan Academy of Science. He was treasurer of that organization from 1944 to 1948 and from 1952 to 1966. He also worked on the Research Endowment and publications committees of that organization.

In 1956 Jones was appointed to the Governor's commission on the social, economic, health and educational status of the Indians of Michigan and was chairman of the committee on education. In Ann Arbor community affairs he had seven years of activity with the Boy Scouts organization, was prominent in Ann Arbor school affairs and became president of the Ann Arbor High School Parent-Teacher Association as well as serving three years on the Council of that organization. In the 1950s an organization was formed called the Memorial to the American Indian Foundation with an Ann Arbor sculptor as a dominant member. Jones became president and a member of the Board of Directors. The Memorial was to be established in the Gallup, New Mexico area, but the many difficulties inherent in such a project could not be overcome.

More than any other member of the Museum of Anthropology staff, Jones was involved in the Exhibit Museum program. As Curator of Ethnology and the resident expert on the American Indian, more requests for assistance in conceiving, designing, and selecting materials for exhibit came to him than to the other curators. The ethnographic specimens on exhibit are a part of the collections of the Ethnology Division, and Jones was responsible for their care and preservation. The present set of highly acclaimed dioramas of American Indian life in the Exhibit Museum owe much of their quality and validity to Jones' knowledge of American Indian life. The accuracy of the Museum's pictorial culture area map also reflects his command of North American ethnology.

In addition to his association with the Exhibit Museum, Jones was called upon by other campus and community units. The Department of Botany requested him to select specimens and participate in the installation of Indian food and narcotic materials for their display, and the College of Pharmacology had an exhibit of narcotics installed with his assistance. Ethnology division materials were borrowed for a Latin American festival and by similar groups until it became clear that even normal treatment by such well-intentioned organizations resulted in too much wear and tear on the specimens. Classroom use of ethnology specimens, while valuable as a teaching device, also had to be curtailed because of similar inevitable deterioration of non-replaceable items.

Jones was called upon to give lectures in the Botany courses of Professors K. L. Jones and Elzada Clover for many years and also on Indian economics in an Economics course of Professor Z. Clark Dickinson. He was called upon by student organizations, dormitory and fraternity groups, and various research clubs. Astronomy, Architecture and Journalism students beat a path to his office for assistance on such subjects as Indian knowledge of astronomy, housing and current Indian affairs. His knowledge and bibliographic references were of great aid to many students with these assignments. In the Ann Arbor community he was in demand by service organizations, women's clubs, church groups, the Boy Scouts and similar organizations.

During World War II Jones was granted a leave of absence in January 1943 to accept an appointment as Production Superintendent in Haiti with the Société Haitiana-Américaine de Développement Agricole. The project was to determine if *Cryptostegia* rubber, an African tropical plant of the milkweed family, could be produced in order to supplant the rubber source from Southeast Asia which was then under Japanese control. While high hopes were held for this experiment, it was not successful. It was abandoned in favor of Government support for synthetic rubber production. Jones returned to the campus in April 1944.

An unusual sidelight of Jones' career was his value as an expert witness during the 1940s to 1960s in the identification of native plants which had been used illegally by some members of American society. In the University fiscal year 1952-1953 he examined eight lots suspected of being marijuana, and four of them were. The analyses were made for the State Police Vice

Squad, the Washtenaw County Sheriff's Department, and the Ann Arbor Police Department. He appeared as an expert witness in the Ann Arbor Municipal Court, the Washtenaw District Court, and the Saginaw District Court. Convictions were obtained in each instance. Despite this record, his position was suspect in some quarters after a prosecuting attorney asked him if in his personal opinion smoking marijuana was a bad thing to do and he answered "no." This terminated his appearances in court.

In the early days of the Radiocarbon Laboratory, Volney took curatorial care of the specimens submitted for dating and set up the record system that was essentially followed to the present time. With the rapid increase of specimens sent in and the pressure of his other activities this responsibility was allocated to graduate students supported by the Laboratory, but he continued as a member of the Radiocarbon Committee. I believe it was Jones who first called my attention to the potential of W. F. Libby's and his colleagues early work on Carbon 14 in the late 1940s. Our interest in the applicability of radiocarbon dating to prehistoric research aided the eventual establishment of the Michigan Radiocarbon Laboratory under the direction of H. R. Crane.

From his arrival at the Museum of Anthropology until his retirement Jones prepared over 350 reports on specimens submitted to the Ethnobotanical Laboratory. While some of these studies required only a short statement others took up months of study and analysis. The specimens submitted usually remained with the laboratory so that over the years the collections grew and provided their invaluable and extensive resource base for continuing research. The curatorial activity on these and on the ethnological specimens was outstanding in its care of the specimens and in the records pertaining to them. His reports to the Director on the activities of his division each year were full and accurate statements and included suggestions for improvement of his program and that of the Museum as a whole. Jones' stability and sound judgment were valuable assets to the Museum program, particularly during the long period when our development was frustrated by a number of University administrative difficulties.

An unfortunate professional practice was that of individuals who received or used reports carefully prepared by Jones and who then abstracted his statements with a nod in his direction. While some which were returned to archaeologists were identifications of single species, others were large enough or important enough to have been placed as a short section of a publication under his name. The Ethnobotanical Laboratory files largely produced by Jones were a valuable research source for students such as Howard, Yarnell, and Nancy and David Asch. Some of their publications made extensive use of these data. Again it is unfortunate that in many instances the fund of ethnobotanical research reports which Jones produced and made available to students and visiting scholars were never adequately acknowledged.

When the editors of professional journals needed a competent individual to review a paper on Ethnobotany they regularly turned to Jones, and he has published 21 reviews. Most of these have been in *American Antiquity* and the *American Anthropologist*. Some of his most important reviews are of George F. Carter's "Plant Geography and Culture History in the American Southwest" (Jones 1946); J. D. Sauer's "The Grain Amaranths: A Survey of their History and Classification" (Jones 1953) and Maitland Bradfield's "The Changing Pattern of Hopi Agriculture" (Jones 1972). All of his reviews have been marked by fairness, comprehension and helpful evidence and citations designed to improve the understanding of the subject matter dealt with in the publication. The reviews were significant additions to the problems rather than scathing indictments or ironic comments on the capabilities of the author; they were truly professional.

## JONES AND THE DISCIPLINE OF ETHNOBOTANY

In an article in *Chronica Botanica* Jones (1941) defined Ethnobotany as the study of the interrelations of primitive man and plants. It was emphasized that the ethnobotanist should be concerned with the entire range of relations between man and plants as they were used in different societies for food, clothing, shelter, implements, utensils and medicines. Plants are also portrayed in art, are important in folklore, religion, music and literature and are organized into a conceptual structure which varies from society to society. Ethnobotany is concerned with those areas of behavior and constructs of a society and the plant world. The difference between the data which could be obtained from living societies and that from archaeological deposits was recognized and the need for competent excavation techniques was stressed. This broad definition of Ethnobotany was improved upon in a paper at the Eighth International Congress of Botany in Paris (Jones 1957) and reprinted in the publication of papers of the Ninth Pacific Science Congress (Jones 1964). His view then was that Ethnobotany is a unit of an ecological study specializing in the interaction of man and the plant world. His approach was, to some degree, a reflection of an interest in ecology which had been a dominant theme in the biological studies of the University Museums since the beginning of the twentieth century.

As a result of Jones' fieldwork among the Great Lakes Indians, he published a number of short papers on the manufacture of wooden brooms, the preparation and uses of basswood fiber, the use of sweet grass, and the manufacture of mats. Each of these is marked by careful field observation, integrated with examples from the literature and reasonable conclusions on the probable age of the technique in the culture and the traditional importance of the several products. He also recognized that these manufactures were rapidly being abandoned under continuing pressures from our contemporary society. His last paper on ethnohistoric plant use was a review of a famine food, bittersweet bark, by the Indians of the Great Lakes and upper Midwest. In this paper he presented the printed evidence that the similar-appearing woodbine vine was also used, but after sampling that, concluded that the authors of the references probably had misidentified the vine actually used as a food source.

For some years W. S. Webb, W. D. Funkhouser, and their associates at the University of Kentucky had pursued a program of investigations at prehistoric sites in Kentucky. In their early publication which dealt with Cave Dwellers and Cliff Dwellers among other subjects, one would never know that such sites contained much botanical material (Funkhouser and Webb 1928). In a 1929 publication on the so-called "Ash Caves" of Lee County, Kentucky they did discuss textiles, skin, animal bone, gourd "shards," and had some references to vegetal material but its significance was not recognized. The same lack of recognition of the importance of an analysis and interpretation of vegetal remains is displayed in the report on "Rock Shelters of Wolfe and Powell Counties, Kentucky" (Funkhouser and Webb 1930). As a result of Webb's association with Carl E. Guthe, specimens from Kentucky were sent in to the Ethnobotanical Laboratory for a report. In the early 1930s Jones prepared five laboratory reports on specimens from Crittenden County; Red Eye Hollow and Buckner Hollow, Lee County; and DeHart Shelter in Powell County, Kentucky, the last three sites being rock shelters. The reports influenced Webb, and in the fall of 1932 he submitted a large amount of preserved vegetal remains from a rock shelter in Powell County and in December 1935 specimens from Newt Kash Hollow in Menifee County. The University of Kentucky excavations had found large quantities of preserved vegetal material in, or associated with, storage pits and "beds" which were interpreted as "aboriginal sleeping places" and from the general digging (Webb and Funkhouser 1936:116).

Jones' report on this material was a major study which has formed the basis for much of the subsequent ethnobotanical work done on Eastern United States archaeological specimens. It was a model of clarity and scholarship. He relied on other botanists for identifications of mosses, fungi and some of the seeds. He also received help from Gilmore, but the work on most of the materials and the comparative data from the literature and from the laboratory were his own effort (Jones 1936). In the shipment there were only two corn cobs, one of 10 and one of 13 rows of kernels and one earstalk. They were identified as probably flint corn. Since he was unaware of the much larger number of corn remains found by the University of Kentucky and because of previous statements that corn only appeared in the upper levels of shelters, he suggested that the Newt Kash corn might be from an occupation subsequent to that of the rest of the botanical material. The other tropical cultigens were of squash and gourd. The squash seed and shells were identified as a small variety of summer squash (*Cucurbita pepo* var. *melopepo ovifera*) and the gourds, seeds and shells as the common or bottlegourd (*Lagenaria vulgaris* [*siceraria*]). Native plants identified included pigweed or goosefoot (*Chenopodium* sp.) sunflower (*Helianthus annuus*), marshelder (*Iva* sp.), giant ragweed (*Ambrosia trifida*), canary grass (*Phalaris caroliniana*). Other plant materials included were acorns, hickory nuts, walnuts, chestnuts, wild grape, pawpaw, honey locust beans and pods, green-briar or cat-brier, compass plant roots, pawpaw bark, Indian hemp and milkweed, the inner bark of leatherwood or moosewood, linden or basswood bark, rattlesnake master leaves, big blue-stem grass; leaves from honey locust, American holly, chestnut, magnolia and oaks; wood specimens from hickory, oak, chestnut, ash, box elder and sumac; several species of moss; large and small varieties of cane; four species of lichens and two fungi and two insect galls. For many of these Jones reviewed the known function from historical accounts and their archaeological distribution. He suggested that the *Chenopodium* might have been a Mexican domesticate or perhaps introduced into eastern Kentucky as a domesticate. He also thought the giant ragweed might well also be a cultivated species.

He identified a small textile fragment as having been made of cotton but viewed it as an historic introduction, and considering the absence of cotton from other eastern sites and the extensive operations in the shelter to produce nitrate for gunpowder, this was the correct interpretation. He also identified three tobacco capsules as *Nicotiana rustica*, the only species cultivated by Indians east of the Mississippi in aboriginal times. This was and still is a puzzling, though certain, identification because of the scarcity of other secure identifications of prehistoric tobacco in the east.

In his summary statements, Jones made a number of major points and hypotheses. The spread of tropical domesticates into the east was not the result of a single or short time spread and the order of their appearance in the east differed significantly from that of the southwest. The four potential North American domesticates appear to precede the appearance of corn and pottery at Newt Kash in the Kentucky region and "seem earlier than any of the tropical plants except possibly the gourd and the squash" (Jones 1936:163). Tropical American agricultural plants at Newt Kash Hollow were of little importance. His identification of considerable prairie flora caused him to suggest that prairie was in existence at the time of the occupation, along with woodland flora and that the present preponderance of forest in the area means that a possible climatic change or floral succession was responsible. It was difficult or impossible when he wrote his paper to identify the time at which the prairie vegetation was present. He also emphasized the similarity of textiles, cultivated and local plant materials from Kentucky and the Ozarks although the latter area had a greater abundance and variety of materials, particularly of cultivated species. His identifications of the importance of vegetal materials and scarcity of faunal material led him to believe that the Newt Kash shelter people were chiefly vegetarian. He might have called this primary prairie and forest flora sufficiency but did not do so.

A further important innovation was Jones' identification of plant food materials from the dissection of human feces and this important tight association of man and his food intake has since been developed both by Jones and other ethnobotanists. It has become a most valuable asset in the study of prehistoric nutrition.

Subsequently a number of Eastern archaeologists almost every decade, have proposed an independent origin of Eastern native domesticates primarily based on Jones' Newt Kash report but often without his caution. At the present time the recovery of squash and gourd from open sites in Kentucky, Missouri, and Tennessee dated between 2500 and 2000 B.C., seems to negate a priority for the domestication of any native domesticates.

Jones' first important paper on maize in the Eastern United States was "Maize from the Davis Site: Its Nature and Interpretation" (Jones 1949). In this he reviewed earlier judgments on eastern maize, improved the definition of the Eastern Complex, and pointed out its striking uniformity over a broad area from the Prairie area to the Southeast and Northeast. At that time he followed some of his peers in suggesting that the morphological similarities of the Eastern Complex and highland Guatemala and Chiapas corn implied a direct connection between the two. His interpretation was strengthened by a date of A.D. 500 on corn from the site, which made it not only the most southern Eastern Complex but also the earliest. He accepted the view that the earlier Basketmaker corn in the east had been brought in from the Southwest instead of directly from Mesoamerica and that one of the distinctive features of Mississippian was their Eastern Complex maize brought in from the Guatemala area.

However, Jones' confidence in a direct Guatemala derivation soon changed to a recognition that the considerable variability of early Southwestern corn could have resulted in the eventual segregation of the Eastern Complex (Jones and Fonner 1954; Jones 1968, 1977). Whether the appearance of Eastern Complex maize took place only in the Southwest or also in the east is still, I gather, a matter of conjecture for there is not much corn known from eastern sites for the period between A.D. 300 to about A.D. 900.

Fig. 3. Volney Jones returned to Newt Kash Hollow shelter with Jimmy Griffin in June 1978. It was Griffin's first visit to this famous site.

In 1952 and in February 1953 the Ethnobotanical Laboratory received archaeological plant material from what was then called the T. O. Fuller Site, now Chucalissa, south of Memphis which was sent by Kenneth L. Beaudoin. Jones prepared a six-page report which was published in mimeograph form in Beaudoin's paper on his excavations (Jones 1953). The statement by Jones was the most complete to appear on a Mississippian site up to that time. There is a full discussion of the characteristics of the maize, which Jones concluded had characteristics of both the Hokokam Basketmaker and Eastern Complexes. Similar material had been identified by Hugh C. Cutler from the Knight Site as probably Guatemalan Tropical flint, but there is some question as to its temporal position at Knight. Jones also identified a cultivated bean, *Phaseolus vulgaris*, persimmon seeds (which were the first he had examined from an archaeological site), hickory and walnut or butternut, the wild plum (*Prunus americana*), and cane. He did not identify the charcoal from deciduous or hardwood trees to species; no conifers were present, for that skill was later added to his identifications. It is unfortunate that this report did not receive wider dissemination.

One of Jones' relatively few papers on living Plains tribes was in association with Gustav G. Carlson (Carlson and Jones 1940) on Comanche knowledge of the plant world. This study, and his M.A. thesis on the ethnobotany of Isleta Pueblo, were his two ventures into tribal ethnobotany beyond Great Lakes material culture. Carlson was a graduate student in Anthropology at the University of Michigan and in the summer of 1933 was a fellow of the Laboratory of Anthropology field school studying the Comanche in Oklahoma under the direction of Ralph Linton. The basic data were made by Carlson, aided by Waldo Wedel, and the identification of the plants, the documentation and comparative study were primarily those of Jones. Some 58 species were identified along with 15 others which were known by the Comanche but were not collected or not identified. Since in the historic period this tribe had been a nomadic group and customarily characterized as typical of the Plains hunting societies, it was somewhat surprising to have obtained such extensive knowledge in rather a short time. The plants and their utilization are discussed under such rubrics as foods, medicines, smoking, ceremonial, hunting and warfare, and in such household activities as tipi seats, soap manufacture, and so on. Comanche ethnobotany reflected their participation in plant world knowledge not only of Plains tribes but also of some Plateau and Southwestern groups.

Beginning in the 1930s Jones compiled an extensive record of Indian uses of the red bean or mescalbean *Sophora secundiflora*. Over the years several of the students in Ethnobotany benefited in their publications from these records. His named participation in publication does not appear until in this decade (Hatfield, Valdes, Keller, Merrill and Jones 1977; Jones and Merrill n.d.). In the first of these papers it is stated that there does not seem to be in the chemistry of mescalbeans any substance that is an hallucinogen and that the reported hallucinogenic effect of consuming mescalbeans by some Indians could well have been because the Indians believed it would produce such a result. The ritual followed in the ceremonies was strenuous, and the combination of physical and mental stress could have brought about an apparent contact with the supernatural.

While Jones made substantial contributions to Eastern ethnobotany, his major works and majority of identifications were of Southwestern materials. An early example was his report on specimens from Jemez Cave in Sandoval County, New Mexico, published in 1935. This report was unfortunately not published in its entirety under his name but was quoted and also abstracted by the archaeologists. Four types of corn were recognized and the presence of a considerable number of abnormalities was stressed. The other domesticated plant was one species of squash. In addition, he commented on the presence and use in the Southwest of 13 native plants. He mentioned cotton, but if he gave much of a statement on it in his report it was not printed.

An excellent statement by him on aboriginal cotton in the Southwest was published in 1936 as a part of symposium papers on Prehistoric Agriculture (Jones 1936). This review cited archaeological sites where cotton had been found; ethnographic data on the growth and use of cotton was included as well as comments on the origin and varieties of New World cotton. The 80 bibliographic citations in the report are typical of Jones' scholarship and ability to integrate the then known references into his interpretations.

As a result of the Chaco Canyon Field Conference in 1938, a list of "Authorities for the Identification of Archaeological Material" was published in the *New Mexico Anthropologist* (Anon. 1938). Jones was identified as an expert on general botanical material, maize, cotton, beans, agaves, and seeds.

In 1941, Jones reviewed the evidence up to that time for the definitely identified presence of tobacco in prehistoric Southwestern Pueblo sites (Jones 1944). He concluded there was no direct evidence for the presence of tobacco, but that it was possible tobacco was used with other materials in pipes and reed grass "cigarettes." Shortly thereafter a publication by Haury (1945) reported an identification of *Nicotiana tabacum* by Dr. H. W. Youngken which Jones has told me should almost certainly be *Nicotiana attenuata*. The specimens were from Double Butte Cave in the Salt River Valley of southern Arizona. Prehistoric evidence for native tobacco (*Nicotiana attenuata*) in the Southwest has appeared since and it was only just that Jones,following his own strictures calling for specific identification,was responsible for the results (Jones and Morris 1960).

The soundly established identifications of tobacco from pipes and particularly dried specimens was published in the paper cited above. Five out of eight pipes with smoked residue from a Basketmaker III site of the early seventh century found by Earl H. Morris in northeastern Arizona had nicotine remains. These identifications by microscopic and chromatographic examination and analysis were made by the research staff of Philip Morris, Inc. Other plant material similar to corn silk was also found in the dottle. From another site explored by Morris, well preserved dry remains of *Nicotiana attenuata* Torr, were contained in a Lino Gray jar. In other caves, small packets of Yucca fiber-bundles contained seeds and fragments identified by Jones as *N. attenuata*. The interpretation was that tobacco and lime (which was also in the packets) were used in a manner similar to southern California Indian practices for medicinal or other purposes. The conclusion was that the tobacco was probably not really cultivated in our sense but had been propagated by scattering tobacco seeds in the ashes of a burned over area, a technique that has been called "tobacco tending."

While working with the Peabody Museum–Harvard University expedition at Awatovi Northeastern Arizona in the late 1930s, Jones noticed that the adobe building blocks of the Franciscan mission included considerable vegetal material. After initially recovering specimens from fragmenting the dry bricks, he developed flotation techniques to remove the earth and recover the lighter vegetal material. The knowledge of these techniques became widely disseminated and adopted in California mission studies and in open sites in the Eastern United States. He was able at Awatovi to identify such early imported plants as apricot or plum, chili pepper, cantaloupe, watermelon, and wheat. The native cultivated plants were maize, pumpkin, cotton, bean and gourd. Additional materials recovered included various trees and shrubs, grasses and weeds. There were also fragments of cordage, fibers, sherds, and chips of obsidian and flint.

In 1954 Jones published a major report in the Carnegie Institution volume on Basketmaker sites in southwest Colorado. The report had been finished in June 1952, and Jones shared authorship with Robert L. Fonner. The discussion of Basketmaker II maize is unusually full and provided an excellent research base for later studies. In commenting on the then current state of knowledge about maize the authors concluded:

> The races of corn are not sharply differentiated entities but are rather peaks which rise from greatly varied populations. Almost certainly the corns reaching the Southwest already had some heterogeneity and it is not necessary or wise to postulate a separate aseptic arrival of each influence. (Jones and Fonner 1954:115)

The only other tropical cultigen identified was pumpkin since neither beans nor cotton had been introduced at the time of Basketmaker II. In this report there are extensive discussions of native plants used for food and cordage and with extensive consideration of probable uses from the literature and his own studies. This study has been a standard reference for Southwestern ethnobotany for years.

A major ethnobotanical report on Southwestern material is that of Vorsila L. Bohrer on the Tonto National Monument Collection. This material was the largest to be received at the Ethnobotanical Laboratory. The identification of the materials was aided by a number of specialists and the comparative data was largely supplied from a punch card file identifying the uses of some 7000 plants by Southwestern Indians. The file was prepared by Dr. Bohrer and Jones with financial support that he had obtained from the Horace H. Rackham School of Graduate Studies. In addition, Jones provided expert guidance during the preparation of the manuscript.

In a joint paper, Robert C. Euler and Jones dealt with hermetic sealing of storage jars in the Southwest. Euler had come across a small storage jar from near Kingman, Arizona resembling the pottery called Pyramid Grey. This vessel contained slabs of mescal and was sealed with a circular sherd and then an organic substance which was identified as lac from a species of insect that lives on the creosote bush. This rare example of effective sealing had been made during the first half of the twelfth century. The several examples from the literature of prehistoric and historic sealing with clay and other substances was reviewed along with an assessment of their efficacy. This provided a single source for such examples of food storage in pottery vessels.

In an earlier report on Tonto National Monument Charlies R. Steen, who had found the specimens, and Jones, discussed the presence of lima bean pods from this fourteenth-century ruin (Steen and Jones 1944). Up to that time very few examples were known or claimed in the Southwest.

The several papers in this volume by former students and colleagues are a testimony to Jones' influence on them in their professional development and careers. He has been a major contributor to the development of Ethnobotany in his Laboratory reports, his publications, his identifications, and in his guidance of students and collaborators. He has been both innovative and sound. He has corrected his own mistakes as well as those of others. Jones did not become involved in published professional argumentation or rank order diatribes. His contributions to the development of the Anthropology program at Michigan as well as those to Anthropology throughout his career are reflected in the body of this paper.

## REFERENCES

Funkhouser, William D., and William S. Webb

1928 Ancient Life in Kentucky. Frankfort: The Kentucky Geological Survey.

1929 The So-Called "Ash Caves" in Lee County, Kentucky. Department of Anthropology and Archaeology, University of Kentucky, Reports in Archaeology and Anthropology 1(2).

1930 Rock Shelters of Wolfe and Powell Counties, Kentucky. Department of Anthropology and Archaeology, University of Kentucky Reports in Archaeology and Anthropology 1(4).

Gilmore, Melvin Randolph

1932 The Ethnobotanical Laboratory at the University of Michigan. Occasional Contributions of the Museum of Anthropology of the University of Michigan No. 1. Ann Arbor: The University of Michigan Press.

Haury, Emil H.

1945 The Excavation of Los Muertos and Neighboring Ruins in the Salt River Valley, Southern Arizona. Papers of the Peabody Museum of American Archaeology and Ethnology, Harvard University 24(1).

Jones, Volney H.

See *Publications of Volney H. Jones*, this volume.

Webb, William S., and William D. Funkhouser

1936 Rock Shelters in Menifee County, Kentucky. Department of Anthropology and Archaeology, University of Kentucky, Reports in Archaeology and Anthropology 3 (4).

# VOLNEY HURT JONES: BIOGRAPHICAL BACKGROUND

*Karen Cowan Ford*
Ann Arbor, Michigan

Since the 1930s many anthropologists and botanists have appreciated the innovative, thoughtful, and detailed ethnobotanical work of Volney H. Jones. Students and colleagues at the University of Michigan have enjoyed his teaching–informally in the Ethnobotanical Laboratory and at coffee breaks–as well as formally in lecture, seminar, and independent reading courses. Little do they realize that from birth Jones was destined to become a world authority on the American Indian and on ethnobotany.

On April 30, 1903, Volney Hurt Jones, the second son and sixth child of Ella Jones Jones and Arthur Willis Jones, was born in Comanche, Texas. A younger sister arrived twelve years later to complete the family. Although Comanche was little more than a sleepy farm town, it bore the name of one of the most famous Indian tribes of the plains whose exploits were still told in Jones' youth and about whom he would some day become an authority. The family lived at various times in Comanche, Arlington, Dallas, El Paso, and Grosbeck, Texas, and Jones attended school in these locations, eventually graduating from Grosbeck High School in 1923.

Jones' early schooling and work experience varied, as the family moved about the state of Texas because his father, a Church of Christ minister, was frequently transferred from one church to another. Since itinerant preaching did not produce the material rewards needed to maintain a large family, the senior Jones pursued numerous entrepreneurial and farming interests as well. From the latter efforts, Volney received his first practical experiences, some of which later would serve him well in the field of ethnobotany. Once, for instance, the Joneses planted a forty-acre pecan grove in Arlington, Texas. Unfortunately, though they lived in the area for four years, they never harvested because pecan trees require eight years to produce a marketable crop. Each work experience seemed to prepare him further for a career in ethnobotany. In addition to planting crops such as watermelons, while in El Paso Volney worked in an ice plant where he claims to have learned how to eat piñon nuts: tossing a handful into the mouth, cracking the shells, retaining the nutmeats, and expectorating the hulls in rapid fashion. These experiences became grist for the many stories he fondly tells at his own expense.

After graduation from high school, Jones enrolled in a junior college course in agriculture at what was then known as North Texas Agricultural College[1] in Arlington, Texas. Before graduating in 1927 with an Associate in Science certificate, he was a student assistant in biology and participated in Little Theater, the Shorthorn Staff, and the Hoof and Plow Club as well as being a key halfback on the football team. He was described in the 1927 yearbook as a shifty back who specialized in passing, off-tackle smashes, and open field running. Now in the late 1970s Jones faithfully watches the charges of the University of Michigan's Bo Schembechler at Michigan Stadium, two city blocks from the Jones' Ann Arbor residence. An avid sports fan, he can empathize with the players as well as appreciate the intricacies of the games. In addition, his courses were popular with athletes at the University of Michigan, and many an All-American has enjoyed Jones' Indians of North America course.

From Arlington, Jones moved to Texas A&M in College Station to earn his B.S. degree in agriculture (horticulture) in 1929. While there, he was student assistant and manager of the Horticultural Demonstration Farm. Supported by a fellowship, he conducted experiments on the effects of nitrogen fertilizers on vegetable crops. For one year following graduation in the midst of the Depression, he supported himself by working at the Post Office. At the same time he undertook part-time graduate work in botany with Professor Robert G. Reeves, studying the genetics of the Japanese persimmon.

His youthful experience and college education culminated in 1930, when he received an opportunity to continue graduate studies in ethnobotany at the University of New Mexico. At the time, this was a unique program in the United States, and Volney made the most of it. As a Graduate Fellow in the Department of Biology, he worked closely with Professor Edward F. Castetter and with funding from the National Research Council wrote his master's thesis, "The Ethnobotany of Isleta Pueblo." His M.A., with a major in biology and a minor in English, was awarded in August 1931.

---

[1] Current name is University of Texas at Arlington.

During his student days at the University of New Mexico, he met many anthropologists, including Edgar Lee Hewett, Marjorie Lambert, and Paul A. Reiter. The latter was to remain his good friend and share many an after-hours adventure. Jones' initiation into anthropology was rapid and initially informal. Always the raconteur, Volney draws upon his UNM exploits for many an entertaining story. One of these involves his participation with a colleague and a guide in an expedition to search for an ice cave in which, the guide assured them, was to be found—frozen and preserved—a Spanish conquistador in full armor. Skeptical though they were, the two friends trekked a full day over the treacherous lava flows, aptly named "malpais,"[2] of western New Mexico. Jones' assignment was to carry the 30 pound camera—just in case the guide found the armored gentleman! In addition to this and other student exploits, while in New Mexico Jones became a member of the National Guard, which gave him access to horses, stable chores, and still other moonlit sorties.

Leaving the University of New Mexico in 1931, Jones moved north and east to begin what has become a long and fruitful association with the University of Michigan and the city of Ann Arbor. He began as a special student of Melvin R. Gilmore, working as a student assistant in the Ethnobotanical Laboratory 1931-1933, continuing his studies under a University Fellowship in Anthropology 1933-1934, and receiving a full-time staff appointment in the Museum of Anthropology in the summer of 1934. In the fall of 1938, Jones returned to classes to pursue a Ph.D. in a combined program of anthropology and botany, while continuing as a part-time staff member. However, Gilmore's health failed, and after he died in 1940, Jones returned to full-time Museum staff status, placed in charge of the Division of Ethnology and the Ethnobotanical Laboratory. In 1945, the titles of Curator of Ethnology and Assistant Professor of Anthropology were bestowed on Jones and hence graduate school rules prevented his receiving a degree from the University of Michigan.

Graduate student life in anthropology at the University of Michigan was exciting and the friendships enduring. Under the auspices of Carl E. Guthe, the director of the Museum, anthropology was actively practiced. Jones was fortunate to work with Gilmore, the premier ethnobotanist of his day, but he also found his own habits compromised by Gilmore's more puritanical expectations. Smoking was not permitted when Gilmore was present in the Ethnobotanical Laboratory. However, in his absence Volney could steal a quick puff on his pipe. Predictably, George Quimby would amble down the hallway imitating Gilmore's gait which, of course, would catch Jones unaware, and, in the midst of George's laughter, he would dash to extinguish the last spark.

---

[2] "bad country"

Fig. 1. Volney H. Jones early in his career at the University of Michigan.

During the 1930s in conjunction with his M.A. degree as well as his study and work at the University of Michigan, Jones conducted many months of field research, especially during the summers, in New Mexico, Arizona, and Michigan. During this period he also found the time to meet and marry his companion of the last 45 years, Joyce Hedrick Jones. They were married in August 1933, and have one son, Alan, who was born in 1937. Joyce, a trained botanist, has accompanied her husband on numerous field trips in addition to pursuing her artistic interests, in particular, weaving.

One "field trip" in which the whole family participated occurred during World War II, when many careers were interrupted and many families disrupted. Although Jones received ROTC training in high school and college and he participated in the New Mexico National Guard (1930-1931) and the

U.S. Infantry Reserve (1924-1937), reaching the rank of captain in the latter, his involvement in World War II was vital to the war effort but remote from the front lines. Taking leave for fifteen months (December 1942-March 1944) from the University of Michigan, Jones engaged in rubber production in Haiti as Production Superintendent, Cryptostegia Rubber Program, Société Haitiano-Américaine de Développement Agricole. The experience of living and working in Haiti proved to be interesting to all three members of the family.

Fig. 2. Cryptostegia Rubber Program in Haiti, Volney H. Jones, Production Superintendent.

Soon after returning to Ann Arbor, son Alan's participation in cub scouting brought his father's talents to the attention of the local Boy Scout Council, which Jones has since served in an advisory capacity for many years. In addition to conducting training courses for adult leaders, he has served as a merit badge counselor for numberous badges, utilizing not only his vast general ethnographic knowledge but also his familiarlity with specific techniques, such as basket weaving and chair caning.

As a parent Jones was actively involved with public school affairs. During Alan's high school days (1952-1955) at the old Ann Arbor High School, which was located in the Frieze Building on State Street, Jones served as president of the parent-teacher organization and for three years on its council.

Other civic activities have included the Ann Arbor Citizens' Library Committee and the Citizens' Committee for Charter Revision. Skill at formal lecturing on a wide range of topics has led to many invitations for him to speak to many different organizations and clubs. Requests for information and/or identifications of botanical specimens or ethnographic objects from individuals have been frequent and have been answered carefully whether from third graders, amateur anthropologists, police officers, or fellow professionals.

In addition to formal teaching, Jones has been an indispensible mentor for many students in less structured situations in the Ethnobotanical Laboratory and at coffee breaks. The twice daily coffee break, faithfully attended by Volney and Jimmy Griffin, informally brought news about recent discoveries or professional controversies to staff and students. As if on cue, the time to disband and return to work was signaled by one final and unimaginable pun provided by Volney.

Although Jones' ability as a raconteur has entertained many students and staff members, proof of his tales sometimes has been demanded. His ethnobotanical and ethnographic collections often have provided indisputable evidence. Perhaps two of the most noteworthy items in these extensive collections are (1) a resident peyote cactus plant, which after many years in captivity in the Ethnobotanical Laboratory, began to bloom in full flower regularly in the 1960s when it was moved from a north to an east facing window, and (2) a collection of dressed fleas, i.e., fleas with clothes on, which can be viewed under a microscope. New students predictably pilgrimage to Jones' office to glimpse these and other wonders of the world. More advanced ethnobotany students are privileged to view "Ouch," the prehistoric human coprolite two and one-half inches in diameter.

Retirement and emeritus status have not greatly changed Volney's schedule. Fewer administrative responsibilities are required now, occasionally an extra wink or two is permitted, and he can spend a bit more time gardening and on short trips, but Jones maintains an office in the Ethnobotanical Laboratory to which he faithfully returns to work each day. This routine certainly has benefitted many recent students as well as faculty who, as newcomers to the University of Michigan, never have had the opportunity to take courses with Professor Jones, and who each year discover one of the most generous and gracious people they will ever meet. For 47 years, this multi-talented man has been a fund of information and an inspiration for each generation of students.

# PART I
# THEORETICAL ISSUES IN ETHNOBOTANY

# INTRODUCTION

Ethnobotany lacks a unifying theory, but it does have a common discourse. The historical threads which give texture to contemporary ethnobotany derive from a variety of sources. Although botany, anthropology, archaeology, natural history, linguistics, and herbal medicine have contributed to the discipline in an unsystematic manner, the accumulation of methods, viewpoints, and data from each field has enriched and enlivened ethnobotany.

To understand ethnobotany is to appreciate its catholic composition and the situations in which it has been practiced. Every culture, despite the training of its reporter, has contributed some knowledge about plants. Archaeologists, historians, travelers, missionaries, biologists, and ethnographers have noted something about some plant in the lives of extinct or extant people. The problems that this plethora of curious data present to the ethnobotanist are enormous. In some cases an anecdote must suffice; in others no voucher specimens were collected or have survived. Sometimes a native name is our only clue; at other times exact botanical descriptions are available unaccompanied by a shred of native botanical knowledge. Even the archaeologist, always alert to the inevitable destruction wrought by every excavation, often fails to acknowledge plant remains as by-products of human cultural activities, and the identified fragments are ignored in the site interpretation.

Because plants permeate or represent all aspects of human affairs, they are studied by a number of unrelated disciplines with which the ethnobotanist must be familiar. The history of technology before the era of fossil fuels is essentially an analysis of plant anatomy and growth. Wood for cooking, housing, and furnishings is selected according to chemical and physical properties. Fibers are used for clothing and bindings. In historic times, the extraction of these raw materials created a lively commerce and an intimate knowledge of the environment in traditional, self-sufficient communities. A background in chemistry, physiology, and anatomy as well as in taxonomy, anthropology and linguistics are requisite for explaining the transformation of one plant or another into desired forms and utilitarian objects.

Feeding people and domesticated animals unites nutrition with ethnobotany. Purposeful protection and cultivation of plants led to genetic changes and dependence upon humans for survival. These domesticated artifacts of human need and ingenuity provided the basis for civilization, but they did so only recently in human history and at the expense of a diversity of gathered plants whose nutritional constituents we are only rediscovering before they are forgotten forever. A meal is more than a biological necessity or a culturally symbolic representation; it is a complex of ethnobotanical knowledge and an epitome of the field's diversity.

The distinction between food and medicine is an artifact of Western specialization. In reality, ethnomedicine as a subfield of ethnobotany may incorporate the study of gastronomy as well. Most cultures, in fact, classify all plants (and many animals) taken internally into a unified taxonomy. Illness may result from overindulgence of one, the exclusion of another, or the consumption of almost any substance under culturally inappropriate circumstances. Yet, restoring a balance to the body or relieving discomfort has been aided by plants that we recognize as medicine, and knowledge of their use and preparation, when extricated from the culture of discovery, has brought new health to people elsewhere. Quinine is a notable example, and the search for other plant medicines continues in the face of rapid culture change and loss of traditional cures.

Plants are indispensible to ritual and religion. They provide a devotional ambience, they symbolize a connection with gods, and they may even transport the believer into the spiritual world itself. Virtually no ceremony in any culture is performed which does not include some plant or reference to growing things. The eternal presence of gods and spiritual beings is continually verified by some plant representing their existence or indelibly marked by their presence. Such beliefs, of course, are not indigenous only to non-Western societies. Certainly in many Christian homes in the United States today Easter could not be observed without the customary white flowered lily (*Lilium longiflorum*), and the presence of Christ is substantiated to believers by the red marks on the four large white bracts of the flowering dogwood (*Cornus florida*) blossom.

Entry into a spiritual world is often facilitated by employing a plant purgative or consuming a psychoactive plant. These plant agents are particularly important in tropical and, to a certain extent, in temperate regions of the New World. Jimsonweed (*Datura inoxia*), peyote (*Lophophora williamsii*), sacred mushrooms (*Psilocybe* sp.), yáhi (*Bannisteriopsis*), and tobacco (*Nicotiana* spp.) are a few that are widespread. Popularization of these plants in Western subcultures should not detract from the profound importance that these and other plants have in the maintenance of many traditional cultures.

No academic interpretation is necessary to realize that plants symbolize and create new social relationships. Sharing food, drinking alcoholic beverages, or smoking together reinforces special relationships. A gift of flowers or fruit has long been a sign of friendship or an unspoken request for a more intimate association. Time devoted to a growing plant calls forth the memory of the giver, and the place where it is planted may represent the closeness of the relative or friend. The preceeding examples are accepted social behavior in U.S. society and are also the subject of ethnobotanical inquiry.

The customary sources of ethnobotanical knowledge and the subject matter appropriate for study have traditionally excluded Western society and have stressed the *uses* of plants. This emphasis, in fact, has provided the most common and misleading definition of ethnobotany—the uses of plants by primitive people. Such an approach denies the intellectual life of non-Western people and excludes inquiry into our own culture.

Volney Jones changed our perspective of ethnobotany by recognizing the reciprocal and dynamic aspect of the human interactions with plants. In this way he anticipated the ecological approach in anthropology, and he provided a framework for the detailed linguistic study of folk plant taxonomy and folk botany, both of which are basic to contemporary ethnobotany. Without a holistic and symbolic study, one cannot understand why certain plants and not others are used, how they are prepared, and when they are significant. In his work, Jones concentrated not only on the methods by which humans exploit plants but also on the consequences of their actions to the plant environment. His definition and his perception of the nature and scope of ethnobotany enable us to continue the dialogue among the heterogenous parts of ethnobotany and to construct a theory explaining the diversity of observations of human-plant interactions in time and space.

The changing quality of ethnobotanical observations within a single culture is most instructive. Not only is there variation in the knowledge about plants among members of a community as a result of age, sex, experience, interest, or a combination of these; in addition, there is also modification through time in the plant lore and even in the organization of plant knowledge itself. Marcus and Flannery have exemplified this principle of dynamic variation through a longitudinal study of the ethnobotany of Zapotec speakers in Oaxaca, Mexico. Their discovery that Fray Juan de Córdova's Zapotec dictionary could provide a baseline for assessing stability and change in a folk classification opens a new dimension to ethnoscience. While they can recognize critical terms for discerning the foundation of Zapotec classification, at the same time they can demonstrate that meanings and even structure itself change over time through contact between neighboring cultures. This lesson illustrates the value of ethnohistorical sources for ethnobotany and provides a warning against assuming that plant lore is immutable, even among seemingly traditional people.

The results of their study are not lost on paleoethnobotanists, whose task is extremely complicated as it is. Archaeological plant remains have long been studied from a materialist perspective, that is, from the standpoint of what people used or ate. Such studies were compatible with the older approaches to ethnobotany but not with the more recent recognition that what was used had a specific cultural meaning and that evolution into new patterns of plant dependency might be unrelated to material needs or environmental conditions.

Recognizing this problem is one thing, but discovering the cognitive system of an extinct culture is quite another matter. Wetterstrom has brought the contradictions of paleoethnobotany and emerging ethnobotanical theory to the surface and is attempting to resolve them by developing cross-cultural models. Her research in the Hueco Bolson in western Texas is predicated upon recognition of the interrelationship between the pattern of plant remains discarded as food debris and the reason for their selection in the first place. A series of analytical questions are designed to examine cognitive mapping and reorientation as the economy of this region shifted from one without agriculture to one virtually dependent upon domesticated plants. Efforts such as Wetterstrom's do not invalidate other approaches for interpreting archaeological plant remains, but they do aid in the quest for a general theory that will be applicable to past cultures as well as to living peoples.

The implications of other aspects of interactions between plants and peoples will be explored in later parts of this volume. The survey of ethnobotanical concerns includes the classificatory relationship between plants and world view and well being, the management of the ecosystem, the impact of human populations on plant dispersal and genetic changes, and the importance of botanical determinations in archaeological land use patterns.

# ETHNOBOTANY: HISTORICAL DIVERSITY AND SYNTHESIS

*Richard I. Ford*
Museum of Anthropology, The University of Michigan

On December 4, 1895, in the course of an address to the University Archaeological Association, a dynamic, young University of Pennsylvania botanist, Dr. John W. Harshberger, first applied the term *ethnobotany* to the study of "plants used by primitive and aboriginal people . . ." (anonymous 1895). His address was published in 1896, and ethnobotany instantly replaced other designations for this realm of botanical knowledge (cf. Fewkes 1896) which had experienced almost half a century of scientific attention and an even longer history of casual observations.

Although several European botanists were interested in the uses of plants by present-day and prehistoric people (e.g., de Candolle 1885; Heer 1866), ethnobotany originated as a result of exploration of the New World. Columbus returned not with the spices of the Orient, but with maize, tobacco, and other cultivars which were quickly assimilated into European cultures. A rapid progression of expeditions came to North America to discover and to colonize, and the chronicles of adventure are a record of the utilitarian value of an unfamiliar landscape and the use the indigenous people made of it. Its economic potential certainly had priority to any interest in the attitudes about the land. The observations of Cartier, Charlevoix, the Jesuit Fathers, Le Page du Pratz, Hariot, Morfi, Kino, and others were mostly happenstance. Nevertheless, they provided the first natural history of North America, and the bases for the beginnings of ethnobotany.

## NATURAL HISTORY AND BOTANICAL OBSERVATIONS

The sources of information about the uses of plants by North American Indians for the pre-Revolutionary War period are scattered in historical chronicles, diaries, and nature studies by European adventurers and botanists who collected specimens for herbaria and museums in Europe. Many foreign born botanists established their reputation by recording the flora of the poorly known new continent. While conducting their collecting trips Peter Kalm, Andre Michaux, Frederick Pursh, Constantine S. Rafinesque and others noted interesting plant uses by the Indians whom they encountered. Although this information was incidental to their primary objectives, by the mid-eighteenth century such observations were rather commonplace, and economic themes about the New World's plants and their benefits for the White man had emerged.

The most significant was medicine. Many colonists adopted herbal medicine of Indian origin, and both botanists and physicians sought to introduce new remedies into the east coast colonies. The accumulation of random observations was sufficient that by 1672 John Josselyn could publish his *New-England's Rarities Discovered*, and almost a century later the distinguished Philadelphia physician, Benjamin Rush, published a major treatise on the *Natural History of Medicine among the Indians of North America . . .* (Rush 1774), and a second edition of this influential reference appeared in 1794. Typical of the practical focus of plant collection, John Bartram, a botanist, in a summary on the uses of plants in 1751, emphasized a newly found cure for venereal disease (Bartram 1751).

Other usages were not denied. Native plants were adopted for food and raw materials, especially since imports to the fledgling colonies were unpredictable and expensive. A practical study on plant dyes, for example, appeared in 1793 (Martin 1793).

With the establishment of the United States, government sponsored exploration became a new source of information about the value of the land westward to the Pacific. President Thomas Jefferson's instructions to Lewis and Clark voiced the expectation that the habits and customs of native people were not to be overlooked while they were assessing the economic potential of the country. Those who followed included scientists who observed the natural history along the route. Thomas Shay, J. T. Rothrock, C. C. Perry, and J. S. Newberry were among the pioneer naturalists who described the flora west of the Mississippi River. Although most of the published reports lacked a systematic appraisal of Indian uses of plants, they expanded knowledge about economic plants.

By mid-century, the great number of army officers, government officials, missionaries, and teachers living with or in close proximity to Indians enabled

Henry R. Schoolcraft to circulate a questionnaire requesting information about the condition of the local people, and many of these replies are a source of material on the environment and particularly on medicine (Schoolcraft 1851).

As the nineteenth century drew to a close, a significant number of systematic descriptions of the flora were available for most of the United States. Botanists began intensive studies of local areas and even of Indian plant uses. At the same time employees of the government, especially the United States Geological Survey and the Bureau of American Ethnology, both directed by Major John Wesley Powell, obtained accurate identifications of the plants used by the tribes they were studying. After almost 400 years, natural history gave way to the scientific study of botany.

## MEDICINE MEN AND "INDIAN DOCTORS"

Before reviewing the history of modern ethnobotany, a sidelight which still influences the field must be mentioned. Not every American in the nineteenth century was touring the prairies or heading by wagon train to California. Most stayed behind and received their vicarious experiences with Indians from newspapers, novels, and exhibitions of paintings. On the frontier, Indian curers occasionally treated whites, and native medicine was adopted by settlers (Vogel 1970). Tales of miraculous healing fed the romantic side of the Indian stereotype in the East, which when combined with the drama of Indian wars, gave the Indian a potency he never deserved. Ethnocentrism fostered an accentuation of cultural differences to create a new Indian, the Medicine Man, and belatedly a large volume of literature with implications for ethnobotany.

The majority of general works on North American ethnobotany appearing before 1896 (Table 1) are herbals published in the East at a time when testimonials of firsthand Indian lore certified the Indian doctor and when any Indian, almost, was endowed with the power to cure. The Native American may have been accorded little other respect, but medicine was his forte. The distortion caused by the medicine wagon and sideshow barker did little to transmit accurate knowledge about Native American beliefs as they pertain to health and illness and built a legend about the efficacy of herbal medicine that we still contend today.

## ABORIGINAL BOTANY AND NATURAL SCIENCE

By the post-Civil War era the taxonomic biological sciences were well established and supported by universities and learned societies. Exploration of an uncharted wilderness was essentially over, and general maps for

TABLE 1
SUMMARY OF ETHNOSCIENCE STUDIES IN NATIVE NORTH AMERICA

| Discipline | Culture Area (following Murdock and O'Leary 1975) | | | | | | | | | | | | | | | | |
|---|---|---|---|---|---|---|---|---|---|---|---|---|---|---|---|---|---|
| | Arctic Coast | Mackenzie-Yukon | Eastern Canada | Midwest | Northeast | Southeast | Plains | Gulf | Southwest (excl. Mexico) | Basin | California | Peninsula (incl. Mexico) | Northwest Coast | Oregon Seaboard | Plateau | General North America | Total |
| *Ethnobiology* | | | | | | | | | | | | | | | | | |
| Before 1896 | | | 1 | | 2 | 4 | 2 | | | 1 | | | | | | | 10 |
| 1896-1920 | | | | 1 | | 1 | 1 | | 2 | | | | | | | 2 | 7 |
| 1921-1950 | 4 | 7 | 1 | | | | 1 | | 7 | 2 | 1 | | 2 | | | 5 | 30 |
| 1951-1977 | 7 | 3 | 8 | | 1 | 3 | 4 | 1 | 13 | 4 | 4 | 3 | 6 | | 1 | 4 | 62 |
| *Ethnobotany* | | | | | | | | | | | | | | | | | |
| Before 1896 | | 3 | 5 | 2 | 15 | 14 | 8 | | 9 | 2 | 5 | 1 | 3 | 3 | | 28 | 98 |
| 1896-1920 | 1 | | 11 | 11 | 15 | 7 | 17 | | 27 | 3 | 9 | 2 | 4 | 2 | 2 | 30 | 141 |
| 1921-1950 | 2 | 2 | 35 | 13 | 32 | 23 | 31 | 1 | 99 | 13 | 20 | 7 | 12 | 10 | 8 | 45 | 353 |
| 1951-1977 | 11 | 2 | 22 | 11 | 22 | 29 | 31 | 3 | 60 | 18 | 19 | 22 | 9 | 5 | 10 | 38 | 312 |
| *Ethnozoology* | | | | | | | | | | | | | | | | | |
| Before 1896 | 5 | 2 | 4 | | 3 | 3 | 6 | | 9 | | | | 3 | 3 | 2 | 4 | 44 |
| 1896-1920 | 11 | 3 | 8 | | 8 | 4 | 8 | | 11 | 2 | 2 | | 4 | 2 | 2 | 3 | 68 |
| 1921-1950 | 22 | 6 | 17 | 3 | 18 | 15 | 19 | | 30 | 7 | 9 | 6 | 17 | 6 | 9 | 8 | 192 |
| 1951-1977 | 36 | 15 | 26 | 4 | 9 | 8 | 12 | | 33 | 7 | 8 | 11 | 8 | 5 | 7 | 17 | 206 |

TABLE 1 (Continued)

| Discipline | Culture Area (following Murdock and O'Leary 1975) | | | | | | | | | | | | | | | | |
|---|---|---|---|---|---|---|---|---|---|---|---|---|---|---|---|---|---|
| | Arctic Coast | Mackenzie-Yukon | Eastern Canada | Midwest | Northeast | Southeast | Plains | Gulf | Southwest (excl. Mexico) | Basin | California | Peninsula (incl. Mexico) | Northwest Coast | Oregon Seaboard | Plateau | General North America | Total |
| *Ethnoastronomy* | | | | | | | | | | | | | | | | | |
| Before 1896 | | | | | | | 4 | | 1 | | | | 1 | | | | 6 |
| 1896-1920 | 1 | | 1 | | 1 | 1 | 4 | | 4 | | | | | | | 1 | 13 |
| 1921-1950 | | | | | 1 | 1 | 6 | | 4 | 1 | | | | | | 1 | 14 |
| 1951-1977 | | | | | | 2 | 3 | | 2 | 1 | | | | | 1 | 2 | 11 |
| *Ethnogeography* | | | | | | | | | | | | | | | | | |
| Before 1896 | 2 | 1 | 2 | 1 | 5 | 1 | 1 | | | | 1 | | 2 | | 1 | 3 | 20 |
| 1896-1920 | 3 | | | 1 | 4 | | 3 | | 14 | 1 | 4 | 1 | | 3 | 3 | 2 | 39 |
| 1921-1950 | 3 | | | 4 | 6 | 6 | 8 | | 15 | 1 | 5 | | 4 | 1 | 1 | 4 | 58 |
| 1951-1977 | 9 | | 2 | 10 | 13 | 4 | 9 | | 16 | 1 | 9 | 2 | 7 | | | 4 | 86 |
| *Natural Phenomena* | | | | | | | | | | | | | | | | | |
| Before 1896 | 2 | | | | | | 3 | | | | | | 1 | | | 1 | 7 |
| 1896-1920 | 2 | | 1 | | | 1 | 2 | | 3 | | | | | | 1 | 4 | 14 |
| 1921-1950 | 2 | | 2 | | 1 | 1 | 3 | | 3 | | 2 | | 1 | | 1 | 2 | 18 |
| 1951-1977 | 5 | 3 | 1 | | | 4 | 2 | | 2 | 1 | 1 | | | 3 | 1 | 6 | 29 |

most of the country were available. Upon the foundation laid by the pioneer observers of nature, trained scientists could build up collections through planned expeditions and could provide accurate taxonomic descriptions of the flora. The study of plants used by Indians reflected now scientific objectivity and taxonomic rigor.

While American botany matured into a science, anthropology was in its infancy. The best studies of Indian uses of plants came from botanists. If any scientist epitomized the new trend, it was Edward L. Palmer, who not only described the vegetation of the West but also made inquiries about plants useful to the Indian people he deliberately visited. His "Food Products of the North American Indians" written in 1870 demonstrates the quality of his work and the copious information of topical value which was available. This seminal paper was followed in 1878 by "Plant Products Used by the Indians of the United States" (Palmer 1871, 1878).

Another influential scientist was Stephen Powers, whose research in California led him to define the field to which he, Palmer, and other botanists were now actively contributing. This he labeled "aboriginal botany" in 1874 to include "all forms of the vegetable world which the aborigines used for medicine, food, textile fabrics, ornaments, etc." (Powers 1873-1875: 373). This term was readily accepted for the next 25 years (cf. Dunbar 1880; Mason 1886; Coville 1895). Palmer and Powers brought botanical exactness and a high standard of fieldwork to aboriginal botany and the quest for new applications of the North American flora.

Following the lead of Palmer, summaries of plant usages became a trend in aboriginal botany. Newberry (1887) compiled food and fiber plants, Trimble (1888-1891) introduced additional edible plants, and Havard gave attention to food (1895) and drink plants (1896). Although diverse literary sources underscored the necessity for these compilations, they also revealed serious gaps in knowledge and the unreliability of previous observations. As a result, exacting studies of specific tribes began to emerge as the dominant form of ethnobotanical investigation.

Although B. R. Ross (1862) and Robert Brown (1868) initiated general studies of plant uses within particular tribes, the impetus for this approach actually came from Palmer. These detailed inquiries were undertaken by government botanists, by long-time residents such as Dr. Washington Matthews, a physician among the Navajo (Matthews 1886), and by the nascent anthropological investigations of Frank Cushing at Zuñi in 1884-1885 (Cushing 1920) and James Mooney with the Cherokee (Mooney 1891).

## ETHNOBOTANY AND ANTHROPOLOGY

Ironically, when Harshberger introduced the immediately popular term ethnobotany in place of aboriginal botany, he was actually recapitulating the theoretical and methodological contributions of botany and natural science to the field. During the 1890s ethnologists replaced botanists as the students of the American Indian's natural resources, and with their entry into the field, the theoretical emphasis of ethnobotany changed. To the botanist, ethnobotany was the study of uses of scientifically identified environmental data. The questions and focus were utilitarian, an old American tradition, and the organization of these data followed scientific classifications. To the anthropologist, however, the focus was the native's point of view and his rules and categories for ordering the universe. These differences in perspective affected the objectives of these complementary disciplines, both practicing ethnobotany.

Perhaps the most important event in the history of this reorientation was the World's Columbian Exposition held in Chicago in 1893. In preparation for this World's Fair, Frederick W. Putnam, Curator of the Peabody Museum, Harvard University, his ethnological assistant, Franz Boas, and his archaeological assistant, George A. Dorsey, deployed anthropologists to assemble collections of the useful products of various world cultures, particularly the American Indian, and even to bring to the Fair the people themselves to live in ethnographic replicas of their homes (Collier 1969). This activity created interest among anthropologists in collections of plants, animals, and other resources, and through the archaeological exhibits to the formalization of ethnobotany itself. While analyzing desiccated plant remains from Mancos Canyon, Colorado, which had been displayed at the Chicago World's Fair, Harshberger conceptualized ethnobotany. But the adoption of the term by anthropologists emphasized the cultural importance and significance of plants in the lives of the people.

Ethnobotany was practiced under the sponsorship of museums, government agencies, and in the twentieth century, universities. The U.S. National Herbarium and Department of Agriculture were crucial to furthering scientific studies of Indian usages of plants. Several of their employees and collaborators conducted field studies and provided ethnologists with plant identifications. Frederick V. Coville published "Directions for Collecting Specimens and Information Illustrating the Aboriginal Uses of Plants" as a guide for standardizing ethnobotanical information (Coville 1895). Coville and J. N. Rose from the National Herbarium identified plants for government anthropologists such as John Peabody Harrington, Walter Hough, and Matilda Coxe Stevenson. Furthermore, the Division of Botany in the Department of Agriculture sponsored field research by Coville (1897, 1904) among the Klamath in Oregon and by V. K. Chestnut (1902) in California. Meanwhile, the Division of Plant Industry of the same department from 1905 to 1926 supported the activities of the economic botanist W. E. Safford (1917, 1927).

At the turn of the century museums actively encouraged research by ethnobotanists. The New York State Museum published the works of W. N. Beauchamp (1905) and A. C. Parker (1910). The Bureau of American Ethnology published a number of outstanding ethnobotanical studies by its employees and collaborators (e.g., Jenks 1900; Stevenson 1915; Robbins et al. 1916; Densmore 1928; Teit 1930). The most ambitious (and unfulfilled) program, however, was initiated by the Department of Ethnology in the U.S. National Museum when it determined "to secure the identification of *every* plant used by North American Indians for any purpose whatever" (Mason 1898:175 [emphasis added]).

The entry of university research programs into ethnobotany during the first three decades of this century soon surpassed other institutions. The first doctoral dissertation in ethnobotany was awarded by the University of Chicago to David P. Barrows in 1900. Later, from 1930-1950, the University of New Mexico, under the direction of Edward F. Castetter, established a masters degree program in ethnobotany within the Department of Biology. Also in 1930, the University of Michigan Museum of Anthropology created an Ethnobotanical Laboratory with Melvin R. Gilmore in charge, and it soon served as "a clearing house and identification center for botanical specimens discovered in the course of archaeological excavations" (Guthe 1930). Gilmore further defined the purposes of the laboratory in a separate publication (Gilmore 1932). Elsewhere, the Botanical Museum of Harvard University pursued the study of economic botany in North America (Vestal and Schultes 1939; Taylor 1940) as an important facet of its worldwide research. Today ethnobotany is recognized; ethnobotanical degree programs and research activities are centered in several American universities.

## ETHNOBOTANY AND OTHER ETHNOSCIENCES

Interestingly, even though ethnozoology never had a theoretical formulation prior to or distinct from ethnobotany, the first "ethno-" prefix to a science was entitled "Ethno-Conchology: A Study of Primitive Money" and published in 1889 (Stearns 1889). Nevertheless, ethnobotany is accorded priority by most scientists, and deservedly so, because it is well defined as a separate domain and has an established intellectual tradition behind it. Castetter (1935, 1944) attempted to emphasize its logical affinities with zoology by submerging it into ethnobiology, but historical precedent and the structure of university training militated against adoption of this association. Recently, however, the term "ethnoscience" has prevailed, and relegated ethnobotany to a subdiscipline.

Ethnoscience in the United States has a history which is coterminous with ethnobotany. Despite efforts to restrict the meaning to "the system of knowledge and cognition of a given culture" (Sturtevant 1964:99), the contemporary definition of the term is the use, importance, and perception of the environment in its most general sense by the original inhabitants of the North American continent or by aboriginal peoples elsewhere. Its allied fields of study are biology, botany, zoology, astronomy, geography, and natural phenomena characteristic of local weather or climate.

Although the figures summarized in Table 1 are relative—for example, native explanations of natural phenomena may have been overlooked because they are found in folktales and not separate studies as such—nonetheless, ethnobotany certainly accounts for the overwhelming majority of the studies for each time period and for each culture area. The two exceptions to the latter generalization are the Arctic Coast and Mackenzie-Yukon culture areas where ethnozoological studies quite expectedly are more numerous. The dates are based upon major theoretical changes in the ethnosciences. The year 1896 was selected as a baseline because of the history of ethnobotany which was just outlined. From 1896 to 1920 all of the ethnosciences were pursued by persons with some natural science training and the overall quality of the results reflects their background. The year 1920 marks the emancipation of anthropology from the natural sciences and its affiliation with the social sciences. Less emphasis is given to natural resources, and the percentage of anthropologists contributing to the ethnosciences decreases. Since 1950 ethnoscience studies have attracted renewed interest, due in part to the impact of both linguistic studies and ecology on anthropology. Throughout these periods ethnobotany has received emphasis because of the importance plants have in the material and mental culture of most people.

## THEORETICAL DEVELOPMENT

Volney H. Jones encapsulated the trends inherent in twentieth-century ethnobotany with his definition: the study of the interrelations of primitive man and plants (Jones 1941:220). It expanded the utilitarian scope of ethnobotany promulgated by Harshberger, Powers, and others, and it brought a dynamic aspect to the field which anticipated ecological approaches in anthropology. It was inevitable that with his background Jones would formulate such an insightful and comprehensive definition. He was trained by Castetter and Gilmore, he had firsthand experience at Isleta Pueblo and with Chippewa and Ottawa Indians in Michigan and Canada, his Newt Kash Hollow archaeobotanical report remains exceptional (Jones 1936), and he continues to be a respected bibliophile. Until Jones assessed the current status of ethnobotany, almost a half century of unresolved debate about the nature of the discipline had elapsed.

Despite his adoption of Harshberger's term, Fewkes' (1896) initial application of ethnobotany in the anthropological literature emphasized Hopi Indian plant names and their etymology. Walter Hough, Fewkes' co-worker, extended the meaning of ethnobotany even further by discarding emphasis on non-literate societies as had been customary and simply calling it "the study of plants in their relations to human culture" (Hough 1898:127), including psychological importance and mythological references.

The departure from a strict concern with practical uses was emphatically underscored by Barrows in his *The Ethno-Botany of the Coahuilla Indians of Southern California*. He instructed ethnobotanists to go beyond the importance of plants in the economy and to investigate their religious significance, their place in folklore, and "so far as he recognizes . . . these plants in his (Indian) language, that is, has named them and understands them" (Barrows 1900:3).

Following a similar line of reasoning, Robbins, Harrington, and Freire-Marreco's *Ethnobotany of the Tewa Indians* (1916) is one of the most detailed studies of Indian plant classification and perception of plants that has ever appeared. The interaction between Harrington, a gifted linguist and indefatigable field worker, and Robbins, a botanist, is obvious. Harrington emphasized the conceptualization of the plant world as revealed through the native language and Robbins called for recognition of the impact of the plant environment in a deterministic sense on the thought, livelihood, and well-being of the people. In later works Harrington continued this theoretical perspective, and his monographic study of tobacco use by the Karak in California remains a classic examination of a single plant in a culture (Harrington 1932).

The next advance in ethnobotanical theory resulted from Gilmore's extensive familiarity with the plants used by Plains and Prairie Indian tribes (Gilmore 1919). He recognized that they were not passively employing what the environment offered, but were actively modifying the plant world by introducing plants to new habitats, eliminating others, and changing the quantity available. As a result of their dependence upon local flora the Indians had an impact on their surroundings.

A. L. Kroeber published a major critique of ethnobotany in his review of Gilmore's monograph. He felt that much ethnobotanical data were collected without a problem for investigation. In particular, he objected to the emphasis on what plants people used when we can learn about the limitations on cultural creativity if we study why people do *not* use many plants available to them. Furthermore, quantitative information about plants that are important would be invaluable for understanding the role of plants in the whole of a culture (Kroeber 1920).

Both Castetter and Whiting were cognizant of these and other criticisms emanating from academic anthropologists, and they addressed them when formulating their research programs. Castetter (1935) defined a series of general problems to which ethnobotany could make a contribution: primitive science, economic botany, environmental impact, origins of agriculture, etc. Whiting (1939) demonstrated the importance of plants in Hopi culture by noting their significance throughout the culture.

When Jones' article appeared, ethnobotany was headed in several directions simultaneously. By calling for an examination of interrelations, he was able to accommodate Robbins' and Gilmore's contributions. By recognizing more than practical concerns for plants, he met the criteria of Barrows and Harrington. By defining specific applications to ethnobotany, he faced Kroeber's objectives. The one traditional position he maintained, in contrast to Hough, was limiting the field to non-literate ethnographic people and prehistoric archaeological material. Finally and most importantly, his definition anticipated and accommodated the ecological and linguistic studies which characterized American anthropology more than a decade later.

Jones recognized that an examination of man-plant interactions would require the tools and methods of the plant ecologist (Jones 1941:220). These would include quantitative techniques to determine mutual impact. George Carter (1950) explicitly argued this position and went so far as to envision ethnobotany as an ecological science capable of forging a link between geography, botany, and ecology.

Jones also emphasized the necessity to study the concepts of primitive science, folk taxonomy, and beliefs about the plant world. Despite Harrington's work, anthropologists, even those without interest in ethnobotany, were slow to do this. In the 1950s and 1960s however, these studies began in earnest and ethnobotany soon became equated in some minds with linguistic concepts and classifications at the expense of understanding human behavior and the reasons why plants were important in the first place.

Conklin's research among the Hanunóo in the Philippines (1954) accomplished what other ethnobotanists had neglected. He not only demonstrated the importance of understanding folk classifications in ethnobotany, but he also showed that even Kroeber's critique missed the mark because he assumed a Western plant taxonomy when, in fact, there are important distinctions between any two systems of classification. In the case of the Hanunóo, they name many more plants than the Western scientist does, and consequently their botany would be unintelligible without recognition of this basic fact.

Berlin and his associates (Berlin et al. 1966, 1973, 1974) have examined plant taxonomies in an effort to demonstrate underlying principles of organization and structure. Their efforts have given ethnobotany a new standard for linguistic competence that had been missing in many ethnographic studies of the past. Without a mastery of a language, much valuable information about plants in daily life, in the form of puns, metaphors, and symbolic referents will be ignored.

In historical retrospect, ethnobotany has become decidedly more cultural than Harshberger envisioned it. To the botanists of his day, and now as well, plants were the independent variable in any study. Emphasis on uses was a natural outcome, particularly if linguistic competence in an Indian language was lacking and national interest had long sought information about resource utilization. Ethnobotanical reports by botanists were ordered according to Linnean categories and not by those of the natives. From this perspective, economic botany can be distinguished from ethnobotany. The latter is concerned with the totality of the place of plants in a culture and the *direct* interaction by the people with the plants. Economic botany, on the other hand, emphasizes the uses of plants, their potential for incorporation into another (usually Western) culture, and that their benefactors have *indirect* contact with the plants through their by-products. In the botanical tradition ethnobotany is subsidiary to economic botany, and Harshberger's definition remains applicable.

## TOWARD A NEW SYNTHESIS

Ethnobotany is at the interface of human needs and thought with nature. Since plants permeate every aspect of human existence, the ethnobotanist must be able to identify what plants are significant; to discover how the people of a culture classify, identify, and relate to them; and to examine how their perception of the plant world actually guides their actions and concomitantly structures the floral environment. With these objectives serving as minimal requirements, Jones' definition of ethnobotany must be modified to accommodate them.

Ethnobotany is the study of *direct* interrelations between humans and plants. One omission is immediately apparent and an additional term requires explanation.

By recognizing that ethnobotany concerns those who actually manipulate or think about the local vegetation directly, the field is immediately demarcated either to the study of self-sufficient societies or to an examination of the conduct of specific activities. Those who must gather or raise their food and build their shelters are in actual contact with plants. They must know how to recognize certain plants, when to gather them, and what to expect from them. Nevertheless, people dependent upon a market economy for provisions are not removed from making decisions about plants. For many Americans the maintenance of a yard places them in direct contact with a variety of plants. These they select for cultural, social and utilitarian reasons; they classify them; and they have rules for manipulating them. The difference between these two points on a continuum of plant interactions is the duration of the contact, not social complexity. At the subsistence level a society's interaction with plants is continuous, while in middle class America it may be no more than several hours per week. A second difference is the relative importance of the plants to the people; however, this is actually an empirical question, and by restricting a definition with this criterion, we may miss an important contribution of ethnobotany toward understanding a particular society, including our own.

Consequently, deletion of "primitive" from the definition is deliberate. Jones did not intend it to have the pejorative connotations it has today; he employed it in the sense of non-literate or aboriginal to restrict the study of ethnobotany to less complex societies. Such an assumption is based upon a belief that a marked difference in botanical knowledge and dependence upon plants separate societies. This is indubitably true in quantitative perspective but not necessarily in qualitative terms. Virtually all members of a complex society possess botanical lore. It may not, however, correspond with that of a professional specialist; yet, the existence of a folk knowledge cannot be ignored. Americans' folk beliefs about plants are important if we are to understand their attitudes and practices regarding personal property, public parks, science education, etc. Hough, of course, recognized and attempted to rectify this logical fallacy in the last century.

To recognize that the world is rapidly changing is a second reason to justify broadening the study of ethnobotany to include all cultures. Most of the societies studied by ethnobotanists in the past are now part of a world economic system. They may have only occasional direct contact with plants, for example, as medicine, even though they continue to speak their native tongue. Unless ethnobotanists are prepared to study cultures that are in the process of change or that have become westernized, the field will soon become a literary tradition.

Basic to any ethnobotanical or ecological study is the recognition that humans form biological populations and are dependent upon culture. A human population may be characterized according to age and sex groupings, and the nutritional requirements of each can be determined. These needs can be assessed for the entire population, and how they are met will dictate the quantity of certain plants to meet them. Similarly, a population's use of fuel for warmth and materials for shelter can be quantified in order to determine how procurement of these materials will affect the floral environment.

The ability to use symbols and the full-time use of language are hallmarks of humans. In ethnobotanical terms, this includes the classification of plants in the environment and man's psychological disposition toward them. Again these qualities determine how vegetation will be manipulated and what the consequences of human utilization will be. Behavior toward plants is not random; it is prescribed by a series of rules that the ethnobotanist must learn in order to determine which plants are used, when and what alternatives will be obtained, and how they will be procured. Cultural beliefs determine the conditions for human existence, and the biological properties of the human population define the quantity of plants that must be obtained. Together they form the human ecology of ethnobotany.

At the same time, human activity and natural environmental disturbance determine the distribution and availability of plants. The actual quantity of a particular plant present in the landscape at any given time may reflect either human gathering practices or a particular environmental variable. Rainfall, temperature, soil properties, insects, etc. will affect the condition of a plant population. The human population must adjust to this dynamic variability. From one time to the next alternative plants may be sought, or new plants may even be introduced to reduce risk from uncontrollable or unpredictable losses. By recognizing phytogeography, plant ecology, and the population biology of each species defined as important by the human culture under study, an understanding of the processual interrelations between humans and plants that Jones emphasized will finally be realized.

## RETROSPECTIVE

Even when the discipline lacked a name, a genre was recognized and was enhanced by the thoughtful observations of explorers, missionaries, naturalists, botanists, physicians, government officials, hobbyists, archaeologists, and ethnologists. Through time each contributed invaluable data that are impossible to collect today. Without this heritage ethnobotany would be unappealing and would lack a comparative framework both for cross-cultural studies of regularity and for an appreciation for culture change.

By continuing to redefine the field in the tradition of Powers, Harshberger, and Jones, ethnobotanists will keep the discipline vital and capable of addressing contemporary concerns and anticipating new horizons.

## REFERENCES

Anonymous

1895 Some New Ideas. Philadelphia Evening Telegraph. December 5, 1895.

Barrows, David P.

1900 The Ethno-Botany of the Coahuilla Indians of Southern California. Chicago: University of Chicago Press.

Bartram, John

1751 Descriptions, Virtues, and Uses of Sundry Plants of These Northern Parts of America, and Particularly of the Newly Discovered Indian Cure for Venereal Disease. Philadelphia.

Beauchamp, William M.

1905 Aboriginal Use of Wood in New York. New York State Museum Bulletin 89:87-272.

Berlin, Brent, Dennis E. Breedlove, and Peter H. Raven

1966 Folk Taxonomies and Biological Classification. Science 154:273-75.

1973 General Principles of Classification and Nomenclature in Folk Biology. American Anthropologist 75(2):214-42.

1974 Principles of Tzeltal Plant Classification. New York: Academic Press.

Brown, Robert

1868 On the Vegetable Products Used by the Northwest American Indians, As Food and Medicine, in the Arts, and in Superstitious Rites. Transcripts of the Botanical Society of Edinburgh (1866-68) 9:378-96.

Carter, George F.

1950 Ecology–Geography–Ethnobotany. Scientific Monthly 70(2):73-80.

Castetter, Edward F.

1935 Ethnobiological Studies in the American Southwest I: Uncultivated Native Plants Used as Sources of Food. University of New Mexico Bulletin, Biological Series 4(1):1-62.

1944 The Domain of Ethnobiology. American Naturalist 78:158-70.

Chestnut, V. K.

1902 Plants Used by the Indians of Mendocino County, California. U.S. National Herbarium Contribution 7(3):295-408.

Collier, Donald

1969 Chicago Comes of Age, the World's Columbian Exposition and the Birth of a Field Museum. Field Museum of Natural History Bulletin 40(5):2-7.

Conklin, Harold C.

1954 The Relation of Hanunóo Culture to the Plant World. Ph.D. Dissertation. Yale University.

Coville, Frederick V.

1895 Directions for Collecting Specimens and Information Illustrating the Aboriginal Uses of Plants. U.S. National Museum Bulletin 39:3-8.

1897 Notes on the Plants Used by the Klamath Indians of Oregon. U.S. National Herbarium Contribution 5:87-108.

1904 Wokas, a Primitive Food of the Klamath Indians. U.S. National Museum Report for 1902:725-39.

Cushing, Frank H.

1920 Zuñi Breadstuff. Indian Notes and Monographs 8:1-673.

De Candolle, Alphonse

1885 Origin of Cultivated Plants. New York: D. Appleton.

Densmore, Frances

1928 Uses of Plants by the Chippewa Indians. Bureau of American Ethnology 44th Annual Report (1926-27):275-397.

Dunbar, John B.

1880 The Pawnee Indians, Their History and Ethnology. Magazine of American History 4(4):241-281.

Fewkes, J. Walter
1896 A Contribution to Ethno-botany. American Anthropologist 9(1):14-21.
Gilmore, Melvin R.
1919 Uses of Plants by the Indians of the Missouri River Region. Bureau of American Ethnology 33rd Annual Report (1911-12):43-154.
1932 The Ethnobotanical Laboratory at the University of Michigan. Museum of Anthropology, University of Michigan, Occasional Contributions 1.
Guthe, Carl E.
1930 Identification of Botanical Material from Excavations. National Research Council, Division of Anthropology and Psychology, Circular No. 6.
Harrington, John P.
1932 Tobacco among the Karuk Indians of California. Bureau of American Ethnology Bulletin 94.
Harshberger, John W.
1896 Purposes of Ethnobotany. Botanical Gazette 21(3):146-54.
Havard, Valery
1895 Food Plants of the North American Indians. Bulletin of the Torrey Botanical Club 22(3):98-123.
1896 Drink Plants of the North American Indians. Bulletin of the Torrey Botanical Club 23(2):33-46.
Heer, Oswald
1866 Treatise on the Plants of the Lake Dwellings. In: The Lake Dwellings of Switzerland and Other Parts of Europe, by F. Keller, translated by J. E. Lee. London: Longmans, Green & Co.
Hough, Walter
1898 Houses Built of Fossil Trees. Plant World 1(1):35-36.
Jenks, Albert E.
1900 The Wild Rice Gatherers of the Upper Lakes: A Study in American Primitive Economics. Bureau of American Ethnology 19th Annual Report (1897-98): 1013-60.
Jones, Volney H.
1936 The Vegetal Remains of Newt Kash Hollow Shelter. In: Rock Shelters in Menifee County, Kentucky, by W. S. Webb and W. D. Funkhouser. University of Kentucky Reports in Archaeology and Anthropology 3:147-65.
1941 The Nature and Scope of Ethnobotany. Chronica Botanica 6(10):219-21.
Josselyn, John
1672 New-England's Rarities Discovered . . . London: G. Widdowes.
Kroeber, Alfred L.
1920 Review of Uses of Plants by the Indians of the Missouri River Region, by Melvin Randolph Gilmore. American Anthropologist 22:384-85.
Martin, Hugh
1793 An Account of the Principal Dyes Used by the American Indians. Transactions of the American Philosophical Society 3:222.
Mason, Otis T.
1886 The Ray Collection from the Hupa Reservation. Smithsonian Institution Annual Report:205-39.
1898 Untitled article in: Plant World 1(11):174.
Matthews, Washington
1886 Navaho Names and Uses for Plants. American Naturalist 20(9):767-77.
Mooney, James
1891 The Sacred Formulas of the Cherokees. Bureau of American Ethnology 7th Annual Report (1885-86):301-97.
Newberry, John S.
1887 Food and Fiber Plants of the North American Indians. Popular Science Monthly 32:31-46.
Palmer, Edward
1871 Food Products of the North American Indians. U.S. Commissioner of

Agriculture Report 1870:404-28.
1878 Plants Used by the Indians of the United States. American Naturalist 12: 593-606, 646-55.

Parker, Arthur C.
1910 Iroquois Uses of Maize and Other Food Plants. New York State Museum Bulletin 144:1-119.

Powers, Stephen
1873-1875 Aboriginal Botany. California Academy of Science Proceedings 5:373-79.

Robbins, Wilfred W., John Peabody Harrington, and Barbara Freire-Marreco
1916 Ethnobotany of the Tewa Indians. Bureau of American Ethnology Bulletin 55.

Ross, Bernard R.
1862 An Account of the Botanical and Mineral Products Useful to the Chipewyan Tribes of Indians. Canadian Naturalist and Geologist 7:133-37.

Rush, Benjamin
1774 An Inquiry into the Natural History of Medicine Among the Indians of North America, and a Comparative View of the Their Diseases and Remedies with Those of Civilized Nations. Philadelphia: American Philosophical Society.

Safford, William E.
1917 Narcotic Plants and Stimulants of the Ancient Americans. Smithsonian Institution Annual Report 1916:387-424.
1927 Our Heritage from the American Indians. Smithsonian Institution Annual Report 1926:405-10.

Schoolcraft, Henry Rowe
1851 Historical and Statistical Information Respecting the History, Condition and Prospects of the Indian Tribes of the United States. 6 vols. Philadelphia: Lippincott, Grambo, and Co.

Stearns, Robert E. C.
1889 Ethno-Conchology: A Study of Primitive Money. U.S. National Museum Annual Report 1887:297-334.

Stevenson, Matilda Coxe
1915 Ethnobotany of the Zuñi Indians. Bureau of American Ethnology 30th Annual Report (1908-09):35-102.

Sturtevant, William C.
1964 Studies in Ethnoscience. In: Transcultural Studies in Cognition, edited by A. K. Romney and R. G. D'Andrade. American Anthropologist 66 (Pt. 2): 99-131.

Taylor, Lyda Averill
1940 Plants Used as Curatives by Certain Southeastern Tribes. Cambridge: Botanical Museum of Harvard University.

Teit, James A.
1930 Ethnobotany of the Thompson Indians of British Columbia. Bureau of American Ethnology 45th Annual Report (1927-28):441-522.

Trimble, Henry
1888-1891 Some Indian Food Plants. American Journal of Pharmacy 60:593-95; 61:4-5, 556-58; 62:281-82, 598-600; 63:525-27.

Vestal, Paul A., and Richard Evans Schultes
1939 The Economic Botany of the Kiowa Indians as It Relates to the History of the Tribe. Cambridge: Botanical Museum of Harvard University.

Vogel, Virgil J.
1970 American Indian Medicine. Norman: University of Oklahoma Press.

Whiting, Alfred F.
1939 Ethnobotany of the Hopi. Musuem of Northern Arizona Bulletin 15.

# ETHNOSCIENCE OF THE SIXTEENTH-CENTURY VALLEY ZAPOTEC

*Joyce Marcus and Kent V. Flannery*
Museum of Anthropology, The University of Michigan

It is a pleasure to be able to write a paper that honors not one, but two distinguished ethnobotanists. Our primary intent, of course, is to honor Volney H. Jones, who has been a friend and colleague to the senior author for 6 years and to the junior author for 12 years. Our selection of sixteenth-century Zapotec ethnoscience as a topic, however, brings us directly in contact with another pioneering ethnobotanist: Fray Juan de Córdova (1503-1595), whose studies of the Zapotec language have never been surpassed. As these lines are being written, we have reached the four hundredth anniversary of the publication of Córdova's two major works (1578a, 1578b).

Córdova was born in Spain nearly two decades before Cortez landed in Veracruz. After serving Carlos V in the battles of Flanders and Vienna, he traveled to New Spain as part of Vásquez Coronado's Cíbola expedition. He joined the Dominican order in Mexico City during 1543-1544 and was assigned to their convent in Oaxaca a few years later. There he served under Fray Bernardo de Albuquerque, who had already mastered Valley Zapotec (Jiménez Moreno 1942:9).

During this period Cosijopi, the last Zapotec *coqui* or hereditary ruler, had been baptized "Don Juan Cortés" by the Spanish and was living in Tehuantepec. Between 1561 and 1564 he was discovered practicing "idolatrous rites"–presumably the same rituals Zapotec lords had always been expected to perform for the welfare of their subjects–and was denounced by the Inquisition. Bishop Albuquerque sent to the investigation his two best Zapotec language experts, Fray Juan de Mata (author of the *Relación de Teozapotlan* [1580], an account of Zaachila which contains

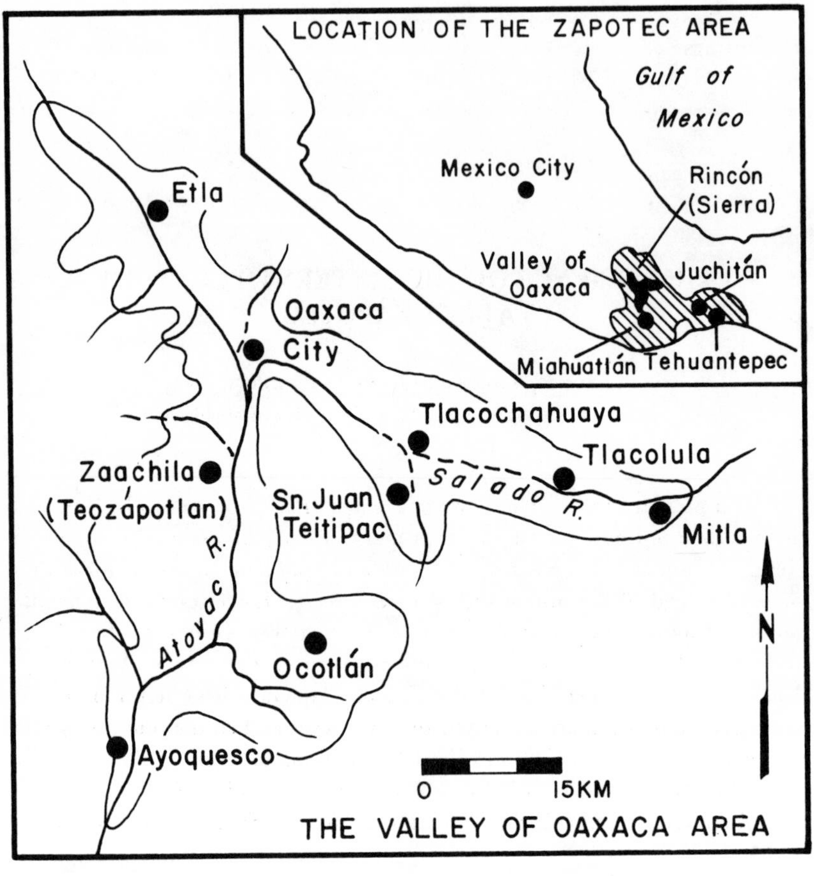


Fig. 1. Map of the Valley of Oaxaca and the Zapotec area of Mexico, showing localities mentioned in the text.

important ethnohistoric data) and Fray Juan de Córdova. Perhaps it was in the course of questioning Cosijopi (who was eventually freed) that Córdova obtained the first glimpses of Zapotec cosmology, religion, and ritual which are reflected in his later writing. While he never understood them fully, preconditioned as he was by a Greco-Roman model of idolatrous polytheism, some of his insights are remarkable.

In 1570, Córdova was assigned to the convent of Teticpac (now San Juan Teitipac) in the eastern, or Tlacolula arm of the Valley of Oaxaca (Fig. 1). Dividing his time between Teticpac and Tlacuechahuaya (now Tlacochahuaya de Morelos, 5 km northeast of Teticpac), he plunged into his writings on Zapotec grammar and vocabulary. His dictionary was meticulously edited by Fray Domingo Grijelmo of Ocotlán and Fray Juan de Villalobos of "Taneche y Zapotecas del Rincón [Sierra]," using four Zapotec informants (Jiménez Moreno 1942:11-12); in addition to increasing its accuracy, this editing may well have broadened the book's coverage of Zapotec dialects. Córdova's works were published in 1578; during 1579-1581, while many other friars were engaged in answering the questionnaires that led to the famous *Relaciones Geográficas* (del Paso y Troncoso 1905), he was vicar of Teitipac. In 1595 he was buried in the Convento de Santo Domingo in Oaxaca City—a building now, appropriately enough, converted to a museum of anthropology.

Fray Juan de Córdova and Volney Jones share a number of characteristics. Both spent many of their formative years out of doors in a land of semiarid valleys, rocky canyons, and rugged mountains. Both contributed to our understanding of the American Southwest, Córdova on the Cíbola expedition and Jones with the ethnobotany of Isleta Pueblo and Jémez Cave. Both made trips to tropical America early in their careers, Córdova to southern Mexico and Jones to Haiti. Both showed an early fascination for the way American Indians classified the world and its vegetation. And Jones, as did Córdova some 400 years ago, continues to do vigorous, important, and original research at an age when most anthropologists have long since settled back into their orthopedic deck chairs at Leisure World. We dedicate this paper to both of them.

## THE ZAPOTEC AND THE VALLEY OF OAXACA

Today there are an estimated 200,000-300,000 speakers of Zapotec (Nader 1969), the generic term for a series of at least seven mutually intelligible "dialects" or "languages" spoken in the state of Oaxaca, Mexico (Pickett 1967:292). Currently-available archaeological data make it seem likely that the Precolumbian Zapotec state had its origins in and around the Valley of Oaxaca, which is where most of Córdova's work was done (Flannery and Marcus 1976). However, the Zapotec also colonized the Sierra de Juárez (relatively early in their history) and the Isthmus of Tehuantepec (relatively late in their history). We will deal only with sixteenth-century Valley Zapotec, but comparisons will occasionally be made with twentieth-century word lists from Juchitán in the Isthmus of Tehuantepec (Pickett 1959, 1967), Ayoquesco in the extreme southern end of the Valley of Oaxaca (MacLaury

1970), and Mitla in the extreme eastern end of the Valley (Messer 1975, 1978).

The Valley of Oaxaca lies in the southern highlands of Mexico at an elevation of 1500 m (Fig. 1). It is a Y-shaped valley produced by the Río Atoyac and its tributary, the Río Salado, and contains some 700 $km^2$ of flat land surrounded by forested mountains which rise to 3000 m. The climate is temperate to semitropical, with 500-700 mm of annual rainfall concentrated mainly in the May-November period.

In considering Zapotec ethnoscience, we should bear in mind two things: their environment and their linguistic origins. Zapotec belongs to the Otomanguean family, a group of languages (Chatino, Mixtec, Cuicatec, Otomí, Chinantec, and others) which have a common origin far back in time (Swadesh 1967). Zapotec may thus share a number of basic classificatory principles with other Otomanguean languages (Marcus, Flannery, and Spores, n.d.). However, the Valley of Oaxaca has its own particular complex of plants and animals which sets it apart from other Otomanguean regions; it is a large, diverse montane valley with a variety of niches created by differences in soil, topography, temperature, rainfall, and human modification. Zapotec taxonomy therefore draws on its linguistic roots but adapts to its specific environmental situation as well.

## RESEARCH OBJECTIVES AND METHODOLOGY

In this paper we will attempt to reconstruct the way the Zapotec Indians of the Valley of Oaxaca described and classified their world during the period A.D. 1578-1581. Since the Zapotec had first been contacted by Europeans less than 60 years previously, this comes as close as we are likely to get to a reconstruction of their classificatory scheme during Precolumbian times, which is the period of greatest interest for the present authors. In addition to the works of Córdova, we will draw on the *Relaciones Geográficas* of 1578-1581 (del Paso y Troncoso 1905).

Clearly, both our research objectives and our methodology are somewhat different from the studies of folk botany and zoology conducted recently in the neighboring state of Chiapas (Berlin, Breedlove, and Raven 1974; Hunn 1977). The goals of those authors were to understand how twentieth-century Indians classify plants and animals, and their method was to interview living Indians. Our ultimate goal is to understand how Precolumbian Indians classified their world, though we are forced to settle for Indians recently contacted by Europeans; our method is to examine the words of the Indians who were interviewed 400 years ago by those same Europeans.

The use of sixteenth-century dictionaries and ethnohistoric documents has both rewards and limitations. One of its limitations is the fact that the recorded data are finite. If an author forgot to collect a word, we cannot go back and get it, nor can we ask about subtle nuances of meaning. Above all, we have to deal with not one, but two, foreign cultures: although our aim is to discover how the Zapotec classified an item, we first have to discover how a sixteenth-century Spaniard would have classified it. A knowledge of modern Mexican Spanish helps, but is not enough; one must also know archaisms such as "adive," "raposa," and "leonado," and realize that the bird he seeks may not be listed as a "paloma" but as a "torcaz" or a "tórtola." One must also be alert to errors introduced by the different structure of Spanish and Zapotec, as evidently happened when Córdova asked one of his informants " ¿Como dicen Vds. 'una red'?" Since Zapotec has no indefinite article, the informant evidently misunderstood and answered "*tòo quixe*"–not "a net," but "one net" (Córdova 1578a:346).

Some of the greatest rewards of using the sixteenth-century documents include ritual, religious, and cosmological data which were more intact in the 1570s and 1580s (Marcus, in press). Today the Precolumbian religion has all but disappeared, and even the most meticulous ethnologist will find only occasional hints. To be sure, the Zapotec still revere their ancestors, the *binigulaza* or "old people of the clouds," but in the valley today virtually nothing remains of the hierarchy of priests (*uéza-eche*, *cópa pitào*, and *bigaña*), the personal bloodletting, the human and animal sacrifice, the cannibalism, the reciprocal relationship with lightning and the supernatural, the deified royal ancestors, and the ritual use of *Datura*, tobacco, and hallucinogenic mushrooms which fascinated the writers of the sixteenth century. At the same time, we must bear in mind that many of the writers were priests who viewed Zapotec religion through somewhat less than objective eyes, since their assignment was to stamp out the very rituals we would most like to know about.

Any reconstruction is a hypothesis that must be tested against other data. In this case, we have compared our reconstruction of the sixteenth-century supernatural world with the seventeenth-century writings of Fray Francisco de Burgoa (1670, 1674) and Gonzalo de Balsalobre (1656). Our reconstruction of animal terms has been compared with Pickett's dictionary of Isthmus Zapotec and MacLaury's Ayoquesco word list; our reconstruction of plant terms has been compared with Pickett, with MacLaury, and with Messer's ethnobotanical work at Mitla. In addition, we have been greatly aided by Ernesto Martínez and Pablo García of Mitla, two native Zapotec speakers who have worked for the University of Michigan's Oaxaca Project for 10 years.

Obviously, when we find that our reconstruction matches the analysis of another author, it renews our confidence. On the other hand, it would be unrealistic to expect a complete overlap between the Zapotec of 1578 and the Zapotec of 1978. Consider, for example, what has happened to English since 1578, when Elizabeth I was queen and William Shakespeare was 14 years old. As a matter of fact, another of the rewards of using the Colonial documents is the glimpse they afford us of the way the Zapotec language has changed since the sixteenth century, not merely in pronunciation but in the meanings of words as well. Like most languages, it was changing even as Córdova wrote; for example, after giving a series of common terms for "male," he added the less common term *pèxo*, and explained, "este pèxo es nombre antiguo y específico y no le alcançan todos" (Córdova 1578a:252). For similar reasons, the writers of the *Relaciones Geográficas* often stated that they had relied on the oldest Indian informants available because the young Zapotecs no longer knew the old customs.

Consider the term for "river" in Valley Zapotec in 1578, *quégo* or *quéco*. In Juchitán today it is still *guiigu'* (Pickett 1959), but in the Valley of Oaxaca the second consonant has been dropped so that "river" is now *gèw* in Ayoquesco (MacLaury 1970), *gèw* or *yèw* in Mitla. Since the term was still recorded as *kigo* in the Valley of Oaxaca in the late 1800s (Pimentel 1875), this change in pronunciation must literally have taken place within the last century. In a later section, we will suggest how the introduced Spanish concepts "pasto" and "yerba" may have modified the meaning of the sixteenth-century Zapotec concepts *quijxi* and *nocuana* over the last 400 years.

## PROBLEMS OF ORTHOGRAPHY

Obviously, there are great differences in the orthography of the various works involved. Córdova was not a linguist, so he tried to transcribe an unwritten language, and a tonal language at that, in ways that sixteenth-century Spaniards could comprehend. MacLaury, on the other hand, used linguistic transcription, complete with tones and glottal stops; Messer recorded all but the tones. Pickett was preparing a dictionary to be used by Mexican laymen, so her orthography was aimed at speakers of Spanish. Thus the gloss for "weed" or "inedible herb" was written *quijxi* by Córdova, *guixhi* by Pickett, *gîẓ̌* by MacLaury, and *giš* by Messer. We have left all spellings exactly as given by the original author, including Córdova's enigmatic and occasionally inconsistent accent marks. Ideally, we would like true linguistic transcriptions for all words, but if we tried to do this for Córdova or the *Relaciones Geográficas* we would only be guessing.

In our opinion, much of the apparent divergence between sixteenth-century and twentieth-century Zapotec may be a function of these differences in orthography. The sound that a modern linguist would record as "b" (as in *beni* or *binni*, "human being") would almost certainly have been

| English Term | Sixteenth-Century Valley Zapotec (Córdova 1578a) | Twentieth-Century Juchitán Zapotec (Pickett 1959) | Twentieth-Century Mitla Zapotec[1] |
|---|---|---|---|
| human being | péni | binni | behn |
| animal | máni | mani' | mahn |
| wild feline | péche | beedxe' | betz |
| mountain | tàni | dani | dahn |
| river | quègo | guiigu' | géw, yéw |
| tree | yàga | yaga | yahg |
| fruit | nocuana | cuana[2] | $k^{w}an$ |
| maguey | tòba | duba | dohb |
| organ cactus | pichij | bidxí | bidz |
| weed | quijxi | guixhi | giš |
| cotton | xilla | xiaa | žil |
| squash | quéto | guitu | gït |

[1] Animal and environmental terms collected by the authors; some plant terms collected by the authors, others by Messer (1975).
[2] Juchitecos also use the term *cuananaxhi* (cf. sixteenth-century Zapotec *nocuananàaxi*, "sweet fruit").

Fig. 2. A comparison of plant, animal, and environmental terms in sixteenth- and twentieth-century Zapotec, showing the trend toward monosyllabic nouns in Mitla.

written with a "p" by Córdova (as in *péni*, "human being"), since the sixteenth-century Spanish "b" was considered more like a "v." Similarly, the modern linguist's hard "g" (as in *guitu*, "squash") might well have been written "qu" by Córdova (as in *quéto* or *quèeto*, "squash"). When such orthographic differences are factored out, modern Juchitán Zapotec (as recorded by Pickett) looks remarkably similar to the Zapotec of Córdova—much more similar than the Mitla dialect, which seems to have gone its own way, making twentieth-century monosyllables out of two-syllable sixteenth-century words (e.g., *behn* instead of *beni; gït* instead of *guitu*; *mahn* instead of *mani.* See Fig. 2).

## THE CONCEPT OF PÈE

At the root of all sixteenth-century Zapotec classification of the world was the concept of *pèe.* Variously translated as "wind," "breath," "spirit," or "ánima, lo que da vida," *pèe* was the vital force that made all living things move and thereby distinguished them from nonliving matter. This was probably a very ancient concept, for it is shared not merely by other Otomanguean speakers, but by less closely related Mesoamerican peoples. According to Ronald Spores, the Zapotecs' sixteenth-century Mixtec neighbors believed in a similar vital force—*yni* or *ini*—which has been translated "spirit," "heart," and "heat" (Marcus, Flannery, and Spores, n.d.).

While inanimate objects could be manipulated by Zapotec science and technology, anything that possessed *pèe* was deserving of respect—indeed, in many cases it was treated as sacred. All animals, great and small, had *pèe*; it is not yet clear to what extent plants did, although *pèe* (as "wind") had the power to make them move, and certain plants (primarily hallucinogens) had enough *pèe* to put one in touch with the supernatural. The sacred wind would also bring inanimate things temporarily to life: a rushing, flooding river became *quécopèe* ("living river"), and the effervescent foam (*pichijna*) on a cup of stirred hot chocolate was considered alive (Marcus, in press).

The most respected of all living things were a series of great forces which we would call "natural," but which for the Zapotec were supernatural beings. Apart from the wind itself, these included clouds, lightning, and earthquakes, all clearly alive because they moved and could move other things. Lightning (*cocijo*) split the clouds (*zàa*) and caused rain (*niça quie*) to descend; thunder was *xòo cocijo*, "lightning's earthquake." *Cocijo* was one of the most powerful supernatural forces, and the Zapotec frequently addressed him as *pitào cocijo*, an expression the sixteenth-century Spanish erroneously translated as "the God of Rain" (Flannery and Marcus 1976). In fact, *pitào* is *pè* + augmentative, and *pitào cocijo* is more accurately translated "Great Spirit (or Breath, or Wind) within the Lightning."

*Zàa* was important to the Valley Zapotec because they had descended from the clouds and called themselves *peni-zàa*, "the cloud people." Their ancestors, "the old people of the clouds" were revered because they had returned to their place of origin and could intercede with the supernatural on behalf of their descendants. Royal ancestors were particularly revered, with sacred images and sacrifices made to them. Because the sacred images ("idols") were sometimes addressed as *pitào*—and because the Spanish had mistranslated this as "god"—the Zapotec have erroneously been considered to have had a pantheon of gods, when, in fact, their view of the supernatural was an animatistic one (Marcus, in press).

The Zapotec did believe that lightning, clouds, people, animals, plants, and the whole world had been created by one supreme being—"he who is without beginning or end, who created everything but was not himself created"—but this supreme being was incorporeal; no images were ever made of him and no mortal man ever came in contact with him. Man dealt with lightning on a regular basis, as well as with a whole series of minor spirits such as *huichàa* ("ghosts," "duendes") or *xini pitào* ("offspring of the Great Spirit").

The Spanish did their best to eradicate this part of Zapotec cosmology, and today there remain only vestiges. The term *pitào* has survived among the

Isthmus Zapotec as *bidu*, among the Valley Zapotec as *bido'o* or *bido. Cocijo* has survived as *gusíu* or *gusiy'*, "rayo," "lightning bolt." *Pèe* has become *bi* in Juchitán, *be* in the Valley of Oaxaca; it is also a kind of "animal classifier," the most common phoneme employed as a prefix to an animal name, just as it was in the sixteenth century when every animal had *pèe*.

## CATEGORIES OF TIME

The Precolumbian Zapotec, like many other Mesoamerican peoples, had two calendars: one secular and one ritual. In the sixteenth century, the 365-day secular calendar was called *yza*. Although the Zapotec did count months or "moons" (*pèo*), the essential subdivision of the *yza* was the season, called *cocij*, "lightning." In Juchitán, these words have remained very similar to their ancient counterparts–*iza* for year, *beeu* for month or moon, *gusi* for season (Pickett 1959).

The sixteenth-century year could be divided several ways, perhaps the most common being the contrast between a dry season, *cocijcobàa*, and a rainy season, *cocijquije* (from *niçaquie* or *niçaquije*, "rain"). These terms survive in Juchitán as *gusi ba* and *gusi guie* (Pickett 1959), while in Mitla, where the Spanish word "tiempo" has been adopted, the only vestige of the ancient system is the term *tiemp gusgih* for "rainy season" (Messer 1975).

In addition to the basic dry season/rainy season contrast, the sixteenth-century Zapotec used *cocij* in a manner analogous to the Spanish term "tiempo," to indicate informal subdivisions of the year during which important activities took place. The maize-planting season was *cocijxòopa* (from *xòopa*, "corn kernel"), the harvest season *cocijlayña* or *cocijcolläpa.* The Zapotec also made distinctions between a "dry year" (*yzacobàa*) and a "rainy year" (*yzaquie, yzaquije*).

The 260-day ritual calendar was called *pije* or *piye*, a term whose initial phoneme suggests that it had *pèe*; ritual or sacred time was alive, it moved, and its calculations were in the hands of a group of ritual specialists called *colanij*, "diviners." The term *cocijo* was also used for units of the *pije*, but in this case the division was into four *cocijo* of 65 days each. In turn, each 65-day *cocijo* was divided into five units, called *cocii*, of 13 days each (Córdova 1578b:202). Each "day" (*chij*) of the *pije* had its own number and name, usually an animal or a natural force; a *chij* began at midday and ran until the next midday (ibid.:212). Figure 3 depicts the *pije* with its 260 *chij*, 20 *cocii*, and 4 *cocijo.*

The four *cocijo* were also called *pitào* ("great spirits"), perhaps a reference to the four lightnings which resided in the four quadrants of the Zapotec world. The *cocijo* or *pitào* were said to cause all events that occurred (Córdova 1578b:202). To these *cocijo* the Zapotec offered sacrifices, including blood from various parts of their own bodies (ears, thighs, tip of the tongue, and so forth). Each of the 260 *chij* had its own corresponding fortune, benevolent or malevolent. The day names and numbers were used to name newborn children, as well as to determine the feasibility of marriage for a particular couple; the *colanij* decided whether the combination of names and numbers augured well for a wedding.

Most 260-day ritual calendars in Mesoamerica had 20 different day names which combined with the numbers 1-13 (Marcus 1976a). However, there are interesting differences between the Zapotec *pije* and the Maya 260-day ritual calendar. For the Maya (as recorded in the codices Dresden and Madrid), the 260 days were usually divided into five units of 52 days; these 52-day units were in turn divided into "irregular intervals" (Thompson 1950:101). By contrast, the Zapotec division into four 65-day units, each subdivided into five 13-day units, was more uniform.

## THE ORGANIZATION OF THE ZAPOTEC COSMOS

The sixteenth-century Zapotec, like virtually all other Mesoamerican Indians, believed that the cosmos was rectangular; that it was divided into four great world quarters; that each quarter was associated with a color, sometimes with a fifth color for the center; and that the main axis along which the cosmos was divided was the east-west path of the sun.

Among the sixteenth-century Valley Zapotec, for example, "east" was *çooche lani copijcha*, "where the sun rises"; "west" was *çootiace copijcha*, "where the sun sets" (Córdova 1578a). The four cardinal directions were also associated with the four quarters of a day, and time moved clockwise from east (morning) to south (midday), then on to west (afternoon) and north (night). Among the twentieth-century Juchitán Zapotec, "east" is *neza ridani gubidxa*, "road of the rising sun"; "west" is *neza riaazi gubidxa*, "road of the descending sun" (Pickett 1959).

While most Mesoamerican Indian groups viewed the world in terms of four quarters, the association of a specific color with a specific quarter was regionally variable. Perhaps best known is the Maya scheme, in which east was red, west black, north white, south yellow, and the center blue-green (Marcus 1973); among the Nahua speakers the colors—red, black (or yellow), white, and blue-green—were used, but the directional associations varied (Nowotny 1970).

It is very difficult to reconstruct the color associations of the sixteenth-century Zapotec, although some vestiges have persisted to the present day. In the Zapotec sierra south of Miahuatlán, several communities studied by Weitlaner and de Cicco (1962) still believe in four "lightnings" that reside on certain hills oriented to the major world directions. These four directions

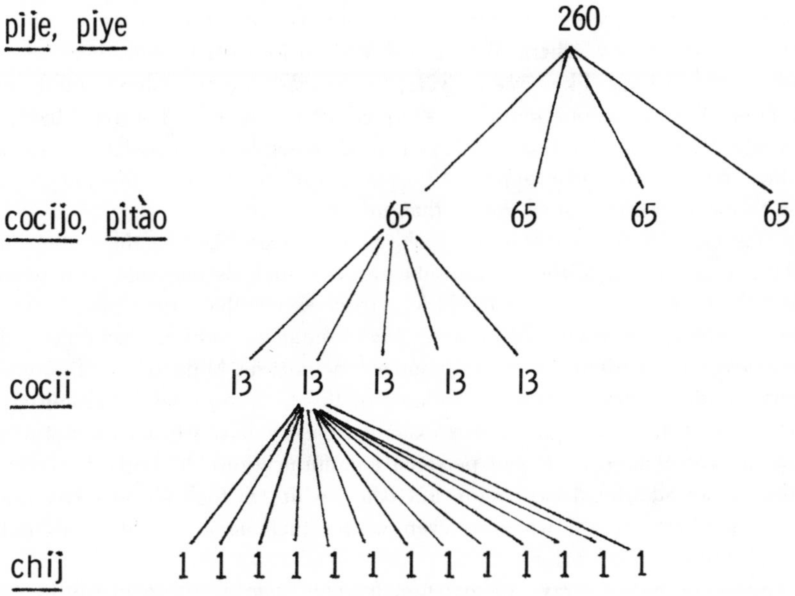


Fig. 3. Diagram of the Zapotec 260-day ritual calendar, with its division into 4 *cocijo* of 65 days, each composed of 5 *cocii* of 13 days (*chij*).

are associated with colors, and these in turn are symbolized during ritual by the use of five kinds of flowers, as follows:

| | | |
|---|---|---|
| east | – | black flower |
| west | – | white flower |
| north | – | green flower |
| south | – | yellow flower |
| midday (center?) | – | blue flower |

While we are happy to have data on this ritual from Miahuatlán, it is nevertheless an acculturated ritual, as indicated by the fact that blue and green are distinguished; the ancient Zapotec, like most Mesoamerican Indians, regarded blue-green as a single color. Let us therefore look briefly at the ancient Zapotec color categories.

Córdova's sensitivity to sixteenth-century Zapotec comes across in his discussion of color terms. Realizing that his informants had no word for "color," he explained to his readers that

> Color como dezimos nosotros de que color es no tienen los yndios sino dizen, de que parecer o presencia es. [Córdova 1578a:80]

We are convinced that Córdova's informants had only six primary color categories, all of which began with *na*: *nagàce*, *naquichi*, *natèe*, *naxiñàa*, *nagàche*, and *nagàa*. There were also shortened forms that could be used as adjectival suffixes like *-yàce*, *-yàte*, or *-yàche*. Some colors could be combined with various modifiers such as *còhui*, "dark," or *huixi*, "light," though it is doubtful that Córdova lists all possible variants. He mentions other terms, but most appear to be attempts to describe the hue of a particular flower or animal, rather than primary colors.

*Nagàce*, "black," survives in Juchitán as *nayaase'* and in the Valley of Oaxaca as *nagase*; Mitleños use subcategories such as *nagaceti*, "un poco negro" or *nagas-bo* "charcoal black" (from sixteenth-century *pào*, "charcoal," which survives in Juchitán as *buu*). *Naquichi*, "white," has remained *naquichi'* in Juchitán but is pronounced *naquitz* in Mitla today. In some parts of the Valley of Oaxaca, variants of this term are used for absolutely white objects; in Ayoquesco, for example, *nàgǎty* is as white as something can get (MacLaury 1970 and personal communication). In Mitla, however, there is an additional expression *nǒl beh*, "white as fog" (from sixteenth-century *pèye*, "fog"), which is even whiter than *naquitz*, but seems not to refer to a primary color.

*Natèe* or *natèo* "gray" (*ceniziento*, *leonado*, *pardo*, *bruno* in Córdova's dictionary) is still *naté* in Juchitán but *natah* in Mitla. *Naxiñàa*, "red" is still *naxiñá'* in Juchitán but has several variants in the Valley of Oaxaca. In Ayoquesco it would be *žnyê* (MacLaury 1970:17), while Mitleños would say *našnyâ* or *šnyâ*. Frequently, "red" will be compared to an object, as in *šnyâ-řehn*, "as red as blood."

*Nagàche* or *nacòche*, "yellow," survives in Juchitán as *naguchi* and in the Mitla region as *nagotz*. In the latter area it may be modified as in *gotz-ya*, "bright yellow" or *gotz-guí*, "yellow as a flower" (their standard may be the bright yellow flowers of *Cassia* sp.). *Nagàa*, "blue-green," was modified in the sixteenth century by the addition of adjectives, as in the case of *nagàa còhui*, "dark blue-green," or *nagàa huixi*, "light blue-green." The modern term is *nagá'* in Juchitán and *nagaa* in Mitla. When asked to describe the color of a Precolumbian jade object, Mitleños use *nagaa-gu*, "strong blue-green."

Other color expressions used by Zapotec speakers today include obvious loan words (*zul*, from the Spanish azul, "blue") as well as native words whose history we have been unable to trace back to the sixteenth century. One example would be the shade the Juchitecos call *mexu* and the Mitleños call *meš* ("güero," "blond," "fawn-colored"). Another would be the Mitleño *bisyao*, "speckled." We consider none of these to be primary color categories.

## THE EARTHLY LANDSCAPE

The rural sixteenth-century Zapotec landscape was defined in terms of several dichotomies—the village vs. the surrounding countryside, the cultivated field vs. the wilderness, the mountain vs. the valley plain. Rivers, roads, and certain named mountain peaks served as landmarks. We also suspect, but cannot prove, that each village would have been surrounded by a zone of several kilometers' diameter where virtually every hill, knoll, arroyo, or rock outcrop would have had a toponym, as Schmieder (1930) demonstrated for Mitla almost half a century ago. Indeed, in the Valley of Oaxaca toponyms can be traced back at least as far as the "place glyphs" of 100 B.C.-A.D. 100 (Marcus 1976b).

The village was *quéche*, and around it were the cultivated corn fields (*quèela*) which had been wrested from the wilderness. The wilderness in turn might be glossed as *láoquijxi* ("countryside") or *quijxitào* ("dense monte"); in either case, the critical term was *quijxi*, "inedible weed," "monte," "yerba en general," suggesting that wilderness was defined on the basis of wild vegetation. Beyond the level plain (*láche* or *làache*) lay the mountain (*tàni*). Both *-táni* and *-quijxi* were also used as suffixes to indicate that a particular race of animal was wild (see below).

Most of these terms persist in modern Zapotec, e.g., *guidxi* (Juchitán) or *giǰ* (Valley of Oaxaca), "village"; *guela* (Juchitán) or *gähl* (Valley of Oaxaca), "cornfield"; *lač*, "level plain"; and *dani*, *dan* or *dayn*, "mountain." However, terms such as *láoquijxi* have generally been replaced by the Spanish loan word, "campo."

Another major landmark for the Valley Zapotec was the river—usually the Río Atoyac, the Río Salado, the Río Mixtepec, or a major tributary. A large river might be referred to as *quégo* or *quéco*, a narrow stream as *quécolàce* or merely *niça*, "water." A river which dried up during the dry season might be referred to as a "seasonal river," *quécopijchi* or *quécolayña* (both *pichijta* and *layña* were also used to specify plants which were available, or "harvested," only during one season per year). By contrast, as we have already mentioned, a full, rushing river was *quécopèe* (alive, moving, possessing *pèe*).

These landmarks gave the Zapotec directional terms which were the terrestrial equivalent of "the road of the rising (or descending) sun." Localities were described as "upriver" (*quiaço quégo*), "downriver" (*quéteçòo quégo*), "in front of the mountain" (*làotàni*), "behind the mountain" (*xìchetàni*), and so forth. MacLaury (1970) reports similar expressions in twentieth-century Ayoquesco Zapotec, e.g., *lō dàyn*, "in front of the mountain" (*lu dani* in Juchitán).

## THE ANIMAL WORLD

The Zapotec divided animals (*màni*) into several broad categories, including creatures walking on four legs (*màni tizàa cotàa*), aquatic animals (*màni niça*) and birds (*màni zàbi* for large birds, *màni piguijni* for small ones). Most names for animals began with a *pe* or *pi* sound, perhaps an animal classifier but just as likely a reflection of the fact that animals had *pèe* because they were alive. Animals were further divided into classes such as *pèche* ("fierce" animals like felines, canids, etc.), *père* (gallinaceous birds), *pèeche* (toads and frogs), and *pèlla* (fish, snakes, worms), which in turn might be divided into still smaller categories by the use of an adjective or suffix.

A glimpse at the most common adjectives or suffixes used communicates the distinctions that were important to the Zapotec. Suffixes such as *-tào* ("large"), as opposed to *-láce* ("thin") or *-huini* ("small"), distinguished related animals by size. Modifiers such as *-quijxi* ("of the monte") or *-tàni* ("of the mountain") identified wild races of animals or plants; in 1578, such races were frequently contrasted with those "from Castille." Interestingly enough, color seems to have been an almost insignificant criterion for distinguishing animals; it was much more important to specify their habitat. Thus *yàga* ("of the trees"), *niça* ("of the water"), or *yòo* ("of the earth") were important modifiers.

A few examples will serve to show how the system worked. A land turtle was glossed *pègo*; a water turtle was *pègo niça.* A dog was *pèco*; an otter was *pèco niça,* "water dog." A female turkey was *père*, "hen," until the arrival of the Spaniards; after that, the native turkey became *père* [*péni*] *zàa*, "Zapotec hen," while the European domestic chicken became *père castilla.* The generic term for dove, *cògo*, has survived in Juchitán as *guugu* (López Chiñas 1937); there it is applied mainly to the white-winged dove, *Zenaida asiatica.* The smaller mourning dove (*Zenaidura macroura*) is referred to as *guugu huini*, "small dove," and the white-fronted dove (*Leptotila verreauxi*) is described appropriately as *guugu yuu*, "ground-dwelling dove." The corresponding terms in Córdova's time would presumably have been *cògohuini* and *cògoyòo.*

A fairly large category was that of *máni péche*, "fierce wild animals" (see Fig. 4). We know from modern zoological studies how many native canines and felines there are in Oaxaca, but unfortunately Córdova did not, and his confusion is all too clear. He did establish that Oaxaca had no wolves, but he lists such a bewildering assortment of "jackals," "vixens," and "foxes" that one can only sort them out by reference to their twentieth-century cognates. The ancient Zapotec did not draw the same line we do between canids and felids, and their Spanish-speaking twentieth-century descendants still refer to the gray fox as a "gato montés."

Largest and fiercest animal in this category was the *péchetào* ("big *péche*") or jaguar (*Felis onca*). The average Valley Zapotec probably never saw a live jaguar, but the animal was important in his ritual, cosmology, and calendrics, and its hide was used by his rulers. He would have been much more familiar with the *péchepiáha* or puma (*Felis concolor*), which was abundantly represented in Zapotec art. Ironically, archaeologists have spent a great deal of time arguing whether a particular Precolumbian representation is that of a jaguar or a puma, when both would have been called by the same generic: *péche*.

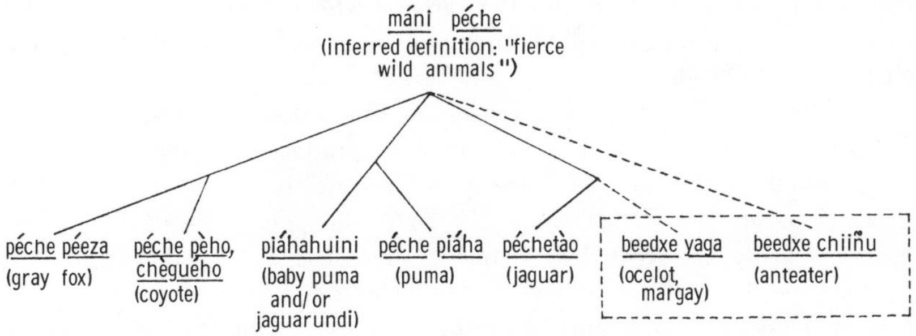


Fig. 4. Subdivisions of *máni péche*. All terms are in sixteenth-century Valley Zapotec except those indicated by dashed lines, which are from twentieth-century Juchiteco Zapotec.

Names of the smaller cats are more ambiguous. *Péche* has survived in Juchitán as *beedxe* (Pickett 1959), where the Zapotec use it to refer to a "tree-dwelling" species, *beedxe yaga*; this could be either the ocelot (*Felis pardalis*), the margay (*Felis weidii*), or both. In Chiapas, zoologist A. Starker Leopold (1959:473) found the margay referred to as *pichigueta*, a name which is distinctly non-Maya and may be a corruption of Zapotec (*péchequéta*?). Córdova also lists the terms *xínipiáha* ("puma's offspring") and *piáhahuini* ("small puma") for an animal he calls a *leoncillo*. This probably refers to a young puma, but in southern Mexico the term *leoncillo* is also used for the jaguarundi (*Felis yagouaroundi*), an animal which at a distance could easily be mistaken for a miniature puma. At any rate, their cat terminology illustrates how the Zapotec treated increasing levels of specificity. We begin with *máni péche*; *máni* drops out as we specify *péche piáha*; *péche* drops out as we specify *piáha huini*.

But we are still not finished with *péche*. The larger of Córdova's two "jackals" is *péche pého* or *chégueho* (possibly a contraction of *péche guého*). This must be the coyote (*Canis latrans*), since it is called *guéu'* in Juchitán and the Valley of Oaxaca today. Thus the smaller "jackal" or "vixen," *péche pèeza*, must have been the gray fox *Urocyon cinereoargenteus*. We have not been able to determine whether any other small carnivores belong to the *péche* or *beedxe* category; it seems not to have included the Procyonids, although more research is needed. At Juchitán today, however, the collared anteater (*Tamandua tetradactyla*) is classified as *beedxe chiiñu*, perhaps because of its very long claws.

Another major category of animal was *máni pèla* or *pèlla*, which included worms, snakes, eels, and fish (Fig. 5). The slithering movement of a snake was described by the sixteenth-century Zapotec as "going *quíli, quíli, quíli*," and it may be that herein lies the variable which united all these diverse animals: each moved through his chosen medium with the undulating motion characteristic of legless creatures.

It is possible that tonal differences, or distinctions between the sounds Córdova recorded as *l* and *ll*, were used to produce subcategories such as "fish" on the one hand and "snakes and worms" on the other. If so, such distinctions are not detectable in the sixteenth-century documents, where *pèla* and *pèlla* are treated as virtually interchangeable (perhaps regional) alternatives. In Juchitán today, a double *e* sound is used to distinguish *beenda'*, "snakes and worms" from *benda*, "fish," although the terms are noticeably similar.

Figure 5 shows some of the other subdivisions of this category–*pèlayòo*, "earthworm" (from *yòo*, "earth"); *pèllatòxo*, "rattlesnake" (from *natóxo*, "fierce," "something that bites"); *pèlatòba*, "maguey worm" (from *tóba*, "maguey"); *pèllaxangàle*, "eel," and *pèlapecòte*, "intestinal worm."

One term, *pèllayòoniça*, deserves special comment. It was described by Córdova as a "water snake," but its name literally reads "*pèlla*-earth-water." It may be the same "ordinary black-and-yellow water snake" described by Parsons (1936:223) at Mitla, a serpent to which her informants attributed the power to increase or decrease the flow of river water. This "culebra de agua" was said to fall from the sky (perhaps as a water-spout), sometimes causing floods. The two names Parsons collected for it, *bil nis* ("water snake") and *bil yuš* ("sand snake") show the same "earth-water" overlap as Córdova's term.

Was the *pèllayòoniça* regarded as a form of earthworm (*pèlayòo*) that lived in water (*niça*), or was it simply a snake that spent time both on land (on earth, or in sand) and in the water? We have no way to resolve this question at the moment. Indeed, a third possibility–metamorphosis–is raised by the ambiguity of the animal's name and its possible supernatural affinities.

The metamorphosis of animals from one life form to another was a widespread concept in Indian Mexico. For example, the Tarahumara studied by Lumholtz (1902:309) believed that the brown ground squirrel which they called a *chipawíki*–an animal that lives among rocks and does not ascend trees like other squirrels—eventually became a serpent. Thus the *pèllayòoniça* may have been a creature which in an earlier incarnation had been a form of *pèllayòo*.

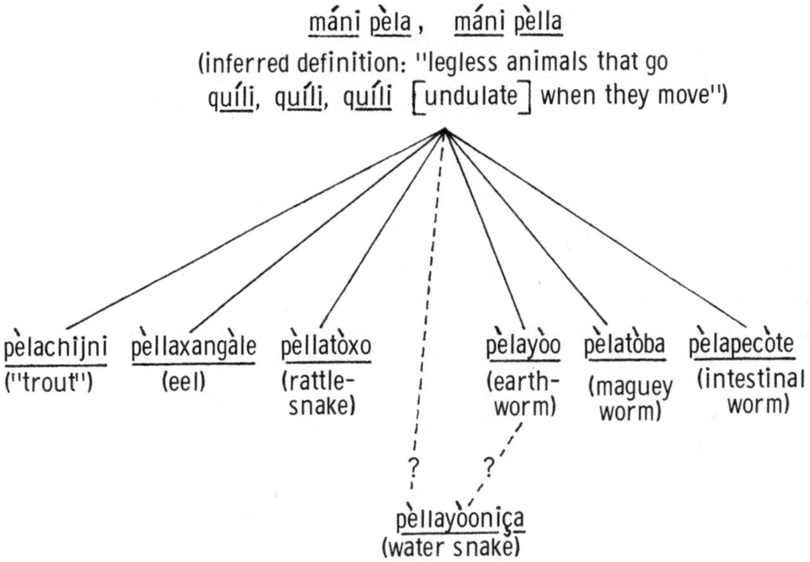


Fig. 5. Subdivisions of *máni pèla* or *pèlla* in sixteenth-century Valley Zapotec. The classificatory relationships of *pèllayòoniça* are unclear; two possibilities are indicated by dashed lines. The *pèlachijni* was an unidentified fish referred to as trucha ("trout") by Córdova; actual trout (family Salmonidae) are not native to Oaxaca.

Metamorphosis was frequently used by Mexican Indians as an explanation for animals that were difficult to classify because they violated one or another rule of the system. Lumholtz's Tarahumara informants believed that bats were rats that had undergone metamorphosis. The Tzeltal Maya studied more recently by Hunn (1977) gave a similar metamorphic explanation for bats' contradictory position in their folk zoology:

> Bats cannot be 'mammals' because 'mammals' 'walk' (*ya šben*). Neither can they be 'birds' though they 'fly' (*ya šwil*), since they lack feathers (*mayuk sk'uk'umal*). Again a folktale provides clarification. Bats are transformed shrews. According to the tale, *ya'al be* 'shrew' attempts to jump across a trail (*be*). If he fails in this attempt he dies. This explains why shrews are so often found dead in the middle of the trail. If he succeeds he is transformed into a bat (*ya šk'ahta ta soc'*). [Hunn 1977:59]

The sixteenth-century Zapotec had the same problem with bats, and resolved the anomaly of their having wings but no feathers by calling them *piguite ziña*, "mouse butterfly" (from *piguite*, "butterfly," and *piciña*, "mouse"). A similar term is used by today's Mitla Zapotec, but in Juchitán the bat is called *biguidi beela*, "meat butterfly" (from *biguidi*, "butterfly," and *beela*, "meat"). The use of "meat" as a modifier to define an unusual animal is also paralleled in Tzeltal, where the armadillo is called "squash" + "meat" (Hunn 1977:87). Bats were important in the cosmology of all these peoples in Precolumbian times, perhaps in part because of their unusual place in the animal world.

Still other animals had supernatural connections. The generic term for frog or toad was *pèeche*, distinguished by a double *e* sound from *pèche*, "fierce animal," and divisible into several varieties. One of these was the *pèeche mao*, a frog that "sang when it rained"; another variety was thought to be produced by spontaneous generation from the rain itself. This frog, which evidently appeared in great numbers after heavy storms, was known as *pèeche xini cocijo* (literally, "the offspring of *cocijo*").

Finally, the Zapotec recognized several growth stages in the life of an animal. A very young animal might have attached to it the adjectives *patào*, *yyni*, or *yyñi* ("newborn," "nursling"), or it might simply be described as *huini*, "small." When adult it might be classed as *pinici* or *pinijci*, "fully grown"; eventually it became *nagola*, "elderly." The Zapotec recognized man's affinity to the animals by using similar terms for human beings of different ages, while employing a completely separate vocabulary for the growth stages of plants.

## THE PLANT WORLD

Botanical data in the sixteenth-century documents are extremely rich; they are also frustrating, because no Latin names are assigned to the plants involved. Fortunately, many sixteenth-century words can be matched with terms collected in Juchitán by López Chiñas or Pickett, in Ayoquesco by MacLaury, or in Mitla by Messer. Still others were recognized by our Zapotec workmen and can be applied to species identified by botanist C. Earle Smith, Jr. (1978 and personal communication). Dozens more, however, were recorded by the sixteenth-century writers only with phrases such as "a tree like our Spanish poplar" or "a tree yielding fruit like our blackberry."

The ancient Zapotec had a single word, *yàga*, which served on several classificatory levels—it could mean "plant," "wood," "tree," "stick," or "firewood," with the exact meaning determined by context. This was by no means unusual among Otomanguean languages; according to Ronald Spores, the sixteenth-century Mixtec term *yutnu* also meant "plant," "tree," or "wood" (Marcus, Flannery, and Spores, n.d.).

It is not clear how many intermediate categories occurred above the level of the specific plant. One informal category was *yàgalayña*, which could be roughly translated "trees which bear seasonal fruit," or "trees which are seasonally harvested." A few of the trees that might occur under this category are given in Figure 6: *yàga piògoxilla* (*Ceiba parvifolia*, the pochote or silk-cotton tree), *yàga làha* (*Leucaena esculenta*, the guaje), and *yàga pichíj*, the collective term for several genera of organ cacti. The sixteenth-century term *pichíj* has become *bidxí* in Juchitán (Pickett 1959) and *bidsz* in the ideolect of our Valley of Oaxaca informants (*bidz* in the ideolect of Messer's [1975] informants). The Zapotec with whom the junior author worked in 1966-1967 divided organ cacti into *bidsz-lats* (*Lemaireocereus*), *bidsz-žob* (*Myrtillocactus*), and *bidsz-top* (*Cephalocereus*), terms which presumably had sixteenth-century counterparts which were simply not collected by Córdova. He did collect terms for *yàga tòba* (the maguey, *Agave* spp.) and *yàgaqui*, a heavy cane used for basketry (probably *Phragmites*, although the term has now been extended to the Old World introduction *Arundo donax*).

It will come as no surprise that some of the important classificatory criteria for the sixteenth-century Zapotec were the utility of the plant ("edible" or "useful" vs. "inedible" or "useless") and its specific resemblance to other plants. The *piògo* in the term for pochote refers to the pod's resemblance to a "corncob," while the *xilla* refers to the "cotton" in the pod; the *žob* in the term for *Myrtillocactus* alludes to the fact that the fruits are the size of a "corn kernel" (called *xòopa* in 1578). The informal category for "plants which are not eaten" was *yàgaquijxi*. This is the same *quijxi* which we saw augmented to mean "dense monte" or "wilderness" (*quijxitào*), or suffixed to an animal name to mean "wild" or "of the monte" (*mániquijxi*); suffixed to *yàga* it seems to have meant "weed," and below it in the classificatory scheme came dozens of species of inedible or useless herbs or grasses. The most frequent use of *quijxi* seems to have been as an adjective or adjectival suffix, rather than a noun.[1]

[1] Some confusion could result from the fact that both in Juchitán and in Mitla, there are two variants of this term. In Juchitán, *guixi* is "basura, zacate"; *gui'xhi'* is "monte" (Pickett 1959). In Mitla, *giš* is "pasto"; *gihš* is "monte" or "silvestre" (Messer 1975). However, Briggs (1961:71) reveals that this is merely a pronunciation difference depending on whether the word stands alone or is an adjectival suffix ("bound morpheme"). Córdova evidently recognized this and made no distinction between *quijxi*, "weed," and *quijxi*, "monte," "silvestre," even though there may have been a slight pronunciation difference even in the sixteenth century. We suspect that just as *quèela* could mean "a corn plant" or "a stand of corn plants" (see below), *quijxi* could mean "a weed" or "a stand of weeds" (i.e., "monte").

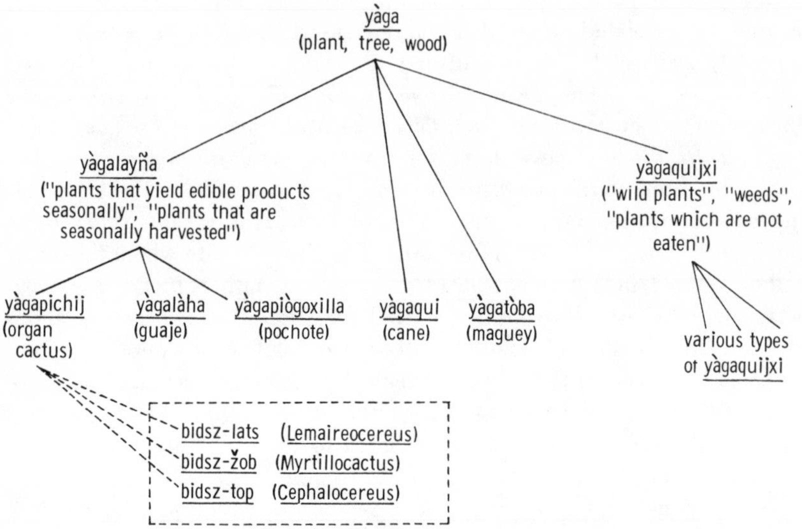


Fig. 6. A few of the many subdivisions of *yàga*. All terms are in sixteenth-century Valley Zapotec except those indicated by dashed lines, which are from twentieth-century Mitla.

This brings us to what we see as a series of differences between the ways some plant terms were used in the sixteenth century, and the way they were evidently used by Messer's informants in present-day Mitla. There can be little question that the sixteenth-century *yàga* (still *yaga* in Juchitán) is Messer's *yahg*; sixteenth-century *quijxi* (still *guixi* in Juchitán) is Messer's *giš*; and sixteenth-century *nocuana* (now *cuana* in Juchitán) is Messer's $k^{w}an$ (Córdova 1578a; Pickett 1959; Messer 1975, 1978). However, while *yàga* seems to have had roughly the same meaning as today's *yahg*, our impression is that *quijxi* and *nocuana* only partially overlap with the definitions given by Messer for present-day *giš* and $k^{w}an$.

For example, the sixteenth-century meaning for *nocuana* was "fruit," or "edible plant part" (Fig. 7). It could include a "sweet fruit" (*nocuananàaxi*) like one of the zapotes, or a "sour fruit" (*nocuananayy*) like the hog-plum (*Spondias* sp.). It could also include a bulb like the local wild onion (*nocuana xijto*). Thus, one of the definitions given by Messer for $k^{w}an$–"the soft, edible part of a plant as opposed to the woody stem"–fits *nocuana* as well. However, her other definitions for $k^{w}an$ do not. We cannot find any evidence that the sixteenth-century Zapotec extended the term *nocuana* to inedible herbs or "broad-leaf herbs" as do present-day Mitleños. Most importantly, *nocuana* seems never to have been used as a "general life form" in contrast to "tree," but was instead used to refer to the edible fruit of a tree. For example, when oranges were introduced to Oaxaca the tree was called *yàga naranjo*, while the fruit became *nocuana naranja* (Córdova 1578a:280).

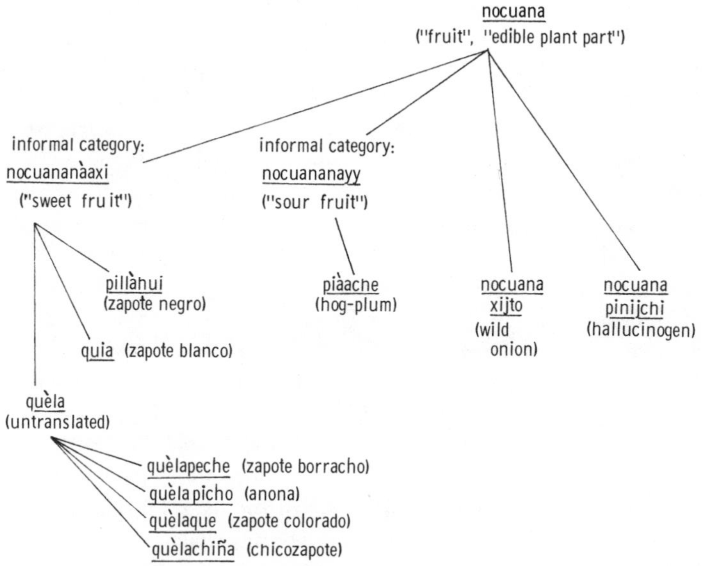


Fig. 7. Some subdivisions of *nocuana* in sixteenth-century Valley Zapotec.

The same could be said for the difference between sixteenth-century *quijxi* and the *giš* of Messer's informants. We can find no evidence that *quijxi* was ever defined as "pasto"; indeed, we find no stated distinction between "grass" and "herb" in sixteenth-century Zapotec. *Quijxi*, as we have already suggested, was used primarily as an adjective meaning "useless," "wild," or "weed," and as such could even be used to describe a tree (*yàgaquijxi*).

One of the most convincing bits of evidence for this difference is Messer's description of the way the terms are now applied to plant growth stages (Messer 1975:100). In Mitla, certain plants may be described as *giš* when newly-sprouted and grasslike, $k^{w}an$ when they reach the size of a leafy herb, and *yahg* when they finally become large and woody. This could never have happened in the sixteenth century. Apart from the incongruity of a *quijxi* turning into a *nocuana* (for which we have no evidence), by the rules of sixteenth-century Valley Zapotec it would have had to be done through a series of adjectival suffixes, e.g., *yàgaquijxi-yàganocuana-yàgayaha.* Not only did *yàganocuana* not exist, even had it existed it would have meant "fruit tree."

How are we to understand these differences between sixteenth-century and twentieth-century Zapotec? Our hypothesis, which obviously should be tested with further research, is as follows. We suspect that the Precolumbian Zapotec did not have the same distinctions between grass, herbs, and trees that they do today. We believe they had one word for plant—*yàga*—while *nocuana* meant "fruit" or "edible plant part" and *quijxi* was an adjective meaning "wild," "of the monte," or "useless." The Spanish arrived with an already-established taxonomy in which grass (pasto), herb (yerba), and tree (arbol, palo) were distinguished as separate life forms. Over the 400 years since the Conquest, these concepts have penetrated so deeply into Zapotec taxonomy that *quijxi* and *nocuana* have been largely redefined, with *giš* approaching Spanish "pasto" and *k*$^{w}$*an* approaching Spanish "yerba." A similar process has taken place in Nahuatl, with the sixteenth-century *zacatl* ("yerba en general") entering Spanish as the loan word zacate, now redefined as "pasto."[2]

With a little detective work, one can reconstruct some of the *nocuananàaxi* or "sweet fruits" known to the Valley Zapotec in 1578. This detective work leads us through two Indian languages (Nahuatl and Zapotec) as well as Colonial Spanish, and therefore serves as an example of the lengths to which one must go to make sense out of ethnohistoric data. Indeed, as tedious and circuitous as our route will be, it is less frustrating than the dead end provided by the terms for which we have no clues at all.

The Valley Zapotec used a wide variety of tropical fruits, some locally grown and some obtained from the coastal lowlands or the Cañada de Cuicatlán. The Aztec guides and interpreters used by the sixteenth-century Spanish were already familiar with these, and had incorporated them into their own Nahuatl dichotomy of *tzapotl*, "sweet fruits," and *xocotl*, "sour fruits." The sweet fruits thus became *çapotes* for the Colonial Spaniards, who went on to distinguish the various genera by color, which is not at all the way the Indians had classified them. The Spanish categories have survived as zapote blanco (*Casimiroa edulis*), zapote amarillo or zapote borracho (*Lucuma salicifolia*), zapote colorado (*Calocarpum sapota*), zapote negro (*Diospyros digyna*), and chicozapote (*Achras sapota*) (Pesman 1962).

According to López Chiñas (1937), in Juchitán many tropical fruits are classed as *enda* (Pickett [1959] gives the alternative *guenda*), with the following subdivisions:

[2]The Spanish Colonial folk theory of "hot" and "cold" foods has had a similar acculturative effect on Mexican Indian ethnobotanical classification. We have found no evidence in the documents to suggest the Zapotec ever employed such a system in Precolumbian times.

*endabedche*, zapote borracho (*Lucuma*)
*endabidchu*, anona or custard-apple (*Annona* spp.)
*endadchiña*, chicozapote (*Achras*)

López Chiñas' suffixes *-bedche*, *-bidchu*, and *-chiña* almost certainly correspond to the sixteenth-century *-peche*, *-picho*, and *-chiña*, and thus are of great use in our reconstruction.

Now let us examine some of the sweet fruits eaten in 1578. *Diospyros*, the black zapote, is native to the Valley of Oaxaca region and was known as *pillàhui*. The white zapote, *Casimiroa* (described by both Córdova [1578a: 104] and Pesman [1962:239] as "resembling a quince"), was known as *quia*. Then came a whole series of tropical fruits classed as *quèla*, a term which may be the sixteenth-century Valley Zapotec equivalent of the present-day Juchiteco *enda* or *guenda*. This category was divided into *quèlapeche* (cf., *-bedche*, *Lucuma* or zapote borracho); *quèlapicho* (cf., *-bidchu*, *Annona* or custard-apple); *quèlaque* (*Calocarpum* or zapote colorado); and *quelachiña* (*Achras sapota* or chicozapote). Note that none of these were classified by the Zapotec on the basis of color. The *quèlapicho* was so called because its seeds were like "cotton seeds" (*pichoxilla*); the *quelachiña* was so called because it tasted like "honey" (*chiña*). For the sixteenth-century Zapotec, as for the Mitleños recently studied by Messer, taste and physical resemblance to other plants were more important criteria than color.

Another category of plant part (perhaps analogous to and contrastive with *nocuana*) was *còo*, "tuber," which survives in Juchitán and Mitla as *gu*. Figure 8 illustrates some of the ways this term was combined with others to define individual species. The jícama (*Pachyrrhizus erosus*) was known in 1578 as *còoyati* ("white root"). In Juchitán today, yuca (*Manihot esculenta*) is called *gu yaga*, "wood-root." The sixteenth-century word for the *Phaseolus* bean was *pizàa*, and the typically Oaxacan small black domestic bean was known simply as *pizàaláce*, "thin bean." However, one species of local wild runner bean, the jicamita (*Phaseolus heterophyllus*), was collected mainly for its edible tuber; this species was called *pizàacòo*, "tuber-bean," a term which has survived in the eastern Valley of Oaxaca as *bisya-gu*.

Finally, we should mention a few of the hallucinogenic plants recognized by the sixteenth-century Zapotec. One of these was a mushroom called *pèyaçòo* (from *pèya*, "mushroom"), probably the well-known hallucinogen *Psilocybe* sp. Its Zapotec name is unclear, unless another term given by Córdova, *còopatào*, is a misspelling of *çoopitào*. In this case, the full name might be *pèya çòo pitào*, "mushroom from the region of the great spirit." (If *còopatào* is the correct spelling, some kind of tuber must be involved.)

From a woodier plant, larger than a mushroom, came the "fruit" *nocuana penèeche* or *pinijchi*, which "caused one to see visions." Two possibilities for this plant are the morning-glory or Nahuatl *ololiuhqui* (*Rivea* sp.), and the jimson-weed or toloache (*Datura* sp.). *Datura meteloides* is still used for curing at Mitla under the name *binij bido* (Messer 1975), which is a likely twentieth-century Mitla Zapotec version of the sixteenth-century expression *pinijchi pitào*. According to Parsons (1936:231), *binij* is a Mitla Zapotec term for "duende" ("ghost"), so *binij bido* might roughly be translated "ghost of the great spirit."

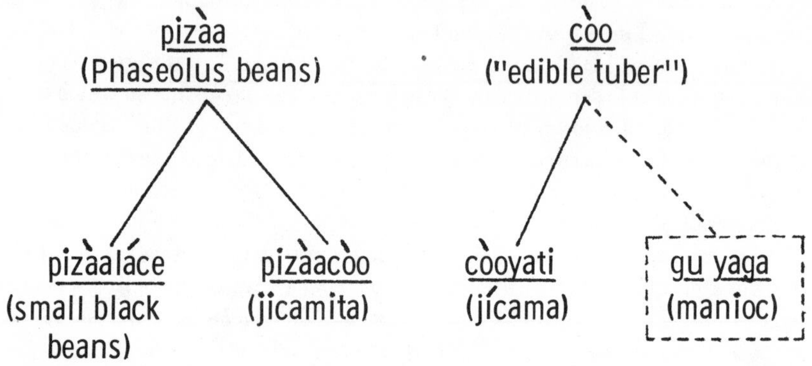


Fig. 8. Definition of one plant species by reference to another plant category. Although *Phaseolus* beans were called *pizàa* and edible tubers *còo*, one species of wild runner bean, the *pizàacòo*, was collected for its edible tuber and named accordingly. All terms are sixteenth-century Valley Zapotec except that indicated by dashed lines, which is from twentieth-century Juchiteco Zapotec.

The problem with identifying *pinijchi* as *Datura*, however, is that Córdova lists another hallucinogen, *nocuana còhui* ("dark fruit"), whose physical description—"a manera de erizo de castaña"—closely fits the spiny seed case of *Datura*. This might suggest that *nocuana còhui* was *Datura*, and *nocuana pinijchi* a different hallucinogen whose name is now applied to *binij bido*, but the question is far from settled.

## RAIN, SACRIFICE, AND AGRICULTURE

The staff of Zapotec life in the sixteenth century was *quèela*, a term that has survived in Juchitán as *guela* and in the Valley of Oaxaca as *gähl* (ideolect of our Zapotec workmen) or *yähl* (ideolect of Messer's informants). Like its modern-day cognates, *quèela* referred to both the maize plant (*Zea mays*) and the field of maize plants ("milpa"); a separate word, *guiñaa* (surviving in Juchitán as *ñaa*), referred to the field itself without the plants.

The Zapotec recognized at least two kinds of cornfields, irrigated (*quèela huizoa*) and non-irrigated (*quèela pichijta*) (Flannery and Marcus 1976:377). The same adjectives, *huizoa* and *pichijta*, were also applied to corresponding varieties of maize: *xòopa huizoa* was "tremesino" maize, maturing in three months and presumably irrigated, while *xòopa pichijta* was "cadañero" or "temporal" maize, grown only once a year by rainfall alone. However, these categories of *xòopa* were only two out of dozens: white corn, red corn, speckled corn, popcorn, and other varieties all had their designations. The harvest of maize was *layña xòopa*, and its season was *cocij layña*, a reference to the division of the year into *cocijo* or "lightnings." In between planting and harvesting there were many growth stages of the milpa. Unfortunately, we have no actual translations for these, only some general Colonial Spanish descriptions:

| | |
|---|---|
| *quèelaxeni* | (growing maize, with as yet no spikes or ears) |
| *quèelatào* | (big, tall maize) |
| *quèelaquèecho* | (maize with still-immature ears [elotes]) |
| *quèelatòça* | (maize with ears 'two palms long' [6 inches?]) |
| *quèelazèhe* | (maize with mature ears [elotes]) |
| *quèelanijza* | (maize fully mature, but not yet harvested) |

Valley Zapotec agricultural techniques include well irrigation, canal irrigation, and floodwater farming, but since water is actually applied to less than 20% of the cultivated land (Kirkby 1974:121), rainfall is a very important factor. For the sixteenth-century Zapotec, rain was simply a form of niça, "water," but it was controlled by *cocijo*, a powerful supernatural who had the power to split the clouds and release it. Apparently, water itself did not possess *pèe* (except in cases such as a turbulent, rushing river, as already mentioned), and could be manipulated by technology once it reached the ground. To bring it down, however, one had to propitiate the sacred lightning by offering drops of one's own blood, or the blood of a quail, dove, turkey, dog, human infant, slave, or captive taken in war. As late as 1578-1581 there was still a rich vocabulary of sacrifice for Córdova and the authors of the *Relaciones Geográficas* to record. A person sacrificed to encourage (or reciprocate for) rain was called *péni guij cocijo* ("man-object of sacrifice-lightning"); the associated ritual cannibalism might include *pèlapeni*, "human meat," or *pèlayyni*, "infant meat." Blood drawn from one's own ears or tongue with a bone, fish spine, or obsidian blade was *tini* or *tini yàa*—fresh, flowing blood which moved and was alive, as opposed to dried blood. Today the only surviving word for "blood" is *rini* (Juchitán) or *r̃ehn* (Mitla), while *tini* (Juchitán) or *tyehn* (Mitla) is used for the fresh, flowing sap of a tree. However, the Valley Zapotec still respect the "rayo" or lightning bolt (*gusiy*); Parsons (1936:211) reports that during 1929-1933 a few Mitla farmers still put down bread for lightning to eat, and "not so very long ago, before planting, a farmer would put down a miniature *tortilla* for Lightning, in the middle of his field."

Should the Zapotec's prayers be answered, rain (*niça quie*) came in many different forms. Indeed, the Zapotec probably had as many terms for rain as the Eskimo are reported to have for snow. In addition to [*niça*] *quie natào*, "gentle rain" and [*niça*] *quiexòo*, "hard rain" (literally, "earthquake-rain," "rain that shakes the ground"), there are expressions which meant "to drizzle," "to rain only in one place," "to rain over a wide area," and "to rain in big drops" (Córdova 1578a:252). The first rain of the year, which the Zapotec still use as a predictor for the remainder of the rainy season (Kirkby 1974:121-23), was called [*niça*] *quiethoce*. Even today, the Mitla Zapotec make note of the "first lightning" (sixteenth-century *cocijo nacòbi*, twentieth-century *gusiy cōb*, "new lightning"), which signals the onset of the rains and sets many things in motion: "the hummingbird revives at first lightning; Lightning embodies himself as a lizard" (Parsons 1936:212). Thus, two animals seen primarily in the rainy season are still associated in the Zapotec mind with the first stirrings of *cocijo*, just as was the frog *xini cocijo* in the sixteenth century.

## SUMMARY AND CONCLUSIONS

Despite the limitations of the sixteenth-century documents, the outlines of an ancient Zapotec system of ethnoscientific classification can be seen. Animals, human beings, clouds, wind, lightning, earthquakes, ghosts, and one's ancestors were considered alive; other things could be brought temporarily to life by wind. The daily route of the sun was a major axis for cosmic orientation. There were four world quarters, each with an associated "lightning"; both the secular and sacred calendars were also divided into "lightnings," and there were certain animals and hallucinogenic plants which were associated with lightning. Animals were classified on the basis of their method of locomotion (walking, swimming, flying), their chosen habitat (earth, water, trees), their behavior (fierce, non-fierce), and their resemblance to other animals. Plants were classified on the basis of their usefulness to man (harvestable, edible, useless), their taste (sweet, sour), and their resemblance to other plants.

We expected to find a lot of Nahuatl loan words in the Spanish vocabulary of 1578-1581, and we did. On the other hand, we were struck by the paucity of loan words in sixteenth-century Zapotec. Considering the length of time the Zapotec had been in contact with the Mixtec and the Aztec, we expected more loan words from those languages than we found. The Zapotec word for "dog," *pèco*, does resemble the Yucatec Maya word for "dog," *pek*, but this is only one probable loan word.

Finally, we remind the reader that this brief glimpse at sixteenth-century Zapotec classification is only a reconstruction. Much of it cannot as yet be considered final, and it doubtless contains some errors which we will need to correct during future research. We therefore offer it not as a completed project, but as a preliminary tribute to two of the most significant ethnobotanists of the last 400 years.

## REFERENCES


Balsalobre, Gonzalo de
1656 Relación Auténtica de las Idolatrías, Supersticiones, Vanas Observaciones de los Indios del Obispado de Oaxaca. Reprinted in: Anales del Museo Nacional de Mexico, Primera Época 6:225-60 (1892).

Berlin, Brent, Dennis E. Breedlove, and Peter H. Raven
1974 Principles of Tzeltal Plant Classification: An Introduction to the Botanical Ethnography of a Mayan-speaking People of Highland Chiapas. New York: Academic Press.

Briggs, Elinor
1961 Mitla Zapotec Grammar. México, D. F.: Instituto Lingüístico de Verano and Centro de Investigaciones Antropológicas.

Burgoa, Fray Francisco de
1670 Palestra Historial de Virtudes, y Exemplares Apostólicos . . . Reprinted in: Publicaciones del Archivo General de la Nación 24. México, D. F.: Talleres Gráficos de la Nación (1934).
1674 Geográfica Descripción. . . . Reprinted in: Publicaciones del Archivo General de la Nación 25, 26. México, D. F.: Talleres Gráficos de la Nación (1934).

Córdova, Fray Juan de
1578a Vocabulario en Lengua Zapoteca. México, D. F.: Pedro Charte y Antonio Ricardo. Reported in: Biblioteca Lingüística Mexicana I. México, D. F.: Secretaria de Educación Pública (1942).
1578b Arte en Lengua Zapoteca. (Reprinted 1886.) México, D. F.: Pedro Balli.

Flannery, Kent V. and Joyce Marcus
1976 Formative Oaxaca and the Zapotec Cosmos. American Scientist 64:374-83.

Hunn, Eugene S.
1977 Tzeltal Folk Zoology: The Classification of Discontinuities in Nature. New York: Academic Press.

Jiménez Moreno, Wigberto
1942 Fr. Juan de Córdova y la Lengua Zapoteca. Introduction to 1942 reprinting of: Vocabulario Castellano-Zapoteco, by Fray Juan de Córdova. Biblioteca Lingüística Mexicana I. México, D. F.: Secretaria de Educación Pública.

Kirkby, Anne V. T.
1974 Individual and Community Responses to Rainfall Variability in Oaxaca, Mexico. In: Natural Hazards: Local, Regional, and Global, edited by G. White. Oxford: Oxford University Press.

Leopold, A. Starker
1959 Wildlife of Mexico: The Game Birds and Mammals. Berkeley: University of California Press.

López Chiñas, Jeremías
1937 Algunos Animales y Plantas Que Conocieron los Antiguos Zapotecas. Neza 3(1):12-24.

Lumholtz, Carl
1902 Unknown Mexico. 2 vols. New York: Charles Scribner's Sons.
MacLaury, Robert E.
1970 Ayoquesco Zapotec: Ethnography, Phonology, and Lexicon. M. A. thesis, Department of Anthropology, University of the Americas, México, D. F.
Marcus, Joyce
1973 Territorial Organization of the Lowland Classic Maya. Science 180:911-16.
1976a The Origins of Mesoamerican Writing. Annual Review of Anthropology 5:35-67.
1976b The Iconography of Militarism at Monte Albán and Neighboring Sites in the Valley of Oaxaca. In: Origins of Religious Art and Iconography in Preclassic Mesoamerica, edited by Henry B. Nicholson, pp. 123-39. Los Angeles: UCLA Latin American Center.
in press Archaeology and Religion: A Comparison of the Zapotec and Maya. World Archaeology, October 1978 issue.
Marcus, Joyce, Kent V. Flannery, and Ronald Spores
n.d. The Cultural Legacy of the Oaxacan Preceramic. To appear in: The Cloud People: Evolution of the Zapotec and Mixtec Civilizations of Oaxaca, Mexico, edited by K. V. Flannery and J. Marcus. School of American Research Advanced Seminar Series. Albuquerque: University of New Mexico Press.
Mata, Fray Juan de
1580 Relación de Teozapotlan, Hecha el 11 de Noviembre de 1580. In: Papeles de Nueva España, edited by F. del Paso y Troncoso. Segunda serie. Geografía y estadística, Vol. 4:190-95. Madrid: Sucesores de Rivadeneyra.
Messer, Ellen
1975 Zapotec Plant Knowledge: Classification, Uses, and Communication about Plants in Mitla, Oaxaca, Mexico. Ph.D. dissertation, Department of Anthropology, University of Michigan, Ann Arbor.
1978 Zapotec Plant Knowledge: Classification, Uses and Communication about Plants in Mitla, Oaxaca, Mexico. Prehistory and Human Ecology of the Valley of Oaxaca, Vol. 5, Pt. 2. Memoirs of the Museum of Anthropology, University of Michigan, 10.
Nader, Laura
1969 The Zapotec of Oaxaca. In: Handbook of Middle American Indians, Vol. 7:329-59. Robert Wauchope, gen ed. Austin: University of Texas Press.
Nowotny, Karl A.
1970 Beiträge zur Geschichte des Weltbildes: Farben und Weltrichtungen. Vienna: Ferdinand Berger & Söhne.
Parsons, Elsie Clews
1936 Mitla: Town of the Souls and Other Zapoteco-speaking Pueblos of Oaxaca, Mexico. Chicago: University of Chicago Press.
Paso y Troncoso, Francisco del
1905 Papeles de Nueva España. Segunda serie. Geografía y estadística, Vol. 4. Madrid: Sucesores de Rivadeneyra.
Pesman, M. Walter
1962 Meet Flora Mexicana. Globe, Arizona: Dale S. King.
Pickett, Velma B.
1959 Castellano-Zapoteco, Zapoteco-Castellano: Dialecto del Zapoteco del Istmo. Serie de Vocabularios Indígenas 3, edited by Mariano Silva y Aceves. México, D. F.: Secretaria de Educación Pública.
1967 Isthmus Zapotec. In: Handbook of Middle American Indians, Vol. 5: 291-310. Robert Wauchope, gen. ed. Austin: University of Texas Press.
Pimentel, Francisco
1875 Cuadro Descriptivo y Comparativo de las Lenguas Indígenas de Mexico ó Tratado de Filología Méxicana. Reprinted in the 1886 edition of Arte en

Lengua Zapoteca (1578), by Fray Juan de Córdova. México, D. F.: Pedro Balli.

Schmieder, Oscar

1930 The Settlements of the Tzapotec and Mije Indians, State of Oaxaca, Mexico. University of California Publications in Geography 4:1-184.

Smith, C. Earle, Jr.

1978 The Vegetational History of the Oaxaca Valley. Prehistory and Human Ecology of the Valley of Oaxaca, Vol. 5, Pt. 1. Memoirs of the Museum of Anthropology, University of Michigan 10.

Swadesh, Morris

1967 Lexicostatistic classification. In: Handbook of Middle American Indians, Vol. 5:79-115. Robert Wauchope, gen. ed. Austin: University of Texas Press.

Thompson, J. Eric S.

1950 Maya Hieroglyphic Writing: Introduction. Carnegie Institution of Washington, Publication 589. Washington, D. C.: Carnegie Institution of Washington.

Weitlaner, Robert J. and Gabriel De Cicco

1962 La Jerarquía de los Dioses Zapotecos del Sur. Proceedings, 34th International Congress of Americanists: 695-710.

# COGNITIVE SYSTEMS, FOOD PATTERNS, AND PALEOETHNOBOTANY

*Wilma Wetterstrom*
Massachusetts Institute of Technology

Some anthropologists recognize a fundamental distinction between two notions of culture: (1) culture that they as scientists can objectively observe and measure including behavior, objects, events, and phenomena and (2) culture as the native perceives it (Goodenough 1961; Keesing 1976: 138-42). The latter is an "organized system of knowledge and belief whereby people structure their experience and perceptions, formulate acts, and choose between alternatives" (Keesing 1976:138).

Archaeologists deal primarily with culture as the realm of things rather than ideas. Most of their data come from material remains, which offer rather limited insights into beliefs and values. Other sources of information about the past, such as paleoecological studies, also inform archaeologists about aspects of the material world rather than systems of knowledge and belief. It is not surprising then that interpretations and explanations in archaeology generally ignore what the native might have thought or believed and view people's behavior as primarily a response to material conditions (e.g., Higgs 1972). One exception is Deetz' concept of the mental template which the prehistoric craftsman used as his guide in making a pot, a basket, or a stone tool (Deetz 1967:45-49).

While it cannot be disputed that the material world sets the conditions in which people must operate and limits the choices that they can make, the way in which people respond is influenced by the way in which they perceive the world and by their systems of belief and values. Many cultural anthropologists believe that what they observe of culture can only be understood with reference to a people's cognitive system (e.g., Bidney 1953:21-53; Frake 1964; Keesing 1976). For example, the drought-resistant corn that the Pueblo Indians raise is an adaptation to the arid conditions of the Southwest. However, the various color varieties that they carefully segregate in different fields can only be explained in terms of their system of color symbolism (Ford 1968:224-27). Undoubtedly, cultures of the past could be better understood if their cognitive systems were considered. When archaeologists narrow their vision to material factors surely they limit their opportunities for understanding and explaining the behavior that created the material remains they study. This paper suggests that culture as an ideational system should form part of the archaeologist's tool kit of concepts and proposes one way in which such a perspective could be incorporated in archaeological research.

Probably all behavior that is "fossilized" in archaeological remains has in some way been influenced by a cognitive system and, therefore, archaeologists might consider the possible role of values and beliefs in all aspects of the past that they study. However, the one area of research that has focused almost exclusively on the material realm and could especially benefit from a cognitive perspective is the development of agriculture.

The materialist approach has been a very productive research strategy in this field. The outlines of agriculture's development are gradually being filled in through more detailed and sophisticated studies of material factors. For example, genetic and ecological research on the cultivars (e.g., Galinat 1970; Harlan and Zohary 1966) and paleoethnobotanical studies of Neolithic plant remains (e.g., Renfrew 1973; Helbaek 1969) have shed light on the origin of domestic plants and their evolutionary pathways. Yet many questions remain. The shift to agriculture involved a remarkable transformation of people's food patterns. How was this accomplished? Why did the early horticulturalists select particular foods and ignore others? What qualities did they look for in their foods? How did insignificant foods such as early maize become the staple of the diet? Food patterns are usually conservative; how can the shift from a hunter-gatherer diet to an agricultural one be explained? Some of these questions can be answered with more detailed studies of material factors, but the answers will be incomplete. As many scholars (e.g., Lee 1959:154-61; de Garine 1972; Richards 1948) have pointed out, people's food patterns are not simply a response to their environment. What people deem fit to eat, how they choose to prepare and eat it, and with whom they share it are all part of a cultural code. What is required in order to answer the perplexing questions is a research strategy that considers these codes as well as material factors. Unfortunately, archaeologists do not have access to the codes and cognitive systems of the past. However, they can draw upon ethnographic sources for models of these systems, and they can use these models to formulate research problems and questions. Such a research strategy could offer new insights into the past and could provide additional ways of interpreting the archaeological record.

I would like to suggest how such research might be done by outlining a method for investigating the dietary changes that accompanied the evolution of agriculture in one marginal area of the Southwest, the Hueco Bolson.

## ARCHAEOLOGY OF THE HUECO BOLSON

The Hueco Bolson is a "broad intermontane lowland, extending from north to south through central New Mexico, western Texas, and into northern Mexico" (Whalen 1977:1). Most of the area is a desert lying between 4000 and 4200 feet above sea level and has no permanent springs or streams. The vegetation, varying with elevation, includes yucca, agave, cacti, creosote bush, mesquite, and snakeweed (Whalen 1977:2-4). Despite the limited resources, the people who inhabited the region shared in the cultural growth of their Pueblo neighbors and evolved from hunter-gatherers into sedentary, irrigation farmers.

A team from the University of Texas at El Paso, headed by M. E. Whalen, has been intensively investigating the cultures of the Hueco Bolson since 1976. Under contracts with the United States Army, Whalen has conducted site surveys and excavations on Fort Bliss, which encompasses a large area of the Hueco Bolson lying north of El Paso, Texas (Whalen 1977). I have been interested in the subsistence patterns of this region since 1977 when I began analyzing plant remains from Hueco Bolson sites. Since the research is ongoing, the discussion that follows outlines methods that might be used in investigating the evolution of the Hueco Bolson diet.

Evidence of human occupation in the Hueco Bolson spans the period from before 6000 B.C. to about A.D. 1400. Almost nothing is known of the earliest inhabitants, the people of the Paleo-Indian period. The Archaic period, beginning at about 6000 B.C., is also poorly understood but some sites do offer evidence of a hunting and gathering life style. At about A.D. 200 tiny villages of subterranean dwellings, pithouses, began to appear. Small quantities of charred corn from pithouse sites suggest that the Hueco Bolson people of this time, the Mesilla phase, were beginning to cultivate plants. Around A.D. 1200 the pithouses gave way to pueblo style architecture along with more elaborate food storage facilities and ceremonial structures (Brethauer 1977:7). At the same time irrigation agriculture apparently replaced hunting and gathering. About 1400, the Hueco Bolson was abandoned, and no one inhabited the area again until the Historic period (Brethauer 1977:8-9).

A number of factors probably contributed to the development of agriculture in the Hueco Bolson. The El Paso phase population was much larger than that of the preceding period and probably demanded more food than the wild resources could have provided. With population growth in the latter part of the Mesilla phase, Hueco Bolson peoples may have begun to intensify their food production. Of course, it should be noted that population growth may have been the result rather than the cause of new subsistence patterns. Environmental factors may have been an impetus for change also. The Hueco Bolson does not offer particularly abundant food resources, and during droughts all of the inhabitants, human and animal, probably suffered. The humans may have tried to compensate for poor years by artificially increasing their food supply through agriculture.

## HUECO BOLSON FOOD PATTERNS

Some aspects of the Hueco Bolson diet can be gleaned from the archaeological record. Plant remains from a small sample of Mesilla and El Paso phase sites suggest that during both periods people were eating seeds of the annual weeds—chenopod, amaranth, purslane and sunflower—as well as the mesquite bean, and cactus fruit. The mainstay of the diet was probably these wild foods and perhaps others that are not represented in this sample. By El Paso times, cultigens had apparently become the major food. Maize was the most common plant material in sites of this period and beans, squash, and bottle gourd seeds were also found (Ford 1977). The settlement patterns appear to confirm the importance of agriculture. All sites from this period are clustered in the foothills region, where rainfall runoff could be channelled and used to irrigate fields (Whalen 1977:140-42). During the El Paso phase, as well as earlier, small mammals such as rabbits and prairie dogs were the major source of meat.

This outline is very sketchy and must be amplified with more studies of flotation samples, with research on the ecology of the food resources, and with more work on the artifacts used in processing and serving foods. A perspective that considers food as part of a cognitive system can provide a useful framework in which to do this work.

Since the Hueco Bolson data says little about cognitive systems, the research now turns to contemporary food patterns. Are there any common features in people's food patterns and the way they conceptualize foods that may offer insights into the Hueco Bolson diet? Can modern examples of dietary change provide clues to the process by which the Hueco Bolson diet evolved? Such generalizations derived from contemporary food patterns can offer models for understanding the Hueco Bolson diet.

## CONTEMPORARY FOOD PATTERNS

Perhaps the most striking feature of contemporary food patterns is that they vary so widely. In some cultures porridge makes the meal while in others meat is prized as the main course. Some people mix meat and vegetables while others abhor such a combination. Shrimp and lobsters are a delicacy to some palates and an untouchable to others. In some societies food must be taken in a leisurely fashion with polite conversation while in others it is bolted as quickly as possible in complete silence. Despite these enormous differences in taste, etiquette, and values, food patterns throughout the world do share some common features. They all transform eating into a patterned activity and these patterns have a number of structural similarities. First, they divide the daily allotment of food among a series of discrete eating events such as meals, snacks, tea and cocktails. Often these events follow a specified order and occur at particular times in the day. For example, urban Americans begin with breakfast in the morning, take a coffee break at mid-morning, followed by lunch at noon, take another coffee break in mid-afternoon, and conclude with dinner in the early evening. The years's food patterns are likewise patterned. In India the year consists of a succession of religious holidays that are marked either by feasting or fasting by some segment of the population (Katona-Apte 1975:315).

Dietary codes also specify which foods or food categories are appropriate for each "eating event." They often prescribe other attributes of the meal as well, including how the food is prepared and served, who may eat the food, and how people are to behave during the meal. In some complex societies the dietary codes have evolved into elaborate constructs that regulate every aspect of eating.

While many cultures do not have such elaborate food codes, most share the concept of a main meal. This meal is taken daily, is composed of specified foods or categories of foods, and is eaten in a particular manner. Among agriculturalists the main meal is often the staple crop and one other dish. For example, among the Bemba of Rhodesia, a meal consists of two elements: (1) a porridge of millet, the staple, which is cooked to the consistency of plasticine and (2) a relish or stew consisting of vegetables or meat. During the meal the men squat around a pot of relish and a bowl of porridge set on the floor. Each man tears off a piece of dough, rolls it into a ball, dips it in the relish, and bolts it whole. They repeat the process until all the food is consumed. No meal is complete without the relish and porridge. When millet is in short supply, other grains, such as maize, may be substituted but they are considered inferior. The relish category, on the other hand, is very flexible and varies with the seasons. However, the proper ratio of relish to dough is always maintained, and if there is little meat or vegetables for a relish only a small quantity of dough is prepared (Richards 1939:46-47).

Chinese cuisine has a similar pattern. A meal must include appropriate amounts of two elements, *fan*, grains and other starchy foods, and *ts'ai*, vegetables and meat dishes (Chang 1977:7-8). Melanesians also make a distinction between "food," starchy plants and vegetables, and "condiments," comprising vegetables, fish, and meat, that embellish the food and enhance its flavor. No meal is complete without food from both of these categories (Barrau 1973:97).

Among hunter-gatherers the concept of a main meal may not be as rigidly defined since their food resources are usually more diverse than those of agriculturalists and they vary more widely through the year. Unfortunately, there is little information about how hunter-gatherers conceptualize foods and meals. However, it appears that the concept of a core food or food category as the basis of a main meal may be important although hunter-gatherers do not raise a staple crop. Lee (1965:100) found that the Dobe !Kung Bushmen of Botswana were highly selective about foods and depended on only nine species of plants for three-quarters of their vegetable diet. They considered one of these, the mongongo nut, their primary food and believed that "an ideal diet should consist primarily of meat and mongongo nuts" (1965:102). South of the !Kung where the environment is drier and has fewer reliable plant resources, the Kade San Bushmen also concentrate on only a few species of plants, each of which serves as a staple during the season when it is available. For example, from February through April, the *Bauhinia macrantha* bean is the "main element of the diet," "while the other plants are almost never eaten despite the fact that this is the richest season for all food plants in the Kalahari bushveld" (Tanaka 1976:106).

Another common feature of dietary codes is that much of the information and detail about foods is systemized through the process of classification. That is, people place foods, meals, and food related items, such as cooking techniques, into categories and taxonomies in the same way that they classify all other aspects of their world. An example of a classificatory scheme is shown in Figure 1. Most North Americans would probably classify "eating events" according to a scheme similar to this one. As this example illustrates, classificatory schemes are nested and hierarchical. However, they are not necessarily exclusive. To the contrary, they often overlap because the same items may be classified under several systems. For example, the Navaho have a category for "domesticated plants which get ripe" that belongs in their taxonomies for food and for plants (Perchonock and Werner 1969:232-33). A North American's food categories overlap with his classification of meals; starchy vegetable is a class of food as well as a component of the main course for dinner.

The process of classification depends upon the attributes that a people choose to recognize and distinguish. A North American views starchy vegetables as a different category from leafy ones. A Navaho distinguishes "meat from night walkers" and "meat from day walkers," but he lumps together

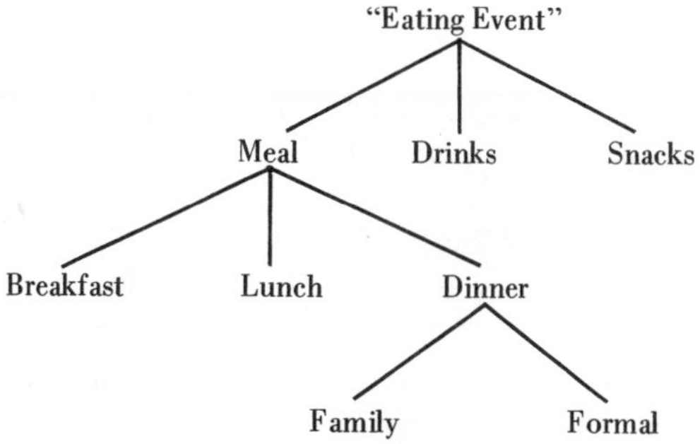


Fig. 1. Part of a North American classificatory scheme for "eating events."

fruits and vegetables (Perchonock and Werner 1960:232). A classification system thus reflects some of a culture's values.

Classificatory schemes serve several purposes. By classifying items as equivalent, they reduce the complexity of the environment. They also offer a way to identify and refer to objects and they provide direction for actions (Bruner, Goodnow, and Austin 1976:178-79). For example, a North American attending a dinner party knows how to deal with a food he has never seen before if his host tells him it is the dessert. The people of San Juan Pueblo, New Mexico, have a hot-cold classification system which tells them how to act in many situations. For example, foods that are classed as "cold" may be used to treat "hot" diseases (Ford 1968:128-34).

The discussion of food patterns and classification schemes thus far has focused on dietary codes in isolation. Yet procuring, preparing, and eating food crosscuts nearly all other domains of a culture including kinship, economy, law, religion, and political organization. Indeed, one of the common features of all food codes is that they are integrated into a cultural pattern and cannot be fully understood without reference to this cultural context.

For example, food patterns may reflect and reinforce religious beliefs or patterns of social and political organization. Each meal for a Hindu reiterates religious doctrine since foods are part of the moral system and are classified according to degrees of purity (Khare 1976:2-5). In the Trobriand Islands, a man must give about half of his yam crop to his sister's husband, thereby reinforcing the tie between affines (Malinowski 1935:189). Hierarchical relationships among the Bemba are reaffirmed at the daily meal; each man draws a portion from the relish pot according to his rank (Richards 1939:76).

Food also conveys meanings about relationships and may be used to create or maintain them. As Mary Douglas (1972:66) points out, people who are not very close may share drinks, whereas meals are reserved for intimate friends. Inviting someone to dinner may be an effort to gain intimacy. In many societies meals are used to bind social groups. For example, the Tsembaga Maring try to win the support of allies for future battles by inviting them to a pig feast (Rappaport 1968:196).

Food is a source of wealth and power, too. On Goodenough Island in Melanesia, the Kaluana people use food to battle for power. Important men, or those who aspire to prominence, invite their enemies to a food exchange and try to shame them by giving them more food than they can return. Their prestige and power are based on the amount of food that they can give away (Young 1971:189-206). A village chief on the Trobriand Islands also bases his power on food. The produce that he collects from his subjects is his primary means for commissioning public works and supporting war parties (Malinowski 1935:47).

## CHANGING FOOD PATTERNS

Although the discussion has presented food patterns as static entities, history amply illustrates that people do develop new food patterns. Hunter-gatherers shifted to an agricultural diet. The New World crops, maize and potatoes, became staples in parts of the Old World. Segments of North American society, traditionally meat-eaters, are converting to a vegetarian diet.

All of these examples involve a unique set of social, cultural, economic, political, and ecological factors. However, some aspects of the way in which new food patterns emerge may be fundamentally similar. It appears that when new foods, new cooking techniques, and new meal patterns are introduced people deal with them in similar ways. The structure of their food patterns and their classificatory schemes play an important role in determining the fate of the introductions. People appear to place the new item within one of the categories they recognize on the basis of attributes that are relevant to them. They then deal with the new item as they would treat any member of that category. For example, when two potentially edible plants were introduced to San Juan Pueblo in the late nineteenth century, the Indians would eat only one of these although they knew that their Spanish American neighbors ate both. The first plant, tumbleweed, was classified as a weed, which is inedible, because it grew in the cultivated fields. The second plant, asparagus, was classified as "thing to eat" because it grew along the ditches and river banks, an area where edible plants are collected (Ford 1968:153).

A new item may be rejected because it does not fit any recognized category. North Americans, for example, had little interest in yogurt as a sour, fermented milk product. It did not fit into any categories of traditional American food patterns. However, when manufacturers added sugar and syrup, yogurt acquired the attributes of a sweet food and could be classified as a dessert or a snack, categories which are not only familiar but very important to Americans. In addition, the manufacturers emphasized yogurt's health benefits in their advertising and thus classified it as a "health food" as well, another category that is dear to Americans.

As this example suggests, new foods or new cooking techniques appear to be most well received when they possess attributes that are important in a culture. When chili peppers were introduced to China in the late nineteenth century, they quickly became an essential ingredient in the cuisines of Szechwan and Hunan, where people were already preparing spicy dishes with brown pepper (Anderson and Anderson 1977:328-29).

While new items are often placed in an indigenous category, new classes also may be borrowed from another culture. For example, the Navaho have a major food category translated as "domesticated edible ripeables" which includes most of the foods that are called fruits and vegetables in English. The Navaho who speak English frequently have added these labels to their classificatory scheme (Perchonock and Werner 1969:235-36). Marathi speakers, a minority community that migrated from North to South India several generations ago, have adopted some of their neighbors' categories. For example, they now eat *tiffin*, one of the two types of meals recognized by South Indians. Unlike a real meal which consists of boiled rice and at least one side dish, *tiffin* is made up of food that the South Indians consider snacks, although it may be more substantial than a real meal. When the Marathi speakers eat wheat in the evening, following a pattern similar to North India, they call the meal *tiffin*. They also eat *tiffin* during fasts as South Indians do and have placed many of their snack and staple foods in this category, although most of these items would have been forbidden as fast foods in the north (Katona-Apte 1975).

Food patterns may also change without the help of introductions and borrowings. These changes seem to evolve as variations on the theme of the original food patterns. They may occur as people begin to recognize new distinctions and create new categories. For example, some North Americans who are particularly concerned with health have added another classificatory scheme to their food patterns. They now classify foods on the basis of health and nutrient attributes and select foods classed as low-cholesterol or low-fat to fill the traditional categories in their meals.

Since foods are integrated into a culture, changes in food patterns cannot be understood without reference to the cultural context. The social, political, and economic roles which foods perform, as well as their symbolic value, influence changes in food patterns. For example, Young (1971) believes that the men of Goodenough Island, described earlier, fight with food because they can no longer wage armed wars. When the British stopped all fighting, the food contests that had once been a minor activity, became the major vehicle for competition and rivalry since food was already associated with power. Deetz (1974) has proposed that food patterns evolved in colonial New England as society was reshaped by the influence of the Renaissance. The original colonists, mostly yeomen from England, followed medieval patterns in their architecture, foods, and world view. In their homes, all daily activities were carried on in one or two rooms, and at meal time they scooped food from a common trencher. As a result of the Renaissance, the Georgian style began to emerge about 1700. The medieval world gave way to a world characterized by "symmetrical cognitive structures, homogeneity in material culture, a progressive and innovative world view, and an insistence on order and balance that permeates all aspects of life from the decorative arts to the organization of space by society" (1974: 22). The common serving bowl of the medieval world was replaced by individual plates belonging to matched sets.

## INVESTIGATING THE EVOLUTION OF THE HUECO BOLSON DIET

The discussion of contemporary food patterns suggests a number of points that may be relevant to the evolution of the Hueco Bolson diet: (1) The food patterns and classificatory schemes of the early Mesilla phase people probably influenced the way in which they first used corn and the other cultigens. (2) Later changes in the diet may have been variations on the theme of the original food patterns. For example, Mesilla people may have added new categories or changed the value of old ones. (3) The changes in the food patterns occurred within a cultural context that was also changing, and the dietary patterns may have reinforced or complemented these new arrangements.

Starting from these points a detailed research strategy could be developed for studying the food patterns of the Hueco Bolson. This research would focus on the following issues:

(1) What were the food patterns of the early Mesilla phase people? Contemporary food patterns suggest that these people did not gather a random assortment of wild foods but probably depended on one or several staples. Since they were sedentary, they may have concentrated on a single staple such as the mesquite bean. This is one of the few abundant, storable resources found in the Hueco Bolson that has been used as a staple by North American Indians such as the Cahuilla (Barrows 1971:307).

This hypothesis, that mesquite, or another plant, was the major food, should be tested with archaeological and ecological data. Ethnographic examples suggest the kinds of food storage and processing facilities that might be found at sites where mesquite beans were an important food. They would also indicate how well mesquite beans might be preserved at archaeological sites. For example, the few tiny fragments of beans that have been found at Mesilla sites may be all that one can expect to find if the beans were prepared by pounding. Ecological studies could be used to determine if the mesquite in fact could have provided a large portion of the Mesilla phase diet.

The research should also be directed at considering the implications of a mesquite bean diet. For example, what kind of social units and settlement patterns would have been compatible with such a food pattern? What kind of problems would the Mesilla phase people have encountered if mesquite beans had been their staple? How would droughts have effected the mesquite crop?

The staple or staples must also be considered in relation to the rest of the diet. Ethnographic examples could suggest models for the way in which the Mesilla people might have conceptualized a diet based on mesquite or another staple. For example, they may have had a concept of a meal similar to that of agriculturalists which required a starchy staple and a relish or a stew of vegetables and meat. Were this the case, which foods filled the relish category? What resources might the Mesilla women have gathered every day in order to have a complete meal? How would such a dietary pattern affect the way in which the Mesilla people used their environment?

(2) How were the cultigens first accommodated in the Mesilla diet? The Mesilla people probably placed them in one of their indigenous categories and treated them as they would treat any member of that category. The early cultigens should be reconstructed and examined with a view toward determining which attributes might have been used to classify them. Ethnographic examples would offer insights into the characteristics that could have been important to the Mesilla people. The cultigens should also be compared with the wild plants since these may have been the model for handling the new plants. For example, the Mesilla people may have classified squashes with the coyote melon, which grows in the area, and they may have collected the plant for the same reason that people in Mexico gather the wild squash—for the seeds (O'Laughlin 1977:182). Viewing the cultigens as the natives might have could offer insights into the attributes that they sought in their foods and the selective pressures that they might have exerted on these plants.

(3) How did the cultigens become staple foods? The new food patterns probably developed as a variation of the old patterns. For example, if the

Mesilla people conceptualized their meals as described above, they may have used maize as a relish. Perhaps maize was a backup food that women turned to when other foods were not available to fill the relish category. Alternatively, maize could have been classified with foods that substituted for staples and may have been used occasionally in place of mesquite. With time maize may have been used more frequently as the main food in a meal and may have eventually become equated with it.

Clearly the value and prestige that Mesilla people associated with maize increased through time. Perhaps this process began when the Mesilla people established a new classificatory scheme that recognized cultivated as a category distinct from wild.

With the help of additional ethnographic data more detailed models of this process could be developed. These would be a useful focus for examining differences among early and late Mesilla sites. For example, they might suggest ways to interpret the sets of plant foods that occur together in hearths and the changes in these sets over time.

(4) In what ways were the food patterns integrated into Mesilla culture? In what ways were they related to the cultural changes in the Hueco Bolson? How did they reinforce them? Foods were undoubtedly significant in social interactions and the cultigens may have become important partially because of the social roles they could perform. For example, the Mesilla people may have found that it was easier to meet kinship obligations by raising more corn than by collecting additional wild foods. During poor years, especially the cultigens may have been substituted for wild products at feasts or food exchanges. As population expanded and the number of villages increased, maize may have become the only food that could be increased for such rituals since the collecting territories were probably growing smaller. Because maize production could be intensified, this cultigen may have become a means for gaining prestige. When families amassed and distribued large quantities of food for marriages, funerals, and other rituals they may have found that growing more maize was the easiest way to produce an impressive display of food.

The evidence for cultural changes in the Hueco Bolson should be examined in the light of such models. Although they simplify the process of change, they may offer insights into some of the relevant factors.

## CONCLUSIONS

People's behavior is influenced both by the material conditions of their world and by the way in which they perceive that world. Although archaeology has focused primarily on the material conditions of past cultures, it could benefit from considering the cognitive systems of these societies as well. I have suggested one way in which this might be done by outlining a research strategy for investigating dietary changes that accompanied the evolution of agriculture in the Hueco Bolson. Since cognitive systems are not preserved in the archaeological record, the first step in this research program is to examine contemporary food patterns for models of the way in which people conceptualize, categorize, and deal with food. These models are then used as a framework in which to analyze the archaeological data. The virtue of such an approach is that it suggests new ways of viewing the archaeological record.

## REFERENCES


Anderson, Eugene N. Jr., and Marja L. Anderson
1977 Modern China: South. In: Food in Chinese Culture: Anthropological and Historical Perspectives, edited by K. C. Chang, pp. 319-82. New Haven: Yale University Press.

Barrau, Jacques
1973 The Oceanians and Their Food Plants. (Roda and Colin Roberts, trans.) In: Man and His Foods: Studies in the Ethnobotany of Nutrition–Contemporary, Primitive and Prehistoric Non-European Diets, edited by C. Earle Smith, Jr., pp. 87-117. University, Alabama: University of Alabama Press.

Barrows, David Prescott
1971 Desert Plant Foods of the Coahuilla. In: The California Indians: A Source Book, edited by R. F. Heizer and M. A. Whipple, pp. 306-14. Berkeley: University of California Press.

Bidney, David
1953 Theoretical Anthropology. New York: Columbia University Press.

Brethauer, Douglas P.
1977 Summary of Cultural Development in the Hueco Bolson and Vicinity. In: Settlement Patterns of the Eastern Hueco Bolson, by Michael E. Whalen. El Paso Centennial Museum Publications in Anthropology 4:4-11.

Bruner, Jerome S., Jacqueline J. Goodnow, and George A. Austin
1976 Categories and Cognition. In: Culture and Cognition: Roles, Maps, and Plans, edited by James P. Spradley, pp. 168-90. San Francisco: Chandler Publishing Co.

Chang, Kwang-chih
1977 Introduction. In: Food in Chinese Culture: Anthropological and Historical Perspectives, edited by K. C. Chang, pp. 3-21. New Haven: Yale University Press.

Deetz, James
1967 Invitation to Archaeology. Garden City, New York: Natural History Press.
1974 A Cognitive Historical Model for American Material Culture: 1620-1835. In: Reconstructing Complex Societies, edited by Charlotte B. Moore. Supplement to the Bulletin of The American Schools of Oriental Research 20: 21-27.

de Garine, Igor
1972 The Socio-Cultural Aspects of Nutrition. Ecology of Food and Nutrition 1:143-63.

Douglas, Mary
1972 Deciphering a Meal. Daedalus 101:61-81.

Frake, Charles
1964 The Structural Description of Subanum "Religious Behavior." In: Explorations in Cultural Anthropology, edited by W. Goodenough, pp. 111-29. New York: McGraw-Hill.

Ford, Richard I.
1968 An Ecological Analysis Involving the Population of San Juan Pueblo, Ph.D. Dissertation, Anthropology Department, University of Michigan, Ann Arbor.
1977 Archeobotany of the Fort Bliss Maneuver Area II. In: Settlement Patterns of the Eastern Hueco Bolson, by Michael E. Whalen. El Paso Centennial Museum Publications in Anthropology 4:199-205.

Galinat, Walton C.
1970 The Cupule and Its Role in the Origin and Evolution of Maize. Massachusetts Agricultural Experiment Station Bulletin 585:1-18.

Goodenough, Ward H.
1961 Comment on Cultural Evolution. Daedalus 90:521-28.

Harlan, Jack R., and D. Zohary
1966 The Distribution of Wild Wheats and Barley. Science 153:1074-80.

Helbaek, Hans
1969 Plant Collecting, Dry Farming, and Irrigation Agriculture in Prehistoric Deh Luran. In: Prehistory and Human Ecology of the Deh Loran Plain, by Frank Hole, Kent V. Flannery, and James A.Neely. University of Michigan Museum of Anthropology Memoirs 1:383-426.

Higgs, Eric S. (ed.)
1972 Papers in Economic Prehistory. Cambridge: Cambridge University Press.

Katona-Apte, Judit
1975 Dietary Aspects of Acculturation: Meals, Feasts and Fasts in a Minority Community in South Asia. In: Gastronomy: The Anthropology of Food and Food Habits, edited by Margaret L. Arnott, pp. 314-26. The Hague: Mouton.

Keesing, Roger
1976 Cultural Anthropology: A Contemporary Perspective. New York: Holt, Rinehart, Winston.

Khare, Ravindra S.
1976 The Hindu Hearth and Home. Durham, North Carolina: Carolina Academic Press.

Lee, Dorothy D.
1959 Freedom and Culture. Englewood Cliffs, New Jersey: Prentice-Hall.

Lee, Richard B.
1965 Subsistence Ecology of The Kung Bushmen. Ph.D. dissertation, Anthropology Department, University of California, Berkeley.

Malinowski, Bronislaw
1935 Coral Gardens and Their Magic, Vol. I: Soil-Tilling and Agricultural Rites in the Trobriand Islands. Bloomington: Indiana University Press.

O'Laughlin, Thomas C.
1977 Excavation of Two Caves in the Mountain Zone of Fort Bliss Maneuver Area II. In: Settlement Patterns of the Eastern Hueco Bolson, by Michael E. Whalen. El Paso Centennial Museum Publications in Anthropology 4:169-89.

Perchonock, Norma, and Oswald Werner
1969 Navaho Systems of Classification: Some Implications for Ethnoscience. Ethnology 8:229-42.

Rappaport, Roy A.
1967 Pigs for the Ancestors: Ritual in the Ecology of a New Guinea People. New Haven: Yale University Press.

Renfrew, Jane
1973 Palaeoethnobotany. New York: Columbia University Press.

Richards, Audrey
1939 Land, Labour and Diet in Northern Rhodesia: An Economic Study of the Bemba Tribe. London: Oxford University Press.

1948 Hunger and Work in a Savage Tribe; a Functional Study of Nutrition Among the Southern Bantu. Glencoe: Free Press.

Tanaka, Jiro
1976 Subsistence Ecology of Central Kalahari San. In: Kalahari Hunter-Gatherers: Studies of the !Kung San and Their Neighbors, edited by Richard B. Lee and Irven DeVore, pp. 98-119. Cambridge: Harvard University Press.

Young, Michael W.
1971 Fighting with Food: Leadership, Values and Social Control in a Massim Society. Cambridge: Cambridge University Press.

Whalen, Michael E.
1977 Settlement Patterns of The Eastern Hueco Bolson. El Paso Centennial Museum Publications in Anthropology 4.

# PART II
# NATIVE EPISTEMOLOGY AND ETHNOBOTANY

## INTRODUCTION

Time, space, being, and becoming are the essence of world view and the basis of ethnoscience. As a folk science, ethnobotany is particularly sensitive to explanations of life and changes in the world order of corporeal and non-corporeal beings. The classification of living objects in space includes the entities that we call *plants*. The primacy of plants, as exemplified by the tree of life in many cultures, is an inalienable concept for understanding how knowledge about the world is attained.

In the creation myths of many cultures, plants (not necessarily in their present form) preceded animals and humans. While humans became the beneficiaries of the learning and meaning of these other living things, they also are dependent upon them for knowledge about life giving plants.

Plants provide a metaphor for life: the oak in ancient Europe, maize in Native America. The germination, growth, and death, especially of annual plants which all can observe, are a metaphor for the stages by which humans pass through this world as well. The power inherent in plants, if understood, can treat illness. In some cultures, the origin myth is a charter for knowledge and the restoration of order in the universe. Since plants preceded animals, they are more powerful and can alter the sickness or misfortune brought by animals. Humans can disrupt this order, but they cannot change it; their knowledge is power, but with it is the seed of their own destruction. In short, harmony is maintained among living beings and other forces in the universe through the proper use of plants.

The place of living things in the world is codified by local folk classifications. Each culture names its own elements; yet, as Berlin and others have shown, there is a similar hierarchical and inclusive classificatory structure that is found in all cultures. The taxonomy of plants may also indicate the relative importance of certain categories of plants to a culture without specifying particular information about them. For example, plants which are not used by anyone in any cultural situation are commonly lumped into a single name. Culturally significant plants such as cultivated crops may be recognized and called by dozens of names even if only a single species is recognized according to Western botanical standards. Understanding folk plant taxonomy as well as the Western botanical systematics is fundamental to ethnobotany. The exact relationship between the two systems of classification—folk and Western science—must be examined empirically.

One could argue that by learning such a system of thought, continually reinforced by legend and ritual, new knowledge of plants would be difficult to obtain. Often it is; random experimentation with plants is rare and risky. Yet, as Marcus and Flannery have described, changes do occur. These may result from a special relationship between a plant or its by-product and its use by a culture.

Merrill has examined the meaning of drinking a fermented corn beverage among the Rarámuri (Tarahumara). To explain the behavior that results from drinking *batári*, Merrill found that "the Rarámuri combine their ideas concerning the nature and operation of their souls with their observations of the events that mark this transformation" from sobriety to inebriation. The analysis of Rarámuri drinking behavior reveals general Rarámuri beliefs about the soul and its relationship to health and maturity.

Among the Aguarunas, Brown discovered an equally complex relationship between psychoactive plants and the human soul. But such a reciprocal act as planting and consuming is not intelligible simply by learning the physical attributes that distinguish the plants in a folk taxonomy. The critical issue is the cultural classification, which orders each according to its function in the society. Once this is recognized, then meaning and knowledge in a particular society are more comprehensible.

Folk medicine also illustrates the complexity of knowledge required to effect a cure. Messer examines all facets of herbal medicine: cultural setting, plant habitat, differential knowledge of the participants. Primary to a successful cure is a classificatory system which subsumes both a disease's causes and its cures. And while many Zapotec Indians share certain principles of this classification—hot and cold—the specifics may vary enormously within the community.

This differential sharing of information raises another problem that ethnobotanists must face: the loss of knowledge. External pressures from the dominant Mexican culture, Messer discovered, are producing new sources of rapid change. The acceptance of alternatives, such as patent medicines, even when they are interpreted according to a traditional classification, leads invariably to the reduction in plant lore. The loss of plant knowledge is a singular crisis throughout the world. Medicines and foods which sustained countless ancestors are converted to memory in a single generation and forgotten entirely by the next.

Discovering the processes through which information about the world is gained and lost is an important contribution that ethnobotany can make to anthropological theory and to the understanding of native epistemologies.

# THINKING AND DRINKING: A RARÁMURI INTERPRETATION

*William L. Merrill*
University of Michigan

The consumption of alcoholic beverages is a particularly interesting phenomenon from an ethnobotanical perspective because the availability of alcohol, unlike that of most other psychotropic substances, depends not on the distribution of a particular plant species but on the natural process of fermentation. As a result, people almost the world over have independently discovered alcohol or at least have experienced its effects and have developed subsequently their own attitudes and approaches to it. Not surprisingly, these reactions and accommodations run the gamut from vitriolic condemnation and total abstinence to near adoration and encouraged indulgence. While this entire range of beliefs and practices sometimes is maintained within a single community, it is more often the case that people who live together and especially people who drink together tend to agree generally about how one should view alcohol and its consumption. Nonetheless, despite such agreement and the fact that alcohol presumably affects the physiological systems of all individuals in basically the same fashion, individuals who drink together frequently display marked variations in the manner in which they behave when drunk.

In the spring of 1977, my wife and I entered the Sierra Madre Occidental of southwestern Chihuahua, Mexico, to live and study among the Rarámuri, or Tarahumara as they are more commonly referred to by non-Rarámuri (for general ethnographic treatments of the Rarámuri, see Lumholtz 1902; Bennett and Zingg 1935; Pennington 1963; Fried 1969; Kennedy 1970a, 1978). The Rarámuri dedicate much of their time, energy, and resources to brewing and drinking a fermented beverage that they call *batári* or *sugí* and known in Spanish as *tesgüino*. Being devoted drinkers with centuries of experience, the Rarámuri have developed their own theories about why individuals vary from one another in their behavior when they are drunk and how one gets drunk in the first place.

In this paper, I will first briefly describe Rarámuri alcoholic beverages and some of their drinking practices and then provide a general account of Rarámuri interpretations of the process of getting drunk and the behaviors associated with a state of inebriation. I believe such an approach to Rarámuri drinking is well in keeping with Volney Jones' (1941) conception of ethnobotany as the study of the interrelationships between humans and plants. Of crucial importance to understanding these interrelationships is a comprehension of the meanings people impose both on the plants (or plant derivatives) with which they interact and on their relationships with them. This is particularly true in regard to psychotropic plant substances in general and alcoholic beverages in particular. The interaction between an individual and alcohol that leads to his inebriation is subjected to constant interpretation by the drinker. It is through such interpretation that the individual attempts to render his drinking experiences meaningful. Similarly, by understanding the ideas upon which the individual draws in making his interpretations, the nature of the interaction becomes more meaningful and understandable to the ethnobotanist.

The information I provide here is derived from my experiences primarily in the Rarámuri community of Rojogochi and secondarily in other Rarámuri communities that, together with Rojogochi, constitute the pueblo of Basihuare (Municipio of Guachochi). The section dealing specifically with the Rarámuri theory of the inebriation process is based principally on conversations with five men, all residents of Rojogochi. In addition, I have discussed various aspects of this theory with ten other men and one woman, of which four reside in Rojogochi and seven in nearby settlements. These individuals varied from one another in specific details of their respective theories of the inebriation process, but all concurred on the general points presented here. Nonetheless, given the rather marked differences that exist among the Rarámuri of different areas in such things as settlement patterns, dress, and vocabulary, it would be presumptuous to claim that this presentation is generally valid for all Rarámuri. Therefore, the following discussion should be considered applicable only to the Rarámuri of the Rojogochi area.

## GENERAL BACKGROUND

The Rarámuri draw upon a wide variety of plants as basic ingredients in the preparation of intoxicating beverages. By far the most important of these is *Zea mays*, of which both the kernels and stalks are used. Beer prepared from maize kernels is consumed throughout the Rarámuri area on a year-round basis, while that fermented from the expressed juice of maize stalks is prepared only when these stalks are approaching maturity but are still juicy and sweet, during the last couple of months before the harvest. The residents of Rojogochi also occasionally prepare alcoholic beverages from apples, peaches, and *manzanilla* berries (*wíchari, Arctostaphylos pungens*) when these fruits are mature and from the roasted stems and leaf bases of *Agave* (*me, chawí*) and *Dasylirion* (*siré*) species, but only during the winter and spring months when the plant parts to be used are sweet. Except in the case of maize stalk beer, which they sometimes consume alone, the inhabitants of the Rojogochi area usually mix the beverages produced from these other plants with maize kernel beer before drinking. In addition to the plants used in Rojogochi, the Rarámuri of other areas are reported to prepare fermented beverages from *Sorghum vulgare*, several cacti, sugarcane, wheat, *Randia, madroño* (*Arbutus* spp.), and, in the recent past, mesquite (*Prosopis juliflora*) (Pennington 1963:155-57; Bye 1976:75-76; Bye et al. 1975).

Regardless of the plant or plants selected for fermentation, the Rarámuri of Rojogochi invariably refer to the end product as *batári* or *sugí* (also pronounced *suwí* and *sugwí*), terms that they consider synonymous and use interchangeably. To indicate the plant source of the beverage, they simply place the name of the plant before the term *batári* or *sugí*: thus, maize kernel beer is *sunú batári* (*sugí*), maize stalk beer is *sonó batári* (*sugí*), maguey beer is *me batári* (*sugí*) and so forth. If the *batári* is prepared from more than a single plant, this fact can be conveyed by stating that it is a mixture and listing the plants from which it is derived.

The basic procedure adopted in preparing *batári* is generally the same for all plants employed throughout the Rarámuri region. The chosen plant part is ground, pounded, mashed, or squeezed, and the resultant extract is then mixed with water and boiled. After boiling, the liquid is allowed to cool and then is transferred to fermenting ollas. One or more plants—which usually are ground up first and which are viewed by the residents of Rojogochi as fermentation catalysts—are added, and the liquid is then allowed to ferment. Of course, most plants need some sort of special treatment beyond these general procedures (the preparation of maize kernel beer, for example, requires several additional steps), and there also seem to be minor regional variations in procedure (see Pennington 1963:149-57 for details).

The amount of *batári* prepared on any occasion varies considerably according to the reasons for the gathering, the number of people to be invited, and the amount of maize the host is willing to expend, among many other factors. Ideally, enough *batári* is brewed to enable all the people present to get very drunk. By my calculations, each drinker requires around four liters of strong *batári* to become reasonably intoxicated. For each liter of maize kernel *batári* prepared, approximately one-quarter liter of kernels (weighing around 175 g) is needed. Thus, a small gathering attended by around ten people would require a minimum of 40 l of *batári* (with an expenditure of 7.0 kg of maize kernels) while a large fiesta involving perhaps as many as a hundred drinkers or more would necessitate at least 400 l of *batári* (requiring 70 kg of kernels).

These figures represent the minimum amount of *strong batári* required to intoxicate those present. Quite often, enough *batári* is prepared for everyone to drink six or more liters each, although on rare occasions the amount of *batári* available is insufficient for each person present to consume even four liters apiece. Even if an average of four liters of *batári* per drinker is on hand, the inebriation of all those present is not assured. The strength of *batári* can vary a great deal from batch to batch, some individuals drink much more than others, and drinkers vary among themselves in their tolerance of alcohol. Given all the factors involved, it is not surprising that on some occasions everyone present becomes very inebriated, on other occasions no one gets drunk, and on still others some individuals will be considerably more intoxicated than their fellow drinkers.

From the Rarámuri perspective, it is the plant or plants added to *batári* during its preparation that transform the liquid from non-intoxicating to intoxicating. In Rojogochi, the plants employed most frequently as additives are *basiáwari* (*Bromus porteri*) and *kotó* (*Phaseolus* sp.). *Basiáwari* is used primarily with maize kernel *batári* and is mixed with the cool liquid when it is set aside to ferment. *Kotó* is employed more frequently with *batári* prepared from maize stalks, apples, peaches, *manzanilla*, *Agave*, and *Dasylirion*. The crushed roots of *kotó* are cooked together with the liquid during the cooking stage and also can be placed with the liquid in the fermenting ollas during the fermentation stage. According to the Rarámuri, these plants make *batári* strong because they cause it to "boil," or ferment, and thus to change its flavor from sweet to bitter. The Rarámuri envision a close connection between the taste of *batári* and its strength and often employ taste categories to indicate potency. If *batári* is *we akáame*, "very sweet," it also is very weak; if it is *we chipúame*, "very bitter," it is very strong. Before fermenting or after having fermented only a short while, the *batári* is *akáame*; after it has fermented a number of hours it is *chipúame*, assuming that nothing has interfered with the desired transformation.

The Rarámuri of Rojogochi state that *batári* passes through two major stages in the course of its fermentation. In the first, which usually commences soon after the sweet liquid is poured into the fermenting ollas, the brew "boils" at a fast pace. As the liquid is increasingly transformed from sweet to bitter, the rate of fermentation slackens until the *batári* is bubbling and fizzing only slightly. Although sometimes people will drink *batári* while it is still "boiling" rapidly, usually they prefer to wait until it has reached this second stage before consuming it.

The Rarámuri consider *batári* to be at its optimum for a period of approximately 12 to 18 hours after it has ceased to "boil" rapidly. After this, it tends to lose its strength and is said to be *simírame*, "passed." Quite frequently, a second batch of *batári* is prepared after the first batch has been poured into the fermenting ollas. When the second batch has been cooked and allowed to cool, the sweet liquid is mixed with the first batch of *batári*. Whether or not the first batch is passed, the mixture then begins fermenting rapidly, and the fermentation process is repeated until the *batári* is deemed suitable for consumption. If a second batch of *batári* is not available, a little sugar can be added to the passed *batári* to make it ferment again. Such twice-fermented *batári* is considered to be particularly potent, but sometimes its taste is a little sweet rather than bitter because of the sugar added. To indicate that such sweet *batári* is also very strong, the Rarámuri usually characterize it as *we iwéame hu*, "it is very strong," or *we ka?rá hu*, "it is very good," phrases that they employ to denote strong *batári* in general. In addition, they sometimes will refer to such twice-fermented *batári* as *we chipúame* despite its slightly sweet taste, using this phrase in its more figurative sense of "very strong" rather than "very bitter."

Besides tasting sweet or bitter, *batári* can alternatively be characterized as *ikótame*, "burned," or *cho?kóame*, "sour," both of which conditions are believed to affect the *batári's* intoxicating properties. *Batári* that is scorched during the preparation process has a very unpleasant taste and is not very strong. In addition, even though drinkers eventually will finish off a batch of scorched *batári*, they usually cannot drink much of it at one time because it tastes bad and is considered to fill one up quickly; therefore, one seldom drinks scorched *batári* in sufficient quantities fast enough to become very intoxicated. Similarly, sour *batári* also is unpleasant to drink, is said to be very filling, and is not very strong; if any intoxication is experienced, it will be slight and of short duration.

The consumption of *batári* among the Rarámuri is preeminently a social event. Historical records dating from the late seventeenth century indicate that the Rarámuri of the colonial period imbibed their alcoholic beverages communally (Neuman, in Pennington 1963:153). Today, the Rarámuri consume their *batári* exclusively in communal contexts, and most extra-familial get-togethers include *batári* drinking. On the other hand, some social events, such as gathering to bet before footraces or assembling on Sundays at the local church to pray and discuss community affairs, as a rule never involve drinking. The social activities associated with drinking are innumerable, but a representative list would include: sending a portion of the newly harvested crops to God; planting, weeding, and harvesting the maize crop; building a house; cutting firewood; performing the celebrations of the Catholic ritual calendar; hauling manure; curing people, animals, or the maize fields; sending food to the dead; moving cattle corrals; building stone walls; and so forth. Frequently, several of these activities will take place during the same gathering. On a typical occasion in Rojogochi, a Rarámuri doctor cured the residents of the house in which the *batári* was served, food and blankets were sent to a relative who had passed away a year before, the men hoed maize, and the women prepared food. Rarely is drinking the only major activity to occur.

The general practice in the literature and among non-Rarámuri is to call all Rarámuri social gatherings that involve *batári* drinking *tesgüinadas*. However, the Rarámuri employ no noun comparable to the term *tesgüinada* to label such events, instead juxtaposing verb forms that indicate one or more of the activities other than drinking that occur in each specific gathering and derivatives of the verb *bahíma*, "to drink." Thus, in reply to the question, "What are you doing?" a Rarámuri would never say, "We are having a *tesgüinada*," but would reply, "We are planting maize and drinking," "We are asking God to send rain and drinking," or "We are cutting firewood and drinking." The reference to drinking or alternatively to the accompanying activities may be dropped altogether, since the co-occurrence of drinking and such communal activities generally is assumed.

The Rarámuri view serving *batári* as a way of attracting people to help them and of expressing their gratitude to people for services rendered during the gathering when the *batári* is served or on some previous occasion. This is not to say that the Rarámuri invariably prepare *batári* with the idea of completing some project. However, when a task is at hand that can be performed by a group, inviting people to drink is a frequent way of fulfilling it.

Most if not all the individuals who attend such a get-together expect and are expected to help out the host and hostess. People provide their assistance by performing any of a variety of activities that range, for women, from making food and cleaning wool to assisting their husbands in curing ceremonies and, for men, from hoeing maize and chopping firewood to singing and dancing in ceremonies or providing the musical accompaniment for these performances. At the same time, these individuals expect the host household to reciprocate by supplying them with *batári* to drink, and, if they consider the amount of *batári* available to be insufficient, frequently will express their dissatisfaction openly.

From a Western economic perspective, it appears that the Rarámuri are exchanging their labor for *batári* in these working and drinking sessions. However, the Rarámuri describe the giving and receiving of *batári* in such contexts in terms distinct from those they use to refer to an economic transaction such as obtaining a basket from a neighbor through barter or earning money by working for local Mexicans. All the people who gather to drink except the members of the host household are said to be *kóriga bahí yéna*, which means they are drinking as the recipients of a gift, the gift being *batári*. If an individual arrives after the work is completed, is not feeling well, or for some other reason is unable to assist the host or hostess, he is not denied *batári* nor begrudged the *batári* he consumes. Of course, the host almost always singles out those individuals who helped him the most or to whom he accords a great deal of respect by calling them individually to drink first or by bringing *batári* to them. Nonetheless, the idea is that each person present, regardless of the degree of assistance he provides, should drink as much *batári* as he desires as long as the supply lasts.

In addition to giving *batári* to his guests, the host usually gives each olla of *batári* he has on hand to one or sometimes more people present, often in recognition and gratitude for some service previously rendered, like the loan of oxen during plowing. Receiving an olla does not mean that the recipient consumes the contents alone but that the *batári* in the olla is his to distribute. The recipient of an olla first sends a bit of the *batári* to God by filling a gourd dipper with *batári* from his olla and then, with a second dipper, tossing out small amounts of this *batári* to the cardinal directions. The *batári* remaining in the first dipper is then exchanged in small portions among the recipient and a few of the other people present, often but not invariably including the host. If one or more ollas of *batári* have been consumed prior to his receipt of an olla, the recipient then may instruct all those present to begin drinking from his olla. However, customarily in the case of the first olla consumed on a given occasion and sometimes in the case of subsequent ollas, the recipient and the host together call certain people to the olla to drink with them before the others in attendance are given access to the *batári*. These individuals—usually those who prepared the *batári* and frequently some of the Rarámuri community officials or other respected figures—exchange a few gourd dippers or commercial vessels of *batári* among themselves until they have drunk their fill. The recipient then opens up the drinking of his olla to the others present, sometimes sitting beside the olla to supervise its distribution.

Most social events that include *batári* drinking are not open to the general public. Instead, a member or members of the family or families who prepare the *batári* will go from house to house inviting the people they want to attend or will dispatch someone else as their representative to do so. Quite frequently after the drinking and associated activities are underway, uninvited people will pass by or just show up and stand around at the periphery. Usually the host will invite them to come join in, although occasionally he will tell the uninvited to return to their homes or ignore them altogether. However, such inhospitality is considered bad form and potentially dangerous. For instance, one winter several years ago, a Rarámuri man from the Narárachi area (approximately 40 km from Rojogochi) stopped by a house in Rojogochi where *batári* drinking was under way. Rather then inviting the traveler in to warm himself and drink, the host ignored him. After shivering in the cold an hour or so, the man departed very angry and in the night bewitched the host, who died several days later.

The fact that the people invited to drink by one household tend to differ somewhat from those invited by other households leads to the emergence of what Kennedy (1963:625-26) has termed the "tesgüino network." The loci of this network are individual households that are linked to other households through the overlapping attendance of their members at drinking get-togethers. Thus, the members of two households that usually do not invite one another to drink in their own homes will drink together in a third household with which both exchange drinking invitations. In Rojogochi and probably throughout the Rarámuri area, the interaction network resulting from an individual's drinking contacts contains the sets of social relationships, other than those at the level of the nuclear or extended family, in which that individual participates most frequently.

## THE RARÁMURI INTERPRETATION OF GETTING DRUNK

In interpreting the inebriation process and associated behaviors, the Rarámuri rely heavily upon their ideas concerning the relationship between body and soul. According to the Rarámuri, every individual has many souls, which are distributed throughout the interior of his body and which, as a group, insure the continued existence and operation of the body. At least one soul is associated with every part of the body that is capable of movement, and each soul is responsible for keeping the body hot inside and moving the portion of the body where it is located.

The Rarámuri employ two synonymous terms to label the souls: *ariwá* and *iwihá*. Both terms also can be translated as "breath" because the Rarámuri consider the souls and breath to be one and the same thing. An individual's souls range in size from large to small and, because they live together, are said to be siblings. In fact, the terminology employed to distinguish among siblings according to the order of thier births also is used—followed by either *ariwá* or *iwihá*—to classify a person's souls. However, since all of a person's souls are the same age, they are labeled according to their relative sizes rather than relative ages. Thus, the smallest souls are known by the term for the youngest sibling, the next largest souls by the term for the next-to-youngest sibling and so on up to the largest soul, which is labeled by the term for the eldest sibling. Since all of a man's souls are male and all of a woman's souls are female, the terminology used for a man's souls is that employed when all the siblings are brothers, and the labels marking a woman's souls are those used when all the siblings are sisters.

The vast majority of an individual's souls are relatively small in size and are located in places like the joints and muscles. The larger souls are fewer in number and are found inside the head and heart, with the largest soul living exclusively inside the heart. However, there is some disagreement among the individuals I have consulted regarding the number of large souls a person has. Most people state that each man possesses three large souls and each woman four. Others state that both men and women have three large souls each, while still others say they do not know exactly how many large souls an individual has.

The largest soul inhabiting the heart and the large soul that lives inside the head, frequently vacate the body, entrusting its care during their absence to the other souls, which abandon the body only at death. When these larger souls decide to leave a person's body, he starts to feel drowsy. Upon their departure, he falls asleep and awakens when they return. The activities of these souls during their wanderings, which usually occur at night, are experienced as dreams. Their temporary absence during sleep is not considered particularly dangerous since they usually are free to return whenever they like. However, sometimes the souls encounter difficulties in their excursions outside the body and the largest soul is unable to return. For example, the largest soul can be grabbed, detained, and eventually eaten by a large snake (*waˀlúluwi*) that lives deep below the surface of bodies of water. As a result, the individual sickens and eventually might die if his soul cannot be retrieved by a doctor, who attempts to do so by performing a ceremony and searching for it in his dreams. Death occurs when all the souls depart, leaving the body as a cold, lifeless hull.

An individual's thinking is done entirely by his souls; his body, being just flesh and bone, cannot think. But not all souls are equally adept at thinking. To a great degree, a soul's ability to think is correlated with its size: the larger a soul is, the better it can think. In fact, the smallest souls (for example, those in the finger joints) are said to be incapable of thinking at all; they just move the body and keep it hot inside, activities that the Rarámuri feel do not require thinking.

An individual's body and souls mature together, so that the ability to think is enhanced with age because the souls are larger. Babies are considered incapable of thinking, while young children can think only a little. Only after attaining a certain age, which generally falls around puberty, is a person able to think well. However, the ability to think well does not depend simply upon the size and maturity of a person's souls. Of utmost importance is the nature of the information thought about during the maturation process. A crucial source of the information needed to develop the ability to think well is the advice given a child by his or her elders, usually the parents or older siblings but also other, mature people who can think well. This advice consists primarily of information about how one should conduct oneself in dealing with others: don't steal, don't fight, be faithful to your spouse, etc. If a child does not receive and incorporate such advice during maturation, his souls will grow up "crazy" and, as an adult he will be able to assimilate such advice only with difficulty, if at all. In addition, even after reaching physical adulthood, a person's thinking ability continues to develop; most older people are believed to think better than younger adults because they have thought about many more things.

When the Rarámuri say that someone thinks very well, they mean that the person in question not only performs intellectual operations at an adult level and in general behaves like an adult, but that his behavior conforms to the Rarámuri standards of propriety. The Rarámuri view an individual's actions above the level of the simplest motor activities to be directed by his thinking and a direct expression of that thinking. Therefore, a person's thinking ability is evaluated in terms of his behavior: good behavior indicates good thinking and vice versa.

The Rarámuri employ several different phrases to characterize a person who fails to behave in accordance with what is considered right. Frequently these phrases constitute explicit evaluations of the individual's thinking ability: *tabiré nátame hu*, "he doesn't think at all," *ki richóti hu*, "he isn't smart," or *we lowíame hu*, "he's very crazy." Another phrase, which is heard easily as often as any of these others, is *ke tási riwéri*. This phrase consists of the negation *ke tási* plus the present tense form of the verb *riwérama*. This verb has a number of connotations, more or less adequately expressed by the Spanish *tener vergüenza* but conveyable in English only partially and only by combining the senses of such terms as "to have pride, shame, self-esteem, honor, respect, and a well-developed sense of appropriate behavior." If, while drinking, a person laughs and talks a great deal, shouts and dances and generally demonstrates that he is having a good time, he might say jokingly of himself *tásini namúti riwéri nihé*, "I have no *vergüenza*." However, the Rarámuri almost never use the term negatively in reference to a third person except in dead seriousness and then almost invariably as an insult. Of course, to say that an individual doesn't think, isn't smart, is crazy, or has no *vergüenza* is to impute these characteristics to his souls. That a person's souls should be in such a condition usually is attributed to the failure of the person in question to have received proper advice during maturation. Sometimes, however, it is seen as the result of his having gotten drunk before he is able to think well.

The only exception to these statements of which I am aware is the behavior of people who are considered to be *sukurúame*. The term *sukurúame* is a general one used to describe individuals who know, among many other things, how to induce as well as alleviate illnesses. The Rarámuri assume that anyone who knows how to cure people also knows how to bewitch them and vice versa. A person who maintains such knowledge must be able to think and dream well. However, if and when he directs this knowledge to the detriment rather than benefit of others, his conduct obviously is considered improper. While such a person is said as a result to have no *vergüenza*, it is not believed that he fails to think but instead that he is thinking differently from other good thinkers who are not so inclined or engaged. The Rarámuri say of him: *ki tási me ga*ʔ*ra wachíniga nátame hu*, "he is not thinking very straight," or *we waná nátame hu*, "he is thinking apart, on the other side," that is, with the Devil rather than with God.

From the Rarámuri perspective, getting drunk is the direct result of the abandonment of the body by a drinker's largest soul. When *batári* is consumed, it passes to the stomach where it remains until it is excreted in the form of urine or feces. While in the stomach, the *batári* continues to ferment, or "boil," and to give off its fumes, just as it did when it still was in the fermenting olla. All the souls dislike the way *batári* smells and the fact that it is "boiling" close to them. However, only the largest soul vacates the body when *batári* is present inside the stomach in quantities sufficient to create a context it considers intolerable.

When the largest soul departs from the body, it remains near the drinker, caring for him from outside. As the person consumes more *batári*, the largest soul moves farther and farther away, and the drinker feels increasingly intoxicated. At the same time, the souls that stay inside the drinker care for him and keep his body hot. The souls themselves do not become intoxicated because they find *batári* distasteful and do not drink any. The state of being drunk is due to the absence of the largest soul rather than the inebriation of those present.

After a person has drunk a great deal, the large soul that lives inside the head sometimes will join the largest soul outside the body and the drinker falls asleep. When this soul reenters the body, the individual wakes up, quite frequently to drink again. After he stops drinking, the souls that are inside the drinker guide him on his way home, with the largest soul usually trailing close behind. As the *batári* passes out of his system, his largest soul comes nearer and nearer to him until all the *batári* is eliminated and the largest soul reenters the body.

The souls do not find all *batári* to be equally obnoxious. Their reaction to a particular batch of *batári* varies with its smell, the amount of it that can be retained inside the body at any given time, and its propensity to continue "boiling," or fermenting, while in the stomach. The souls find bitter *batári* exceedingly offensive because it can be consumed in large quantities without overly bloating the drinker, it "boils" a lot while in the stomach, making it feel hot inside, and it has a smell they consider extremely distasteful. Sour and burned *batári* fill the drinker up quickly and tend to pass out of the system soon after consumption, "boiling" just a bit while still in the stomach. The odor of sour and burned *batári* is only slightly obnoxious to the souls, the largest of which might leave the body when confronted with sour or burned *batári* but remains close by and soon returns. The souls are indifferent to the smell of sweet *batári*, which also tends to be very filling and which fails entirely to "boil" in the stomach. *Batári* that has been induced to ferment twice by the addition of sugar also has a slightly sweet smell, but the souls dislike it because a great deal of it can be retained in the stomach with relative ease and it "boils" a great deal while there. Thus, excepting the special case of twice-fermented *batári*, the Rarámuri consider the strength of a given batch of *batári* to be directly correlated with its taste and smell and view both bitter and twice-fermented *batári* to be especially intoxicating because they induce the largest soul to leave more rapidly and go farther away than sour, burned, or sweet *batári*.

As a general rule, the Rarámuri want to get as drunk as possible when they drink. With this goal in mind, the drinkers try to keep their stomachs filled to capacity with *batári* as long as a supply lasts or until they have passed out or are too drunk to drink more. In fact, getting drunk and keeping the stomach filled with *batári* are seen as so closely connected that the phrases *ácha mu bosáre*, "Did you fill up?", *uché bosása*, "Fill up again", and *chéni bosáre*, "How full did you get?", are used interchangeably with *ácha mu bahíre*, "Have you drunk?", *uché bahísa*, "Drink again", and *chéni rikúri*, "How drunk did you get?".

To further insure inebriation, the Rarámuri invariably supplement their drinking by smoking cigarettes, either purchased pre-rolled tobacco or stronger homegrown (in Rojogochi, *Nicotiana rustica*) that they roll in cornhusks, discarded cigarette packages, or other paper. Some individuals claim that they become slightly intoxicated by smoking tobacco alone, but others report no such effects. In any case, there is a general consensus that smoking cigarettes enhances the probability of getting drunk. The smoke is inhaled into the heart where the largest soul resides and hurts it just as smoke from a fire hurts us. When this smoke is combined inside the body with the unpleasant fumes of *batári*, the largest soul will be strongly inclined to leave the body as quickly as possible.

In many ways, the Rarámuri envision the getting drunk process as a reversal of the maturation process. During one local drinking session, a man arrived much drunker than the others present. While he stumbled around, mumbling incoherently, the other men sat watching him, speculating on his age. After considerable discussion and much laughter, they finally agreed that he was around 3 years old, although he was born approximately 55 years ago.

Residents of the Rojogochi community frequently make the observation that drunk people act like children and point to several correspondences between the behavior of drunk adults and children. Neither small children nor drunk adults are remarkably adept at executing such motor activities as walking and talking. Further, small children have a definitely limited memory capacity, and adults after they have sobered up usually are incapable of recalling all the details of their activities while they were inebriated. In addition, when drunk, people laugh, cry, dance, shout, chase each other around, and generally behave in ways that they consider to closely approximate in form if not always in content the behavior of their children. The Rarámuri attribute the existence of all these similarities to the fact that, when an individual is very inebriated, his largest soul is relatively far away, with the souls that remain inside his body approximating in size and thinking ability the largest souls of small children. Thus, a person acts like a child when he is drunk because the souls that are primarily responsible for directing his behavior are comparable to those of children.

The Rarámuri classify a great deal of the behavior that takes place in drinking contexts as "play" and consider it to bear general similarities to the play of children. They employ the verb *riyéma* to refer to play in general, whether the players are children, sober adults, or drunk adults. However, they use a second verb, *ramuhérama*, to denote a special subclass of play that is distinguished by its emphasis on reciprocal teasing or joking between two or more individuals and its frequent and explicit references to sexual matters. This kind of playing takes place more often among adults than among children, although it is not uncommon for adults to engage in such play with children.

There are a wide variety of behavioral acts that exemplify this special kind of play. On a strictly verbal level, a person can disparage the sexual capacity of another, inquire about his or her sexual activities, or simply joke about sexual or other matters. Such verbal playing often is combined with non-verbal acts such as wrestling; grabbing at another person's genitals or the clothing covering the genitals; simulating sexual intercourse; exposing the genitals (in this case, usually a man will expose his penis to a woman, who acts revolted, often spits at it, and then laughs); dancing in front of a member of the opposite sex (sometimes a man will don a discarded dress and dance like a woman); pretending to eat the genitals of a person of the opposite sex; or making a variety of signs with the hands or other body parts that signify sexual intercourse.

This kind of behavior often occurs outside drinking contexts but usually only in a rather muted form. It emerges most elaborately and most frequently when people are drinking. Such interactions can take place between any individuals present, regardless of their sex, relative ages, or relationship. In fact, some people are noted for their propensity to play in this fashion with everyone and are referred to as *ramuhérawame*, a term with approximately the same connotations as the English colloquialism, "a regular joker." However, the Rarámuri of Rojogochi (and at least some other areas as well; see Kennedy 1970b) consider this kind of behavior to be especially appropriate between certain sets of relatives, most particularly grandparents and grandchildren and siblings-in-law of the same and opposite sex. In fact, the Rarámuri almost invariably explain the occurrence of such behavior between individuals so related simply by pointing out that such a relation exists between them. When the people involved are not related as grandparent-grandchild or siblings-in-law, which quite frequently is the case, people usually say, "They're just playing," or "They're just joking around."

This kind of play is expected and encouraged by the Rarámuri and provides a major source of entertainment both for the individuals who engage in it and for those who watch (for a more detailed discussion of the importance of play in Rarámuri drinking contexts, see Kennedy 1970b). However, there are other acts that occur primarily in association with drinking that are universally and strongly condemned. Most important of these condemned actions are getting into fights and engaging in sexual activities with someone other than a spouse. For the Rarámuri, such unacceptable behavior cannot be explained simply by pointing to the inebriation of the people involved, since not everyone who gets drunk engages in such behavior. To account for such individual differences in behavior, the Rarámuri again rely on their ideas about the relation between an individual's souls and his behavior, but in this instance they emphasize the nature of a person's souls rather than their presence or absence.

Both fighting and philandering elicit the comments of others that the transgressor is crazy, does not think well, or has no *vergüenza.* All three characterizations are considered equivalent pragmatically and indicate that the souls of the person in question are to a greater or lesser degree crazy. Manifestations of insanity are believed to occur more frequently in drinking contexts because a person's ability to think well and behave appropriately is significantly diminished when his largest soul abandons the body. In addition, the Devil always shows up where drinking is underway and attempts to induce people to fight and commit adultery by rendering their souls insane.

Despite the fact that instances of inebriation and insanity usually co-occur, the Rarámuri clearly distinguish between being drunk and being crazy. The verb *rikúma*, "to become intoxicated," is never used to describe a person whose behavior is seen as crazy, and *batári* is not believed to make a person crazy, only drunk. A person who thinks well when sober does not engage in undesirable behavior even when very drunk because, before his largest soul leaves, it instructs the ones left behind in how to act and care for the body, much like an elder advises a child in how to behave properly. People whose largest souls are consistently crazy engage in fights and commit adultery when drunk because these souls have no good advice—or, better, only bad advice—to impart upon their departure.

## CONCLUSIONS

From the Rarámuri perspective, their ability to transform a sweet, innocuous liquid into a bitter, intoxicating beverage allows them to transform themselves from sober to drunk. To explain the consistent occurrence of this transformation, the Rarámuri combine their ideas concerning the nature and operation of their souls with their observations of the events that mark this transformation. The result is a theory in terms of which the individual can interpret meaningfully his experience of getting drunk as well as understand the behavior of others who are drinking with him.

One very attractive feature of the Rarámuri theory explaining the getting drunk process is that it deals rather effectively with both similarities and differences in the behavior of different drinkers. From the perspective of this model, the vast majority of an individual's behavior above the level of reflexes is motivated by his thinking, and his thinking is performed exclusively by his major souls. Therefore, an individual's behavior can be related directly to the nature and activities of his major souls. The souls of all individuals react similarly to the proximity of *batári*, so their behavior while drunk also exhibits general similarities. Differences in the behavior of individual drinkers derive from variations in the nature of their souls, variations that are basically reflections of differences in how and what they think.

This theory not only provides the drinker with an explanation of what is happening to him while he is getting drunk but simultaneously suggests the general procedures he must follow in order to achieve a state of inebriation. To become drunk, according to this model, a person must induce his largest soul to abandon his body, and the degree of inebriation he will attain is directly proportional to the distance the largest soul removes itself from the body. Thus, the Rarámuri approach drinking with the idea of rendering the interiors of their bodies as offensive to their souls as possible. Their practices of continuously smoking cigarettes while drinking and of filling up fast with *batári* rather than gradually sipping it are intended to achieve this goal and are logical outgrowths of the more general Rarámuri theory of the inebriation process.

Given the fact that the Rarámuri associate most serious illnesses as well as death with soul loss, it might seem ironic that they intentionally attempt to drive their largest souls out of their bodies by drinking *batári*. However, when a person's largest soul leaves during drinking, it is not lost but only temporarily absent, remaining relatively near the drinker until the appropriate time for its return. An individual's well-being is threatened only when his largest soul is harmed or otherwise molested by malevolent entities who are capable of subjecting this soul to their machinations whether it is inside or outside the body. The risks run by a person's largest soul during drinking are no greater than those to which it is exposed during sleep or, to a lesser degree, during the individual's waking, sober life.

In undertaking the study of the interrelations between humans and plants, it is incumbent upon the ethnobotanist to examine seriously the manner in which the people involved in these interactions view the plants in question and their relations with them. Individuals tend to behave in ways that are meaningful to them—that is, in terms of their views of the world—and this is as true for their behavior in relation to plants as it is for their actions vis-à-vis their relatives or their deities. To begin understanding why the Rarámuri relate in the way they do to *batári*, it is crucial to be acquainted with their theory of the process of getting drunk, a theory that in itself constitutes an important facet of their interaction with alcoholic beverages. This is not to claim that Rarámuri drinking behavior is explicable only or fully in terms of their theories. There are several aspects of Rarámuri drinking behavior for which the Rarámuri have developed no elaborate explanatory models. For example, the individuals I have consulted consider the fact that they drink to get drunk as the obvious approach to alcohol consumption and thus not wanting of explanation. On the other hand, by putting forth an explanation of alcohol intoxication, the Rarámuri isolate and bring to the fore those aspects of the getting drunk process that they feel do require explanation, thereby providing valuable insights into the phenomenon of inebriation. Comparing and combining such insights and explanations with those that derive from comparable theories generated within the context of a Western view of the world will result in a more adequate ethnobotanical understanding of the relationship between humans and alcoholic beverages.

**Acknowledgments**

From 1973 to 1977, during my tenure as a predoctoral graduate student at the University of Michigan, I had the pleasure of spending many hours with Volney Jones discussing the wide range of plants and plant derivatives, including alcohol, that people employ to alter their states of being. My research on and understanding of such psychotropic substances have been considerably enhanced by Professor Jones' insights and encouragement.

This paper is partially the result of graduate study under the auspices of a National Science Foundation Predoctoral Fellowship and of field research sponsored by the National Institute of Mental Health (Predoctoral Research Fellowship number 5-F31-MH05687) and the Rackham School of Graduate Studies, University of Michigan. I gratefully acknowledge the financial support of these institutions.

I am deeply indebted to the residents of Basihuare pueblo and the community of Rojogochi, Chihuahua, Mexico, for allowing my wife and me to live and drink with them. In particular, I want to thank Candelario Martinez, Moreno Gonzalez, Mauricio Aquichi, Ventura Aquichi, and Mauricio Sahuárare for taking the time to explain to me what happens to our souls when we drink. Robert A. Bye, Jr. kindly identified the plant specimens I collected and commented on an earlier version of this paper. Enrique Gonzalez Zafiro graciously loaned me some maize and a set of scales on which to weigh it. Michael F. Brown, Donald H. Burgess, Claus Deimel, Jacob Fried, John G. Kennedy, Cecilia T. Merrill, María Elena Trillo, and Luis Verplancken read drafts of this paper and provided insightful criticisms and suggestions which have been of great assistance to me. I am especially indebted to my wife, Cecilia, whose companionship, daily encouragement, and intellectual stimulation made the writing of this paper and the research on which it is based both possible and enjoyable.

## REFERENCES

Bennett, Wendell C., and Robert M. Zingg
1935 The Tarahumara: An Indian Tribe of Northern Mexico. Chicago: University of Chicago Press.

Bye, Robert A., Jr.
1976 Ethnoecology of the Tarahumara of Chihuahua, Mexico. Ph.D. dissertation, Department of Biology, Harvard University.

Bye, Robert A., Jr., Don Burgess, and Albino Mares Trías
1975 Ethnobotany of the Western Tarahumara of Chihuahua, Mexico. I. Notes on the Genus *Agave*. Harvard University, Botanical Museum Leaflets 24(5): 85-112.

Fried, Jacob
1969 The Tarahumara. In: Handbook of Middle American Indians, Vol. 8, Pt. 2, edited by Evon Z. Vogt, pp. 846-70. Robert Wauchope, gen. ed. Austin: University of Texas Press.

Jones, Volney H.
1941 The Nature and Status of Ethnobotany. Chronica Botanica 6:219-21.

Kennedy, John G.
1963 Tesguino Complex: The Role of Beer in Tarahumara Culture. American Anthropologist 65:620-40.
1970a Inápuchi: Una Comunidad Tarahumara Gentil. México: Instituto Indigenista Interamericano.
1970b Bonds of Laughter among the Tarahumara Indians: Toward a Rethinking of Joking Relationship Theory. In: The Social Anthropology of Latin America, edited by Walter Goldschmidt and Harry Hoijer, pp. 36-68. Los Angeles: University of California Press.
1978 Tarahumara of the Sierra Madre: Beer, Ecology, and Social Organization. Arlington Heights, Illinois: AHM Publishing Corporation.

Lumholtz, Carl
1902 Unknown Mexico. 2 vols. New York: Charles Scribner's Sons.

Pennington, Campbell W.
1963 The Tarahumar of Mexico: Their Environment and Material Culture. Salt Lake City: University of Utah Press.

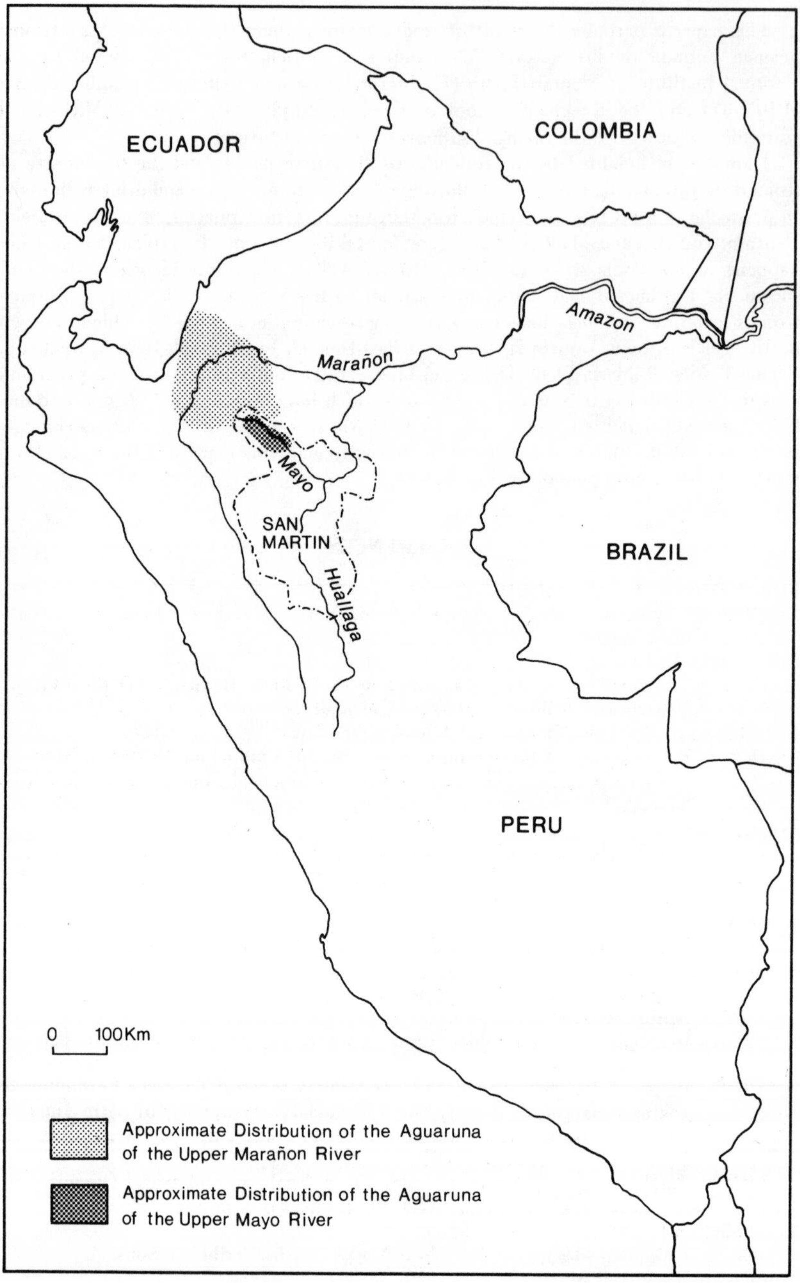


Fig. 1. Map showing localities occupied by Aguaruna groups in Peru.

# FROM THE HERO'S BONES: THREE AGUARUNA HALLUCINOGENS AND THEIR USES

*Michael F. Brown*
University of Michigan

The Jivaroan tribes of Ecuador and Peru have, since their first contact with Europeans, attracted attention for their bellicosity and unwillingness to accept foreign domination. Later research, most notably the work of Michael J. Harner among the *untsuri shuar* or Jívaro proper of Ecuador, has given us an appreciation of Jivaroan peoples' intensely religious view of the world. In Jivaroan thought, the normally invisible world of spirits parallels the visible world in that both are populated by beings which can be hostile or friendly. To insure individual prosperity, and even one's very survival, it is necessary to gain access to the spiritual world through the use of hallucinogenic plants.

In the present article, I wish to describe the varied uses of three hallucinogens which are cultivated by the Aguaruna Jívaro of the Upper Mayo River, Peru. These three plants, all of which are apparently members of the solanaceous genus *Datura* (some species of which are now classified as *Brugmansia* by some taxonomists), form a discrete unit in Aguaruna thought, yet they are attributed widely varying and even contradictory properties. By examining the wealth of detail surrounding the mythical origin, ethnotaxonomy, and uses of these species, which in themselves form only a segment of the inventory of hallucinogens known to the Aguaruna, I hope to draw attention to some general principles regarding the relationship between men and hallucinogenic plants in the Aguaruna scheme of things, as well as provide an example of the fine distinctions which a native people can make between varieties of culturally important plants. Finally, it seems appropriate to make some general observations about the changing role of hallucinogens in contemporary Aguaruna society. My interest in these problems is in many ways a result of contact with Professor Volney Jones, whose pioneering research on North American Indian hallucinogens—for example, his study of the so-called redbean, *Sophora secundiflora*, of the central and southern plains—has served as an inspiration to at least two generations of students at the University of Michigan.

## GENERAL BACKGROUND

Approximately 1000 Aguaruna Jívaro live along the Upper Mayo River and its tributaries in the Department of San Martín, Peru. This population represents the result of continuous emigration from tributaries of the Marañón River which began 30-35 years ago. Like their relatives in the Marañón, the Mayo River Aguaruna are sweet manioc cultivators, hunters, and fishermen who live in relatively dispersed communities. The Aguaruna have received general ethnographic treatment in several works (for example, Stirling 1938; Tessman 1930) and bear a strong cultural similarity to the Jívaro proper (Harner 1972).

The major agent of culture change for the Mayo River communities has been the bilingual schools, which were first established by the Summer Institute of Linguistics in 1972 and which are now controlled by the Peruvian Ministry of Education. The establishment of permanent schools had a profound effect on the use of hallucinogenic plants, and though the present account is written in the ethnographic present, certain practices were terminated soon after 1972. These practices are well remembered, however, and are still a frequent topic of conversation.

The Aguaruna Jívaro, like many indigenous groups of South America, utilize psychoactive or hallucinogenic plants for various ends usually associated with curing, bewitching, and the acquisition of visions. While I have never been able to discover an Aguaruna word which is equivalent to our cover term "hallucinogens" one can formulate a question like "What plants are there which cause visions?" in the Aguaruna language and elicit a list of the various psychoactive plants. This suggests that there is a clear concept of vision-inducing plants even if they are not linguistically marked. In Aguaruna, the verb *nampét*, "to be intoxicated," is used to describe the effects of both manioc beer and hallucinogenic plants. Though both kinds of intoxication are seen to have certain similarities, I have never heard of a case of a person having a vision while drunk with manioc beer. Fermented beverages have predominantly a secular significance; their preparation and consumption is a primary expression of sociability in Aguaruna society. In contrast, intoxication by hallucinogens is rarely considered pleasurable but rather a necessary means to a sacred or, in some cases, magical end.

In free speech, the verb *kahamát* "to dream," may denote dreaming during sleep or while intoxicated by an hallucinogenic plant. In Aguaruna thought, the dreams of sleep and the visions of tobacco, ayahuasca, or *baikuá* (*Datura* sp.) have many elements in common. Both can be omens of success in hunting, victory in battle, or impending danger. However, the dreams of sleep are given less significance than the dreams of hallucinogenic plants because they lack the element of intentionality. Only by the effort of will required to drink the strong-smelling juice of tobacco or *baikuá* and by the suffering associated with prolonged fasting and sexual abstinence, can one acquire a powerful and significant vision.

When asked why a given plant induces visions, an Aguaruna is likely to say that it has this effect because it is "bitter" or "strong" or even "foul-smelling." The association of strong smells or tastes with medicinal or psychoactive properties is an important element in Aguaruna plant use in general. E. A. Berlin (1977:10) suggests that the dominance of a single introduced plant, ginger (*Zingiber officinale*) in the contemporary Aguaruna pharmacopoeia may be due to its exceptional astringent properties. Evidence from the Mayo River tends to support this view and may explain the unexpected inclusion of ginger in the list of hallucinogens provided by some informants, especially since to my knowledge there is no published evidence that ginger has psychoactive properties.

The plants to which the Aguaruna attribute the power to induce visions are listed below.

| COMMON NAME | AGUARUNA NAME | SCIENTIFIC NAME[1] |
|---|---|---|
| tobacco | tsáng | *Nicotiana tabacum* L. |
| ayahuasca | yáhi | *Banisteriopsis cabrerana* Cuatr. |
| ayahuasca | datém | *Banisteriopsis caapi* (Spruce ex Grisebach) Morton |
| –– | baikuá | *Datura* sp. (probably *Datura sanguinea* R & P.) |
| –– | mamabaíkua | *Datura* sp. (probably *D.* |
| –– | bíkut | *suaveolens* H. & B. ex Willd. |
| –– | tsúak | or *D. candida* [Pers.] Safford) |

Also mentioned by some informants but by no means all, were a variety of ginger, *tunchi ajéng* (*Zingiber officinale*), and two herbs *karián pihipíng* (probably a species of *Carex* or *Cyperus*) and *chúchka* (unidentified), specimens of which could not be obtained. *Tunchí ajéng* is reported to be used to cure witchcraft-induced illness, while the latter two are said to be used by people who wish to become witches. The peculiar property of these plants

[1]The plant identifications were made by Drs. W. R. Anderson and B. Gates of the University of Michigan Herbarium, to whom I would like to express my thanks.

is that while they are said to make one "drunk," they are not supposed to cause visions.

Michael J. Harner's works on the Jívaro vividly portray the importance attached to the search for a vision with which to defeat one's enemies (Harner 1972, 1973). His analysis of the Jívaro case holds largely for the Aguaruna as well, with some minor differences. As an Aguaruna youth nears puberty, he is encouraged to begin taking infusions made from the juice of tobacco, *datém*, or *baikuá*. This is done individually or with a group of young men from the same neighborhood. After taking one of these infusions, the youths retire to a shelter in the forest to await their dreams. Principally, they hope for the appearance of an *ahútap*, a fearsome being associated with the spirit of a deceased warrior. Aguaruna accounts of the appearance of the *ahútap* vary greatly; the spirit may come in the form of an old warrior, a fox, a cave bird, a jaguar, or a comet-like flash of red light. The spirit speaks to the dreamer, telling him of his future feats in battle and how he will kill a specific enemy.

Unlike the Jívaro, the Aguaruna do not believe that they acquire an *ahútap* soul. The *ahútap* is an independent entity which returns to the sky after appearing to the dreamer. It is said that the *ahútap* imparts a dream which then becomes an almost physical presence inside the body of the dreamer. A person who has obtained one of these visions is said to be *káhintin* "owner of a dream," or *waímaku*, which can be loosely translated as "one who has had a vision." The *waímaku* vision is preeminently associated with death and warfare. It is a dream which is eternal and transcends the life of its owner, since soon after his death it will leave his body and ascend into the sky to become an *ahútap*. While the *waímaku* vision, through its transformation to an *ahútap*, has some of the properties of a soul, it is never confused with the two "true" souls—one lodged in the eye and the other residing in one's shadow—which each person is said to possess from birth until death.

Closely related to the *waímaku* or killing vision, but distinguished from it, is the vision called *niímangbau*, a term apparently derived from the verb meaning "to see." Often, instead of seeing the *ahútap*, a vision seeker sees a vision of his own future life. These *niímangbau* visions are invested with all the symbols of domestic tranquility: an attractive spouse, many healthy children, abundant chickens and pigs, and fine hunting dogs. By having a *niímangbau* vision, one is assured of prosperity and long life. This kind of dream is obtained by drinking an infusion of tobacco juice or the liquid expressed from the stem of *bíkut*. It is not eternal, however, and it disappears after the death of its owner. But the *waímaku* and *niímangbau* visions need to be periodically renewed during a person's life.

While the desire to acquire visions is particularly associated with young men, women sometimes participate and may even acquire a killing vision.

A female *waímaku* may take a prominent role in the festivities before a vengeance raid by urging men into a killing frenzy, or she may aid her faction by acting as a spy during a social visit to an enemy's house. More commonly, though, women seek life-giving *niímangbau* visions that insure their success in gardening and in raising domestic animals.

Besides the desire for specific visions, the Aguaruna associate hallucinogen use with right or "straight" thinking in general. This is clearly seen in the origin myths of some of the hallucinogens, to be discussed shortly, and in the practice of forcing intransigent children to drink *baikuá* so that they will respect their parents (cf. Harner 1972:90).

Hallucinogens also play a major role in Aguaruna shamanism and witchcraft. The Aguaruna believe that some men, voluntarily and occasionally involuntarily, acquire the power to cure or bewitch others by supernatural means. In the Upper Mayo River, two kinds of shamans are said to exist, publicly recognized shamans, who can cure and bewitch, and hidden shamans, who only bewitch. Hallucinogens play a key role in obtaining witchcraft power, particularly in the acquisition of a saliva-like substance which is the medium for invisible killing darts held in the mouth and upper torso of the shaman. Bewitching conventionally involves sending these darts to one's enemies, causing their illness or death. Curing usually consists of the removal of the offending darts. The hallucinogens tobacco, *datém*, and *yáhi* are used for curing seances and, reportedly, for bewitching.

For curing sessions, most shamans in the Upper Mayo use either tobacco juice or a mixture of *yáhi* or *datém* to arrive at the state of intoxication required for them to see the magical darts inside the patient's body. Even though the leaves of *yáhi* are mixed with the pounded stem of *datém* for curing seances, the mixture is usually referred to as *yáhi*. This reflects the fact that most of the power of the mixture is attributed to *yáhi*, not *datém*. *Yáhi* has no uses other than curing or bewitching, and many people in the Upper Mayo say that a non-shaman who takes *yáhi* will not even become intoxicated, since he does not possess magical darts. So close is the connection between *yáhi* and witchcraft that I found it hard to collect specimens since no one cared to admit that he knew of its whereabouts, nor could anyone imagine why I would want it unless it were to engage in the black arts.

In certain cases of illness for which a shaman is not available, a sick person may drink the juice expressed from the stem of *tsúak*. During the ensuing vision, the "soul" of the plant appears to the dreamer and effects a cure by throwing out witchcraft substance or, in another typical case, by setting a broken bone. The few informants who mentioned ginger as an hallucinogen say that an infusion of a special variety of ginger produces a dream in which doctors appear and render the patient healthy through their ministrations. It is perhaps significant that this use of ginger is reportedly limited to two isolated villages which have no publicly recognized shaman.

Ingestion of tobacco juice or smoke is held to be essential to the acquisition and use of magical songs which guarantee success in courtship, hunting, gardening, and the care of domestic animals. These songs incorporate esoteric language that is said to be impossible to memorize without first consuming tobacco juice. Similarly, such songs can only be "heard" by the intended recipient if the singer is intoxicated by tobacco.

The various uses of hallucinogens in Aguaruna society can be seen as a way of projecting one's will into what I will, for lack of a better word, call a non-ordinary plane or, in some cases, in making oneself receptive to powerful forces which are found in this non-ordinary domain. While I have found few Aguaruna who agree with the Jívaro's statement that the ordinary world is a "lie," and the world of hallucinogens "the real world" (Harner 1972: 134), there is no doubt that the non-ordinary world of hallucinogens is an important component of Aguaruna reality.

## FROM THE HERO'S BONES: BAIKUÁ, BÍKUT AND TSÚAK

I would now like to turn my attention to three solanaceous plants which are closely related both in the western botanical sense and in Aguaruna ethnotaxonomy: *baikuá*, *bíkut*, and *tsúak*. These three plants are members of the genus *Datura* and are morphologically similar in that they have shrub-like or tree-like growth habits, simple ovate leaves, and prominent, elongated flowers which bloom frequently throughout the year (see Fig. 2). *Baikuá*, *bíkut*, and *tsúak* are cultivated in manioc gardens, hidden forest plots, or sometimes adjacent to houses so that they will be readily available when needed. The outer layers of stem tissue of all three plants yield a dark green, strong-smelling liquid which, when taken in sufficient quantities, produce intoxication and hallucinations. The juice is said to be *tsupaú*, "bitter," or *seéseng*, a nauseating smell associated with such things as raw fish or fresh blood. The Aguaruna consider these to be the strongest hallucinogens and potentially dangerous because of their unpredictable effects, but their power makes them indispensible for the acquisition of killing or healing visions.

The taxa *baikuá*, *bíkut*, and *tsúak* seem to correspond to the generic rank in Aguaruna plant taxonomy as described by Brent Berlin (1976). The folk genera *baikuá* and *tsúak* are further subdivided into several folk species (see Fig. 3). One of these species, *mamabaíkua*, is worthy of special consideration because of its importance in the origin myth of the three genera. *Mamabaíkua*, as distinct from the other varieties of *baikuá* is not cultivated. It grows only in the wild, most frequently in damp, sandy soils near rivers and streams. While it is rarely used as an hallucinogen, the Aguaruna state

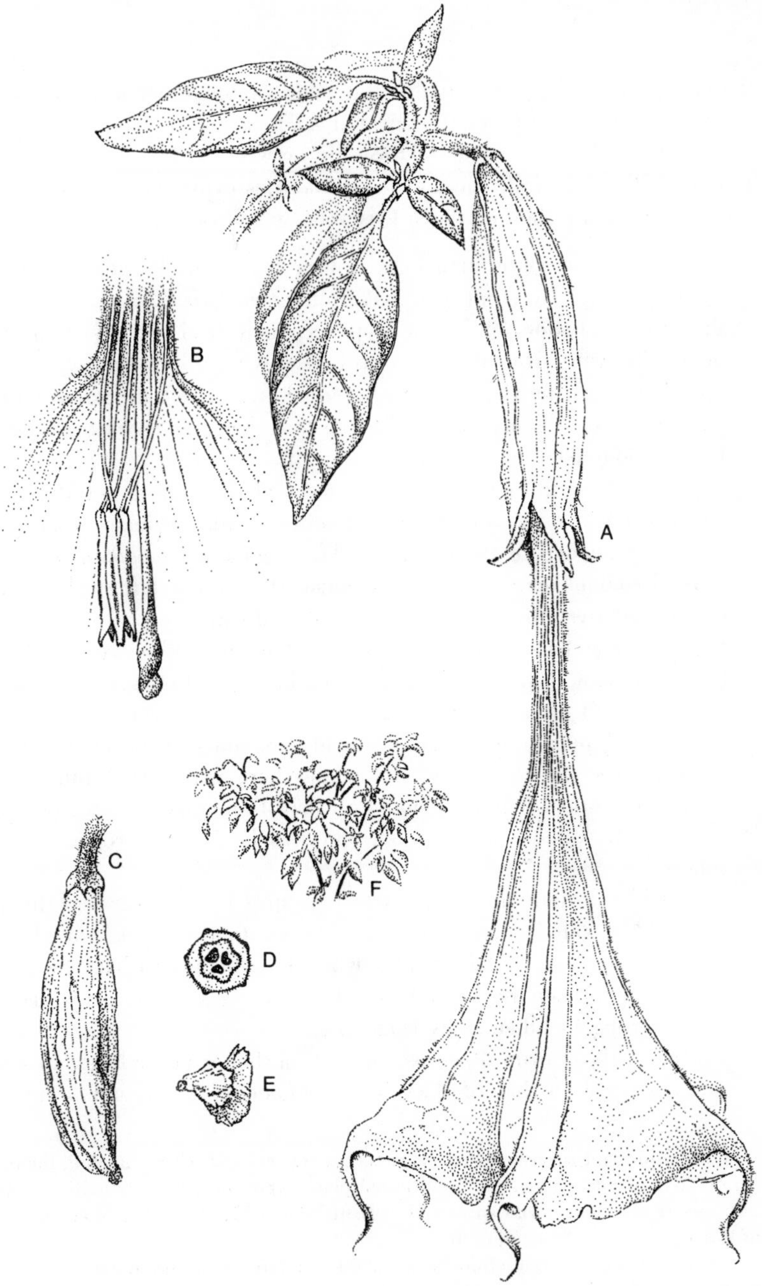


Fig. 2. *Tsúak, Datura suaveolens* H. & B. ex Willd. or *D. candida* (Pers.) Safford. A, flowering branch ca. x 1/3; B, segment of flower, expanded, ca. x 1/2; C, fruit, ca. x 1/4; D, ovary, cross section, ca. x 1; E, seed, ca. x 3/4; F, plant, ca. x 1/60.

that its properties are essentially the same as the other kinds of *baikuá*, and that it can be consumed if cultivated *baikuá* is not available.[2]

The close ethnotaxonomic relationship between *baikuá*, *bíkut*, and *tsúak*, as well as the wild folk species *mamabaíkua*, is expressed in their common myth of origin, a variant of which I recorded as follows:

> Long ago, a young man became very hungry while returning from a hunting trip. He stopped to eat fruits of *mamabaíkua*, not knowing the plant, and soon fell into a deep sleep as if dead. He was found by his family, who brought him to his house. After a long time, he woke up and told of the visions he had seen. He was now a *waímaku*. Other youths began to plant *mamabaíkua* so that they could seek the dreams that would make them *waímaku*.
>
> Another youth [named Bíkut in most versions of the myth] who had never had sexual intercourse or contact with a woman, drank the juice of *mamabaíkua* twenty times. After taking *mamabaíkua* so many times, he could see things like a shaman. If a woman left the house to commit adultery, he saw it and cut her head with a machete. He could tell if a man had come to eat after defecating without washing his hands. He saw if a man had committed incest, and always killed such a person with his lance saying to the others "Can't you see that this was an evil person who had just had intercourse with his sister or someone else in his family? He comes here with his clothing covered with worms. A brave man who eats with him becomes a coward," he said. He killed many people, and was bound with ropes by the others.[3]
>
> During this time, there was a war with those from downriver [Huambisa]. They killed many Aguaruna. The youth Bíkut was untied and given his lance so that he could fight. He bathed, tied back his hair, and dressed. Then he went to meet his enemies. Because his dream was so strong, he killed many Huambisa.
>
> One of those from downriver went to a shelter in the forest to seek a vision equal to that of Bíkut's. He obtained a vision, and wounded

---

[2] One informant suggested that *mamabaíkua* was created when Núngkui, the mythological giver of manioc and other cultivated plants, transformed many cultivated plants into useless but similar wild species. See Brent Berlin (1977) for an interesting analysis of this aspect of the Núngkui myth.

[3] The apparent contradiction between Bíkut's defense of the moral order and his being bound by vines was baffling until Manuel Garcia Rendueles, who has made an exhaustive study of Aguaruna mythology, informed me that the binding of fierce warrior-heroes is a common theme in Aguaruna myths. The power of these super-warriors was felt to be so dangerous that they were bound to keep them from harming members of their family.

Bíkut in battle. Finally, the enemies from downriver killed Bíkut. They left his body on the ground, and it was buried by his family.

From the right femur of Bíkut's body grew *shiwáng baikuá*, and from the left femur grew *tahímat bíkut*. From his spine grew *muntúk tsúak*. From these plants, our ancestors took the plants we have today: *baikuá*, *bíkut*, and *tsúak*.

While all myths are subject to a wide variety of interpretations, this particular one has several features which bear directly on the present discussion:

1) An historical relationship is suggested between a putative wild ancestor, *mamabaíkua*, and three closely related plants which are presently cultivated.

2) While the three cultivated plants have a common ancestor, the myth suggests that they might have opposing or at least different characteristics inasmuch as they are given different names and issue forth from different parts of the decaying body of the myth's hero.

3) There is a relationship of reciprocity between the hero of the myth and the plants mentioned. Bíkut acquires his power through the use of *mamabaíkua*, and it is implied that the three descendants of *mamabaíkua* which come from his bones acquire their power by virtue of his exploits as visionary and warrior. He thus aids the plants by transforming them from a natural to a cultural state, and they reciprocate by transforming him from an ordinary person to a culture hero.[4]

It is difficult to assess the degree to which the botanical relationships expressed in the myth conform to Western taxonomic knowledge, especially since the classification of the species in question, *Datura sanguinea*, *D. candida*, and *D. suaveolens*, is undergoing a revision by taxonomic botanists. A further complication is that the use of these species as ornamentals has resulted in considerable hybridization among them. The group consisting of *baikuá*, *bíkut* and *tsúak* does conform to Linnean taxonomy insofar as all three pertain to the same genus, *Datura*. The various species level folk taxa (e.g., *mamabaíkua*, *tahímat bíkut*, etc.) are the subject of considerable disagreement among native informants, with some people even doubting the existence of some folk species which had been named by Aguaruna from other communities. Figure 3 represents the synthesis of all informants' statements with regard to the taxonomic relationships between *baikuá* and its relatives.

While all informants agreed that *baikuá*, *bíkut*, and *tsúak* are "alike" or "friends" (cf. B. Berlin and E. A. Berlin 1977:8), there was a tendency to say that *baikuá* and *bíkut* are "more alike" than *tsúak*. All three generic

[4] Bíkut is said to be the source of knowledge about personal cleanliness and, according to some informants, knowledge of how to treat childhood illnesses caused by contact with certain species of animals and plants.

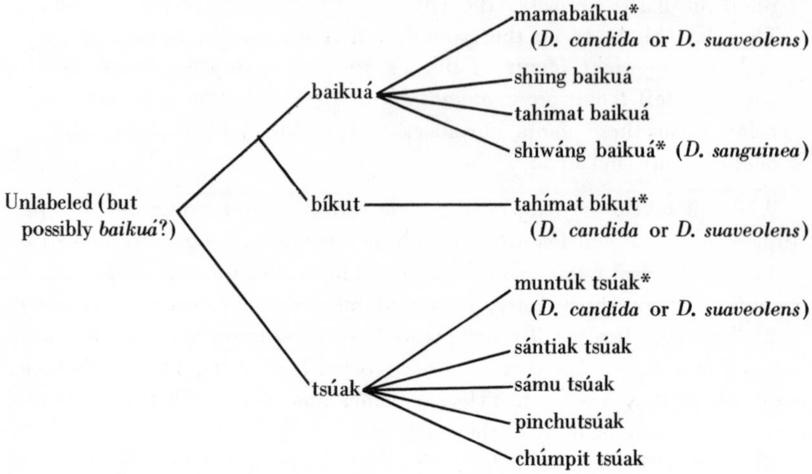


Fig. 3. Synthesis of all informants' statements regarding the taxonomic relationships between baikua and its relatives. Asterisk (*) indicates folk species for which specimens were collected.

level taxa are seen to be more like each other than they are to other hallucinogens such as tobacco and ayahuasca. The group consisting of *baikuá*, *bíkut*, and *tsúak* appears to have no name, though in free conversation I have heard people use *baikuá* to refer to all three folk genera. This use of *baikuá* as a general term for the three may result from the fact that many younger Aguaruna are now unable to distinguish between these plants. Thus they may mistakenly refer to a specimen as *baikuá* when it is really *bíkut* or *tsúak*.

Besides a common appearance, there are other characteristics which cause the Aguaruna to see *baikuá*, *bíkut*, and *tsúak* as a natural group. First, as was mentioned earlier, they are considered to be very strong and potentially dangerous if taken repeatedly. They are all prepared in the same fashion and the preparations have a similar taste and smell. They are also consistently attributed souls or, sometimes, "people" (*áents*). While any animate object, and some inanimate ones, may be attributed souls (which in the present context is simply defined as an enduring, spiritual component), no other plants are so consistently said to possess a soul. The appearance and importance of these souls vary among the taxa under consideration. The soul of *baikuá*, for example, has the appearance of a tall, thin, man who appears to the person who has consumed *baikuá* and whose state of intoxication is beginning to dissipate. This soul or person has no relation to the much sought after *ahútap*, and its appearance is considered to be little more than a peculiar and unimportant side effect. The soul of *tsúak*, on the other hand, has a central role in the cures attributed to the plant. The *tsúak* soul appears in the form of a *wíakuch*—a person of European features dressed in fine clothes—or as one of the mestizo physicians which the Aguaruna have seen when visiting the hospital in Moyobamba. Alternatively, some people say that the soul of *tsúak* is that of the person from whom the plant was obtained. Whatever its appearance, this soul treats the illness of one who has taken *tsúak* and is therefore the key element in the power attributed to that particular hallucinogen.

While *baikuá*, *bíkut*, and *tsúak* thus form a natural group in native thought, each has distinguishing properties that set it apart from the others. There are several morphological characteristics, including size and growth habit of plant, color of flower, and leaf shape, which some Aguaruna say can be used to distinguish between them to a certain extent. Younger informants are not always able to distinguish between them on this basis, but they insist that such morphological criteria do exist. In fact, the problem of identifying the plant rarely arises because the owner and cultivator of the plant will have learned its name from the person who gave or sold him the cutting. From the Aguaruna perspective, what principally distinguishes each plant are the powers attributed to it and its appropriate uses. These are summarized as follows.

## Baikuá

All varieties of *baikuá*, but most especially *shiwáng* ("enemy") *baikuá*, are used for the acquisition of *waímaku*, or killing visions. When the Aguaruna speak of *baikuá*, it is in the context of their attempts to see the *ahútap* during their adolescence. Besides taking the juice of *baikuá*, the consumption of which is in itself seen to be a form of suffering because of its bad taste and powerful narcotic effect, the youth searching for a killing vision has to be sexually pure. By fasting, sexual abstinence, and the continuous consumption of hallucinogens, the vision seeker attempts to arrive at a state of suffering so acute that an *ahútap* will take pity and appear in one of its many forms, thereby conferring to the supplicant the power to defeat his enemies.

The close connection between *baikua* and the killing vision is powerfully expressed in the words of a song which is sung by a man as he prepares *baikuá* for his son:

I am not cutting the stem of *baikuá*
but the bones of my enemy.
I am not scraping the bark of *baikuá*
but the bones of my enemy.

I am transforming the enemy's heart.
The vulture high above, takes the heart
  with its talons
  squeezing it, crushing it
In the bowl is the enemy's blood
The enemy's blood drips into the bowl.
The vampire bat high above, folding
  back its wings
  descends to drink the enemy's blood.
It drinks with great thirst.
High above, the vampire bat never stops
  drinking blood.
With great thirst, it drinks.

Besides causing powerful visions, the juice of *baikuá* is said to prevent sexual arousal for several months when applied to the penis. This secondary use can clearly be understood within the context of sexual abstinence associated with the search for an *ahútap*.

## Bíkut

*Bíkut* emerged from the mythical hero's left side, and it takes on many characteristics which are opposed to *baikuá*, which came from the hero's right side. While *baikua* is associated with self-denial and the acquisition of *waimaku* visions, *bíkut* usually is said to be a source of *niimangbau* or life-giving visions. Indeed, the name *tahímat bíkut* that appears in the myth carries with it connotations of abundance, and good health. *Niimangbau* visions are omens of a successful sexual and family life, and the acquisition of wealth in the form of dogs, pigs, and other domestic animals. A *baikuá* vision deals with blood, vengeance, and the death of an enemy; a *bíkut* vision deals with prosperity, long life, and good health.

## Tsúak

*Tsúak* is, in many ways, the most enigmatic of this botanical trio. As was noted before, *tsúak*—which literally means "remedy"—is used by persons who are afflicted with various sorts of ailments and do not have access to a curing shaman. The two kinds of ailments most frequently treated by *tsúak* are witchcraft-induced illness and accidentally broken bones. In the case of witchcraft, the soul of *tsúak* appears to the patient, examines him, and then removes the intruding witchcraft substance. The patient recovers immediately after he has awakened from his intoxication. Similarly, in cases of bone fracture the soul of *tsúak* is said to manipulate the bones, blow on them, and restore them to their proper location. Informants who have observed people intoxicated by *tsúak* report that the patients themselves are rubbing and blowing on the site of the fracture, but that in their intoxication they see the phantom curer doing the manipulation. Thus it may be that the intoxicating effect ot *tsúak* allows the victim to endure the pain of arranging fractured bones for proper healing.

The peculiar property of *tsúak* is that while it is a formidable source of healing power, the Aguaruna say that by its use one inadvertently acquires the magical darts which shamans use to bewitch. In fact, *tsúak* is allegedly the major means by which hidden or bewitching shamans acquire killing powers without obtaining them from another person and thereby running the risk of discovery. By taking *tsúak* many times in succession, one can acquire sufficient darts to bewitch and kill. A person who wishes to use this plant strictly for curative purposes must rid himself of these bewitching darts after taking *tsúak* by exposing his body to the smoke of burning chili peppers. The strong smell of the peppers throws out the accumulating darts before they become dangerous.

*Tsúak* is considered to be an "enemy" of *baikuá* and *bíkut* because it cannot be mixed with them without threatening the life of the person who drinks the mixture. *Bíkut* and *baikuá* are "friends" because they are mixed in certain cases by someone very anxious to obtain a vision of either the *waímaku* or *niímangbau* variety.

## Reciprocity Between Man and Plant

For the Aguaruna, the power attributed to each of these three plants and their varieties is a kind of statistical norm which may or may not be realized in a given specimen under cultivation. This is a point which returns us to the notion of reciprocity between man and plant mentioned earlier in connection with the myth of Bíkut. To retain and develop its inherent power, each plant needs special treatment by its owner. Ideally, he should fast before planting a stem cutting, avoid contact with the plant for several days after engaging in sexual intercourse, and cultivate it in a place where it will not be disturbed or soiled by children and domestic animals. The inherent power of the plant is further increased by frequent use, and its owner, in turn, acquires ever more powerful visions as his plant's strength grows. It is said that a person who wishes to cultivate *baikuá* will try to obtain a stem cutting from someone who is known to be *waímaku*, possessor of a killing vision, and who has observed the proper precautions in the care and use of his *baikuá*. So close is the identification between a plant and its owner that, in the case of *tsúak* at least, a person who borrows some of another person's plant for a cure may see the owner appear as the soul of the plant during his intoxication.

To review the preceding discussion, I have shown how the beliefs and practices associated with *baikuá*, *bíkut*, and *tsúak* are reflected in their common myth of origin. They are seen as having a common ancestor and certain natural similarities, and that within this similarity there is remarkable diversity. *Baikuá* and *bíkut* stand as paired opposites, the former associated with sexual abstinence, austerity, and death, while the latter is linked to sexual activity, prosperity, and life. Despite these differences, they are "friends" in that they can be mixed together and their use has a common goal, the acquisition of visions necessary to survive in the Aguaruna world. *Tsúak* stands apart as a kind of transforming agent which can change a person from sick to well or from non-shaman to witch. The power of each of these plants is only partially inherent, coming to fullest realization when they are correctly cultivated and used by man.

If the connection between myth and ethnographic reality is fairly clear in this case, there are other problems which are not so easily resolved. First, why is it that there is so much informant disagreement about the varieties of *baikuá* and *tsúak* and the morphological factors which distinguish them? Second, how is it that plants which are so closely related taxonomically can be attributed such different properties?

Informant disagreement is always difficult to assess, and I am willing to concede that it may have been caused by the failure of the ethnographer to ask the right questions. But one remark which I heard frequently in the course of this investigation was that the person who owns and cares for the plant is the only one who can provide a definitive identification of a given specimen, especially at the variety level. Medicinal plants, unlike other cultivars, are given individual treatment in the garden, and different varieties of the same species are usually planted separately so that they won't intermingle. Varieties of plants like sweet manioc, maize, or peanuts, on the other hand, are mixed together in the garden, so that to identify them one must have in mind certain readily observable characteristics such as leaf form, fruit shape and size, etc. The Aguaruna can consistently describe the defining characteristics of varieties of manioc, bananas, and various edible wild plants, but descriptions of the morphological characteristics of varieties of some medicinal plants such as *baikuá*, *tsúak*, and ginger varied significantly from person to person. The characteristics enumerated by older people and specialists (curing shamans) were found to be no more consistent than those of others. It should be restated that I am here speaking of *morphological* characteristics; once a given variety had been identified, people generally did agree about what the plant's medicinal or hallucinogenic properties were. To give a more specific example, Aguaruna women invariably cultivate two or more varieties of ginger in their gardens, each of which has a different medicinal use.[5] They say that the varieties look virtually identical and that one must carefully cultivate each one in a separate part of the garden, keeping in mind where it was planted so that the appropriate variety can be collected when needed. The knowledge of how each variety is to be used was obtained from the woman who originally sold or gave the cultivator her ginger. It thus seems possible that certain medicinal plants may be distinguished by what I would call, with some trepidation, non-observable factors, i.e., properties known only to the individual cultivator and not necessarily to anyone else. If this is the case, it is probably a direct result of the notion of reciprocity between man and plant mentioned earlier, in that only part of the plant's character is intrinsic, the rest being the product of the treatment which it receives from its owner. To put it another way, if the owner of a specific specimen of *shiwáng baikuá* says that his plant is *shiwáng baikuá*, and it is known that by using his plant he has successfully acquired *waímaku* visions, it will hardly matter to someone else who might want to use this man's plant whether its morphological characteristics are similar to or different from, say, *tahímat baikuá*. The important thing is that the plant has been determined to be *shiwáng baikuá* by its owner, and that it has been shown to be powerful in producing visions.

The second troublesome question is the different powers which are ascribed to plants which are so closely related botanically. Harner (1973: 146-47) has suggested that, for the Ecuadorian Jívaro, the difference in use between *Banisteriopsis*- and *Datura*-derived hallucinogens can be related to chemical differences between them, because the dissociating effects of the scopolamine, atropine, etc. present in *Datura* make it unsuitable for the sustained concentration demanded of the curing shaman. It is, however, doubtful that such significant differences exist in the chemical composition of *baikuá*, *bíkut*, and *tsúak* because they are so closely related botanically. One must therefore conclude that the expectations of the user contribute significantly to the sort of vision that he or she will have when consuming one of these plants.

The foregoing comments are not intended to imply that the Aguaruna are unscientific or illogical. The intensive ethnobiological research being carried out by Brent Berlin with the Aguaruna of the Deparment of Amazonas (see, for example, B. Berlin 1976) has given ample evidence that

[5] As is the case with the different kinds of *Datura*, certain varieties of ginger are sometimes attributed completely opposite effects. E. A. Berlin (1977:5-6) reports a variety of ginger (*uchigmátai ajég*), which assures conception, and another (*kága ajég*), which is said to prevent conception. Her observations are also valid for the Upper Mayo region.

their organization and classification of the natural world is systematic and closely resembles our own in its logical structure. I wish only to suggest that in certain classificatory domains, things may be distinguished according to factors which are not directly observable or even empirically verifiable.

## HALLUCINOGENS AND THE PRESENT SITUATION OF THE UPPER MAYO AGUARUNA

Since the introduction of bilingual schools to communities of the Upper Mayo in 1972, Aguaruna opinions about the role of hallucinogens have begun to change significantly. While the use of hallucinogens by curing shamans continues undiminished, the systematic search for visions by adolescents has ceased completely, partly because of the objections of Aguaruna teachers trained by the Summer Institute of Linguistics. All older men, and most men and women over the age of 16, have had some experience with hallucinogens, but a new generation of adolescents is appearing which has had no direct contact with the traditional visionary experience.

One observable effect of this is that young men, who once postponed courtship and marriage to maintain the sexual purity needed to acquire *waímaku* visions, are now inclined to marry at an earlier age than their parents. According to the Aguaruna, this has caused an increase in domestic strife, because early marriages are very unstable and changing economic realities create many hardships for couples who marry young. Older men often complain that young men are now inclined to do "crazy" things at drinking parties, such as fighting with close family members or attempting suicide. This is because their hearts are "soft" and full of clear water, unlike the heart of a *waímaku* or vision possessor who has a hard, red heart. Various elders state that they intend to revive the preparation of *datém* for young boys so that they might acquire life-giving *niímangbau* visions. There is a vague feeling that *waímaku* killing visions are bad because they encourage homicide, though it should be said that men who are known to be *waímaku* are still greatly admired. It seems doubtful, in any case, that the desired revival of the vision quest will take place in the near future.

Several villages of the Upper Mayo River are strongly influenced by Aguaruna evangelists who have had some contact with Protestant missionary groups. Aguaruna Christians are supposed to reject intoxication by hallucinogenic plants as well as fermented beverages. Ironically, traditional theories of witchcraft do not conflict with the teachings of Aguaruna evangelists, and in fact may be reinforced by the new emphasis given to Satan and his omnipresence in the world. It is not possible to say whether witchcraft accusations have increased or declined in the present situation since comparable data from earlier periods are not available, but it is certain that the continuing belief in witchcraft insures the survival of curing shamanism. The ever-growing patronage of Aguaruna curers by non-Indian patients also contributes to the survival of the traditional curing role, albeit in a slightly syncretized form.

We have seen that the relationship of the Aguaruna with vision-inducing plants is a complex matrix of beliefs and practices all directed toward participation in and manipulation of the non-ordinary world. While many New World societies have utilized hallucinogens for shamanistic or curative purposes, the Jivaroan groups can claim some uniqueness in the degree to which their use of hallucinogenic plants is linked to the very foundations of the moral order of society. For the Upper Mayo Aguaruna, the current perspective is one of a simplification and redefinition of their relationship with hallucinogenic plants. Nevertheless, through their use in shamanistic curing, these plants will continue to play an important role in the Aguaruna interpretation of the mysteries of an often confusing world.

## Acknowledgments

The research on which this article is based was conducted between December 1976 and September 1978 in the communities of Huascayacu and Alto Naranjillo, Department of San Martín, Peru. I gratefully acknowledge the financial support of the Doherty Foundation, the Wenner-Gren Foundation for Anthropological Research, and the Centro Amazonico de Antropología y Aplicación Practica. I would also like to thank Margaret Van Bolt and Mary Hodge for preparing this article for publication while I was still in the field. The drawings were prepared by Margaret Van Bolt.

## REFERENCES

Berlin, Brent

1976 Some Evidence from Aguaruna Folk Botany for the Concept of Rank in Ethnobiological Classification. American Ethnologist 3:381-99.

1977 Bases Empíricas de la Cosmología Aguaruna Jívaro, Amazónas, Perú. Language Behavior Research Laboratory, University of California, Berkeley, Studies in Aguaruna Jívaro Ethnobiology, Report No. 3.

Berlin, Brent, and E. A. Berlin

1977 Ethnobiology, Subsistence, and Nutrition in a Tropical Forest Society: The Aguaruna Jívaro. Language Behavior Research Laboratory, University of California, Berkeley, Studies in Aguaruna Jívaro Ethnobiology, Report No. 1.

Berlin, Elois Ann

1977 Aspects of Aguaruna Fertility Regulation. Language Behavior Research Laboratory, University of California, Berkeley, Studies in Aguaruna Jívaro Ethnobiology, Report No. 2.

Harner, Michael J.

1972 The Jívaro. Garden City: Doubleday/Natural History Press.

1973 The Sound of Rushing Water. In: Hallucinogens and Shamanism, edited by Michael J. Harner, pp. 15-27. New York: Oxford University Press.

Stirling, Matthew W.

1938 Historical and Ethnographic Material on the Jívaro Indians. Bureau of American Ethnology Bulletin 117.

Tessmann, Günter

1930 Die Indianer Nordost-Perus. Hamburg: Friederichsen, de Gruyter und Co.

# PRESENT AND FUTURE PROSPECTS OF HERBAL MEDICINE IN A MEXICAN COMMUNITY

*Ellen Messer*
Yale University

Herbal medicines continue to fill an important place in the curing systems of many traditional societies. In spite of the inroads of modern medicines, their prominence in myth, manipulation by specialized healers, and "real" psycho-social-biological effects make them cultural elements which are not easily discarded without major cultural disruption. Volney Jones stressed that ethnobotanical investigations should pursue not only the economic values of plants, but also the "entire range of relations" that exist between human and plant populations (Jones 1941). In particular cases, he searched for biochemical as well as native religious and native scientific explanations for why particular plants were used in particular cultural contexts (Jones 1965). These combined perspectives, which he transmitted to his students and through them to others, have influenced my approach to the topic of herbal medicine in Mitla, Oaxaca, Mexico.

This essay will examine first the biosystematic and functional classifications of medicinal plants in Mitla and then three factors which are affecting the persistence of traditional herbal medicine there. Two routes to the analysis of medicinal plant use in traditional cultures are (1) the pharmacological perspective, which seeks the potential efficacy of indigenous herbs in biochemical terms (e.g., Swain 1972; Lozoya L. 1976) and (2) the symbolic perspective, which views plants as part of the particular cultural system of beliefs and practices surrounding health and healing (e.g., Harvey and Armitage 1961; Kleinman 1973). Principally the latter perspective will dominate this essay, which looks at the role of herbs in Mitla medicine as a cultural system. Data derive from the author's 12 months of ethnobotanical fieldwork in Mitla conducted from 1972 to 1975. During this time, more than 95 Mitleños provided information on medicinal plant identification and use, in participant-observation, home interview, and field contexts.[1]

## SETTING

San Pablo Mitla, a rapidly modernizing Mexican Indian community, is located in the southern highlands of Mexico, approximately 16° N latitude. With elevations ranging from 1600-1850 m, the climate is temperate all year and relatively frost-free. The major environmental markers are a pronounced dry season, which occurs roughly November through March, and the alternate rainy season, which begins in mid- to late April and tapers off in late October. Rains are extremely variable and unpredictable in timing, quantity, and spacing, and make dry farming—the cultivation of maize, beans, and squash—extremely unpredictable. Though approximately half of the town households still farm *milpa* (maize field) lands, most have alternative sources of income. They either manufacture or trade textiles produced for the national and international tourist market or sell them locally to the tourists who flock to the Mitla ruins, the town's archaeological attraction.

Most of Mitla's approximately 4600 residents are bilinguals who speak Mitla Zapotec in their homes as a first language and later learn Spanish. Though many adults never completed primary school, most men and many women are functionally literate, and most children of both sexes now complete the first six grades of compulsory education. Mitla, which lies on a spur of the Pan American Highway, is connected to Oaxaca City, the Tehuantepec coast, and other parts of Mexico by bus, telegraph, telephone, and radio. Modern communication facilities, in combination with government educational and social programs and commercial activities, have greatly affected medical as well as other cultural aspects of the town since Parsons' classic report of Mitla life in the 1930s (Parsons 1936).

## MEDICAL SYSTEMS

The use of herbs must be viewed in the context of Mitla's medical options. In the past, before there were doctors, people cured with "herbs . . . and a little bit of witchcraft," but currently, Mitleños choose from among several alternative medical systems, categories of medicaments, and practitioners.

---

[1]Research was carried on as part of the project, Human Ecology in the Valley of Oaxaca, Mexico under the direction of Kent V. Flannery of the University of Michigan Museum of Anthropology. The author was funded during that period as a Junior Fellow of the Michigan Society of Fellows.

Modern (prescription) medicine, in the form of injections, pills, and syrups, is administered by licensed physicians. Low cost medical care is available through government clinics in Tlacolula, the district capital, 20 minutes away by bus, or in a suburb of Oaxaca City, the state capital, only an hour's bus ride away from Mitla. The government also provides preventive medical care through inoculations and through instructions in nutrition and hygiene, all of which are administered at the local level. Private doctors also provide modern medicine. Two reside in Mitla, but many more practice in Tlacolula, Oaxaca City, and other towns within reasonable distances. In addition, pharmacies and general stores provide modern medicines that can be purchased over the counter without a prescription. The growing popularity of specialized pharmacies is illustrated by the opening of two such stores in Mitla in 1976. Proprietors of modern pharmacies, with the aid of pharmacology texts, provide their clients with medical advice, as do the proprietors of general stores, with or without formal pharmacological training.

Patent medicines, a second general class of medicaments, like prescription medicines, are sold over the counter by pharmacies and general stores. These include pills and powders for stomach upset, worms, and other digestive complaints; ointments and creams for sores; balms for aches and pains; and numerous aromatic rubs for cough and colds. Aspirin and other headache pills, Alka Seltzer tablets, and Vicks Vaporub are familiar parts of the stocks. Apart from these are others that are cross-culturally less familiar and that definitely form part of the local "folk medicine." These include several oils and spirits used to apply leaf poultices for headache (see Table 1), copal incense (the gum of *Bursera* spp.) burned ordinarily on altarplaces and in the cemetery but also burned during curing ceremonies for "fright"; and special curing paraphernalia such as vessels used for "cupping," a traditional folk treatment done to release the "air" which is thought to underlie certain aches and pains.

Besides these is the class of local folk medicines which is predominantly herbal. It includes medicinal herbs, which grow "wild" along the river floodplain, in the fields, and on the hills, herbs cultivated in local gardens, and dried medicinal herbs distributed by local and non-local vendors. Local "wild" herbs are a "free good," most of which come from within two hours walking distance from the town center. Since not everyone ventures out or knows the herbs or their whereabouts, they are occasionally gathered, dried, and sold by those who do. Common remedies are cultivated by most households for home use. Some persons, including the owners of general stores, also cultivate small quantities for sale. Non-local herbs for various remedies are offered by vendors from the regions where they are purchased, either in the Mitla marketplace or occasionally door to door. Such transactions are always on a small scale.

Herbs are also sold by herb vendors in permanent stalls in the Oaxaca City marketplace and by herbalists who frequent the market plazas on particular days within the regional periodic market system. Most of the latter are members of national or international herb syndicates and advertise that "doctors" (*médicos*) serve as their medical advisors. They all sell well-known local garden remedies in dried form as well as non-local species. In addition, homeopathic practitioners and other "doctors" treat patients with herbs, further expanding the variety of herbal medicines available to Mitleños. All available herbs are used by Mitleños to treat some category of physical and sometimes psycho-social illness. Non-local herbalists supplement local herb supplies and herbal knowledge and, at the same time, partake of the reverential prestige accorded to traditional herbal medicine. The syndicated herbalist and herbal "doctor" draw on the respect allotted to modern medical authorities and on the sanctity surrounding traditional herbal cures, which are believed to be given by God. Mitleños continually seek new herbal remedies both out of respect for herbal medicines and because most herbalists communicate about symptoms in terms of traditional illness categories, which people already understand. Herbalists offer continuities with local medical and herbal culture and "modern medicine" at a price significantly below that of most pharmaceuticals.

## MEDICINAL PLANT CLASSIFICATIONS

In spite of the many options available to them, Mitleños still use significant numbers of local herbs for medicine. That most adults can identify and name more than 130 plants used medicinally from their natural environment and gardens demonstrates their biosystematic and pharmacological mastery of the local landscape. Any plant can be described and identified according to a combination of morphological features (root, leaves, stem, flowers, fruit, sap, growth habit), habitat, and botanical associations. Smell and taste are distinctions by which many medicinal plants are identified as well. With one exception, all trees, shrubs, and herbs used medicinally have at least one known name; many have at least two—one Zapotec, one Spanish. In contrast to modern botanical systematic nomenclature, synonymy (more than one label for a single plant category) and homonymy (more than one plant category named by a single label) are permitted, and are even common. This is in part the inevitable result of bilingualism but is also due to the tremendous fluidity of names and plants between regions of Mexico (see Messer 1975:132-33).

In addition to this biosystematic classification, all plants used as remedies belong to the category "medicinal plants" (Zapotec *k$^{w}$an r̃med*, Spanish *plantas medicinales*) and to subcategories of "medicinal plants" defined on the basis of (1) local illness categories and (2) humoral qualities. The medicinal classifications of plants follow the folk illness categories, and group the biosystematic categories into sets of plants that are functional alternatives for treating particular illnesses.

Mitla folk medicine recognizes illnesses of natural ("of God") and for lack of a better term supernatural ("of evil") origin. In the former category are all ordinary aches and pains, e.g., headache, body ache, stomachache, diarrhea, skin eruptions and rashes, colds, eye infections, and ear infections. Local folk science characterizes pains as either from "air" (Zapotec *beh*, Spanish *aire*), a kind of "wind" in the body, in which the pain seems to "move around," or of *pasmo* (Spanish loanword used by all speakers) in which the pain just aches and aches "like pain in the bone." Included with illness of natural cause are certain categories of digestive disorders: *empacho* (Zapotec *yerkáh*)–a feeling of swelling or bloating in which the intestines seem to be "blocked" although the victim has diarrhea around the blockage. It may be brought on from eating particular foods or from being forced to eat when or what one does not care to. "Inflammation" (Zapotec *šliyá*, Spanish *inflamación*) is a kind of pain and swelling, particularly characterizing the "mouth of the stomach" or "the side of the spleen." *Latido* (Spanish loanword used by all speakers) refers to a pulsation around the navel, which also produces diarrhea or cramps of diarrhea. All are diagnosed and treated within the local system by a combination of herbs and store remedies or by a doctor.

Natural ailments are likewise symptoms stemming from disequilibrium of hot-cold. As in other Mesoamerican cultures (see Foster 1953), Mitleños believe that all bodies, herbs, foods, medicines, and other natural things have an intrinsic hot-cold quality. In humans, a healthy equilibrium is maintained through balancing opposing hot-cold influences. Overheating, overexposure to heat followed by sudden chilling, persistent work in a hot or cold environment, or too many "hot" or "cold" quality foods can upset the balance in one direction and result in illness symptoms of that quality. When recognized, the symptoms are redressed by treating the victim with foods and medicines of the opposite quality ("the principle of opposites"). For example, aroused bile, a kind of acid indigestion produced from anger, "overheated blood" and "nerves" is treated by "cool" herbs. Illnesses of either emotional or physical origin are treated within the hot-cold system.

In addition to these ailments, Mitleños also entertain notions of illnesses of evil origin. These include "soul loss" due to "fright" (Zapotec *žahn*, Spanish *susto*) which label a set of generalized symptoms which are believed to stem from social-emotional stress as well as other physical causes (cf. Rubel 1964); "witchcraft" (Zapotec *bi̮jaa*, Spanish *brujo*), "evil eye" (Zapotec *biže?*, Spanish *ojo*), "evil air" (Zapotec *beh da*, Spanish *mal aire*) and "sorcery" (*maldad* Spanish loanword) which label many unexplained or persistent sets of symptoms which do not ordinarily respond to other treatments; "anger" (Zapotec *žahn*, Spanish *muina*) which leads to other illnesses of evil origin and infections from corpses (Zapotec and Spanish *congrena*, from Spanish *gangrena*?). Many of these illnesses though recognized, are not well understood by the majority of people. It is difficult to elicit a scholarly folk etiology except from specialists, but the symptoms are real and people treat them. They are illnesses "which the doctor cannot cure" and therefore *must* be treated by home curers or folk specialists, most of whose remedies are herbal.

All medicinal plants are integrated into the curing system according to these illness concepts. They are primarily assigned according to tradition ("inherited knowledge"), experimentation, and occasional insights, with secondary references to physical properties. (See appendix for Latin scientific names.) In general, sweet, aromatic plants are "hot," bitter tasting plants "cold," and they are used respectively to treat illnesses of the opposite quality. Physiological effects, however, take precedence over taste and smell, as in the classification of *cacahuatón* as "heating," the internal sensation which follows ingestion of the extremely bitter brew. Visual and tactile characteristics are signs for use to some limited extent. For example, the viscous clear to red juice of *suzí* is applied to wounds: the resinous leaves *yerba del aigre* and of other species are applied to external eruptions. While root, leaf, and stem are not known or used for their distinctive shapes, colors are to some limited extent another guide to use, according to hot-cold quality. Generally, red leaves are "hot," white or very fresh leaves "cool." But again, the more consistent reasoning to hot-cold quality is perceived physiological medicinal effect. From their stock of tradition, Mitleños try first one, then another herb until something works.

In addition, some plants are named for the ailments they treat, but their names carry no magical connotations. *Yerba de espanto* ("magical fright herb") produces a "fresh" bright green sudsing beverage when mixed with water and is used to remove the "heat" of illness in magical fright-soul loss. It is perceived to be refreshing. Similarly, *cuanasana* (hispanicization of the Zapotec label "family herb" used to induce fertility and to bathe the mother after delivery) is used because it is perceived to be effective. For "cleaning" (or "sweeping") the body of illnesses of evil origin, strong scented leaves (*ruda, pirú, albahaca*) are preferred because sweet smells "repel the witch." But in general, one must conclude that trial and error produced most of the traditional classifications, some of which signs, such as smell/taste, are still useful guides in explaining pharmacological effects (and occasionally cosmological effects–such as driving out witchcraft). Mitleños look to no "doctrine of signatures" nor "law of planetary correspondences" to guide their use. Nor are concepts of internal soul, soul color, or mythological referents used to direct medicinal constructions (see Vogt 1976 for contrasting examples). Though it is believed now, as in the past, that wedding flowers and *badoo* seeds used in divination have souls (see Parsons 1936:228), for medicinal purposes the only plant approached with ceremony is *toloache* (Zapotec *binij̆ bidoo*, "elf" plant). To gather its leaves to wrap on sprains, one throws three or thirteen stones to "buy" it from the Earth Lord and ask that it provide an effective remedy.

## MEDICINAL PLANTS IN CURRENT MEDICAL PRACTICE

Given symptoms of illness, their possible classifications into one or more illness categories, and the number of herbal and non-herbal remedies from which they can choose, people select herbal medications according to (1) what is available, (2) their knowledge about them, (3) their analysis of symptoms, (4) their preference for and belief in herbs for particular symptoms, and (5) their relative costs. These aspects of herbal medicine in Mitla will now be discussed in turn.

### Availability: Local, Non-Local, Wild and Cultivated

Selection of herbal over non-herbal medications is conditioned in part by absolute availability of the plants. There are many people who know botanical categories but who never leave their sewing machines and other household chores to gather plants. When illness strikes, they go directly to the pharmacy or general store for a patent remedy, in lieu of a natural remedy from the hills. During the course of my fieldwork, there were several medicinal species that were sought for people who had heard about or used them as remedies but who did not have knowledge of their whereabouts or, alternatively, the time to gather them. These included the root of *huesito*, to be boiled and fed to an infant who refuses to nurse, the leaves of *lengua de vaca* to treat *empacho*, the leaves of *salvia* to be boiled into teas for "cold" stomachache, the bitter leaves of *cacahuatón* to be boiled into a brew drunk by men for a kind of "internal massage" and by women for postpartum cleaning and to "strengthen the blood," *pericón* which grows in the higher hills, to remedy infertility, and *espinosilla* to cure magical fright. Though there are still certain individuals who make it a practice to gather wild herbs for drying and sale, particularly in August, when the flowering herbs are believed to be at full strength, they do so in the course of other activities, such as fuel gathering and cultivation and on consignment for others.

Other wild and cultivated herbs are available only intermittently in the marketplace, for example, *monstranza*, boiled as a remedy for "cold" stomach; *poleo*, boiled as a remedy for cough and sore throat, but also a favorite spice in beans; and two kinds of *yerba de espanto*, which are important components in beverages consumed in the magical fright curing ceremony. These all originate mainly outside of Mitla and either are sold directly by their gatherers, who come to Mitla to make household sales, or are bought by Mitla middlemen, who then sell them to others.

Garden cultigens are more generally available in houseyards or through local purchase. They include *yerba buena* and *manzanilla dulce* prepared as a tea for "cold" stomach, *manzanilla amarga* drunk or applied externally for *bilis* and *latido*, *yerba maestra* drunk for aroused bile, lack of appetite, anger and thirst, *ruda* boiled as a remedy for *aire*, to revive victims of faints, and to clean victims of evil eye and witchcraft, *romero* burned against witches, and *sauco* boiled to bathe red, infected eyes. In addition, several herbs, occasionally used medicinally, are easily available since they are more widely used as spices. These include *orégano* prepared as a tea for stomach cramps, infant diarrhea, and menstrual difficulties, *pitiona*, another spice prepared as a tea for sore throat and "cold stomach," and *epazote*, commonly used in the past as a remedy for intestinal worms.

## Sociology of Knowledge

A second factor affecting differential choice and use of herbs is absolute knowledge. Detailed familiarity and use of herbs varies greatly from person to person. Those with minimal knowledge know only a few useful plants, such as spearmint tea (*yerba buena*) for "cold" stomach. Even so, they may prefer to use a patent remedy, such as Alka Seltzer. Others know and use upwards of 30 medicinal plants, and many times that number of combination remedies (i.e., several different classes of leaves boiled into one tea for "cold" stomach). Intensive interviewing revealed that many women could recite the formulas for herbal cures, such as boiling and gargling with the water of *gordolobo* for cough but that they knew neither the botanical category associated with the name nor where to collect it.

In general, there is also a division of knowledge by sex, with women knowing more remedies than men. In many interviews, men deferred to their wives for information and both indicated that curing and cooking form part of the homemaker role. There is also a division of knowledge according to age and experience. This is most evident among young women. Prior to child bearing and rearing, girls show limited knowledge and interest in putting together the various bits of information about identities, hot-cold qualities, and medicinal virtues of plants. However, once faced with curing their own young, they integrate what information they have and solicit more from mothers, mothers-in-law, older relatives, fictive kin, and neighbors. Major sources of medicinal plant information are, not surprisingly, elderly women who have borne and dosed one or more generations of children. "Lay herbalists"—those few individuals who have unusually complete knowledge of local wild and garden herbs—can also be approached by those outside their own kinship and fictive kinship-neighborhood social networks for information and occasionally medicinal plants. They differ from *curanderas* (professional curers) in that they never venture out of their immediate social networks to cure and never cure for money.

Finally, there are the *curanderas* (almost all female), who differ from lay persons if not in the numbers of herbs they know and employ, then at least in their applications of them. *Curanderas*, even if just learning, cure outside of their homes for a fee. Generally they specialize in one or another illness of evil origin, such as magical fright, and have particular herbal and manipulative techniques which distinguish one from another. They also differ in their complexity and effectiveness. Both lay persons and curers learn their cures mainly by imitating experienced curers. They copy herbs, movements, and procedures of those who "know how to cure," (which explains why the same sets of herbs for various ailments are used so widely within the community), but do not, however, learn the "words" which go with the plants and other techniques, which explains why they are perceived to be of varying (and limited) effectiveness. In the words of one 76-year-old curer, there are many who know the herbs and "think they know how to cure but don't." Therefore, despite her advanced age, her services to cure (magical fright) remain in demand.

In contrast, the most reknowned curers command esoteric knowledge of herbs and the nature of evil which most lay persons and even most curers do not share. In Mitla, in 1972, the most reknowned curer was a middle-aged native woman who had learned most of the traditional herbs of the garden and countryside as a child. To these, she later added herbs, perfumes, and procedures which she acquired either through "divine inspiration" and/or trips to spiritualist centers in Mexico. Her skills thus combined traditional specialties, such as diagnostic pulsing and divining with *badoo* seeds, with perfumes, candles, prayers, and rubbing procedures which originated outside of the local tradition. She also proved to be an innovator, who popularized the use of *albahaca*, a common element in spiritualist "cleansing" cures (see Kelly 1961; Madsen 1965) in Mitla curing procedures for evil eye. Before imitating her, people had used mainly the leaves of *pirú* and *ruda*. But inimitable is her unique ability to discourse upon and recognize the signs of witchcraft and other evils and to use herbs systematically rather than by fiat to overcome them. In contrast to others, she does not simply cleanse patients according to "tradition" but reasons according to a logical system which views witchcraft as repelled by certain sweet smells (e.g., *albahaca* or perfumes) and the color red (her reason for using geraniums as the thirteen flowers in cures for magical fright) and gold (the color of mustard seeds scattered around a child to "chase away the witch"). The herbs and flowers also talk to her and she to them, a rapport with the natural world which is not visibly part of the lifeways of other Mitleños.

Finally, people also differ in their faith and belief that the herbs will work. This depends in part on their differential knowledge and experience with them but also on their religious orientation. "With faith" herbs will cure any illness, according to the local herb users. Nevertheless, the continuing use of herbs can be described according to the several types of illness which they continue to treat.

## Analysis of Symptoms

Most illnesses today are treated by the doctor. Even though more expensive, people still seek physicians' remedies for dysentery, fevers, inflammations and other ills for which the doctor "always cures," and for which there are no good herbal alternatives. Even so, herbs are retained for simple complaints (stomachaches, headaches), illnesses which the doctor cannot recognize (magical fright-soul loss, evil eye) and, as a last resort, illnesses which the doctor does not cure. To some extent, the third category overlaps the second, as people rediagnose symptoms into categories of illness which the doctor does not recognize (witchcraft, etc.) or reassess and treat with herbs headaches, stomachaches, urinary disorders, and persistent fevers which prove to be somehow beyond the doctor's healing capacities.

For simple ailments, herbs are an alternative to pharmaceuticals in treating immediate ills. Table 1 displays examples of different herbal and non-herbal remedies for dosing various afflictions in Mitla. (It is a composite of information supplied by more than 95 members of the culture, and no one individual held all of this knowledge in this form, though experienced herb users knew several herbal and non-herbal alternatives for each ailment.) For stomachache, one can drink one or another kind of herbal tea or take some patent remedy. For headaches, teas (for headaches due to colds) and/or leaf poultices applied with one or another grease (oil or wax) are used. Table 1 also indicates the various leaf poultices used to treat inflammations and ingredients combined for purges or enemas which were the common treatments for extreme intestinal disturbances. For natural ailments there were and are herbal plus patent remedies (to which are ascribed hot-cold properties), or non-herbal (e.g., physician) options. When herbs are used, they are viewed as part of the order of nature, and can be combined with non-herbal elements

TABLE 1

HERBS AND OTHER REMEDIES USED IN MITLA

(Key: *Source:* g–garden, w–wild, local, W–wild, non-local, p–pharmacy/general store; *Humoral quality:* H–Hot, C–Cold, T–Temperate; *Preparation:* t–tea, a–external application)

| Illness | Herbs | Botanical Identification | Patent Remedies | Other Remedies |
|---|---|---|---|---|
| Dolor, | yerba buena (g) H (t) | *Menthus* spp. | Alka Seltzer H | Massage |
| Frialdad | manzanilla dulce (g) H (t) | *Matricaria chamomilla* L. | sodium bicarbonate H | |
| (stomachache) | monstranza (g,W) H (t) | *Mentha rotundifolia* Huds. | Picot (Sal de Uvas) C | |
| | salvia blanco (w) H (t) | *Lippia graveolens* H.B.K. | Terramycin H | |
| | salvia amarillo (w) H (t) | *Turnera diffusa* Willd. | milk of magnesia H | |
| | yerba dulce (g) H (t) | *Lippia dulcis* Trev. | salt H | |
| | anís (p) H (t) | | mescal H | |
| | anís del campo (w) H (t) | *Tagetes* sp. | | |
| | orégano (g) H (t) | *Origanum vulgare* L. | | |
| | yerba buena with monstranza, salvia, yerba dulce | | | |
| | yerba buena with ruda | | | |
| | manzanilla dulce (g) H | *Matricaria chamomilla* L. | | Enemas |
| | espinosilla (W) C | *Loeselia mexicana* L. | | |
| | malba (w) C/T | *Malva parviflora* L. | milk C | |
| | rosa de castilla (g) C | | aceites (oils) H/C | Purgatives |
| | hoja sen (p) | | | |
| | hoja lante (p) | | | |

TABLE 1 (Continued)

| Illness | Herbs | Botanical Identification | Patent Remedies | Other Remedies |
|---|---|---|---|---|
| Dolor, Frialdad (cont'd.) | biushito (w) H | *Mollugo verticillata* L. | | Purgatives |
| | mesquite (w) H | *Prosopis laevigata* (Willd.) M.C. Johnst. | | |
| | gordobahn (w) C | *Pedilanthus tomentellus* (Rob. & Greenm.) | | |
| Empacho | yerba buena (g) H (t) | *Menthus* spp. | zarcón H | Massage |
| | biushito (w) H (t) | *Mollugo verticillata* L. | milk of magnesia H | |
| | lengua de vaca (w) H (t,a) | *Buddleia sessiliflora* H.B.K. | Terramycin H | |
| | mesquite (bark, gum) (w) ? (t) | *Prosopis laevigata* (Willd.) M.C. Johnst. | Interovioformo H | |
| | nanches (green fruit) (w,g) ? (t) | *Malpighia* sp. | | |
| | anís (p) H (t) | | | |
| | yerba de empacho (w) H (t) | *Tecoma stans* (L.) H.B.K. | | |
| | linasa (seeds) (p) | | | |
| | rosa de castilla (g) C (t) | | | |
| Inflamación | chamizo (w) C (a) | *Baccharis salicifolia* (R.&P.) Pers | manteca (H/C/T) | Doctor's pills, injections |
| | yak tsun las (w) C (a) | *Montanoa* sp. | aceite de comer C | |
| | gubedundh (w) C (a) | *Anisacanthus quadrifidus* (Vahl) Standl. | aceite rosado ? | |
| | pájaro bobo (w) C/T (a) | *Ipomoea pauciflora* Mart & Gal. | alcohol C | |
| | grilla blanca (g) C (a) | *Ricinus communis* L. | catalán C | |
| | yerba mora (w) C (a) | *Solanum* (*nigrum* group) | aguardiente C | |
| | malba (w) C/T (t) | *Malva parviflora* L. | mescal H | |
| | manzanilla amarga (g) C (a) | *Chrysanthemum parmethium* (L.) Bernh. | | |

TABLE 1 (Continued)

| Illness | Herbs | Botanical Identification | Patent Remedies | Other Remedies |
|---|---|---|---|---|
| Inflamación | shobarobo (w) ? (a) | *Cordia curassavica* (Jacq.) R.&S. | | Doctor's pills |
| (cont'd.) | mostaza (w) C (a) | *Nicotiana glauca* Grah. | | injections |
| | yerba de lartija (w) ? (a) | *Euphorbia maculata* L. | | |
| | collar de lartija (w) ? (a) | | | |
| | susí (w) C (a) | Unidentified | | |
| | ruda (g) ? (a) | *Ruta graveolens* L. | | |
| | lengua de vaca (w) H (a) | *Buddleia sessiliflora* H.B.K. | | |
| | limón (tender fruit) (g) C (a) | *Citrus* sp. | | |
| | miltomate (tender fruit) (g) C (a) | *Physalis* sp. | | |

Note: combination leaf/tomato/lemon/spirit poultices applied early in the morning before the sun is up to "cool" inflammation. The fresh leaves are "cooked" as they draw out the "heat."

| Illness | Herbs | Botanical Identification | Patent Remedies | Other Remedies |
|---|---|---|---|---|
| Dolor de | manzanilla dulce (g) H (t) | *Matricaria chamomilla* L. | cough syrups | |
| cabeza | poleo (g,W) H (t) | *Satureja mexicana* (Benth.) Briq. | aromatic rubs: | |
| (headache) | ruda (g) ? (t) | *Ruta graveolens* L. | Disenfriol C | |
| | pitiona (g) H (t) | *Lippia alba* (Mill.) N.E. Brown | Vaporub | |
| | yerba santa del campo (w) ? (a) | *Marsdenia mexicana* Decne | lard H/C/T | |
| | grilla verde (g) C (a) | *Ricinus communis* L. | chicken fat C | |
| | San Cayetano (w) C (a) | *Solanum laurifolium* Mill. | aceite rosado H? | |
| | patoshiwit (w) C (a) | *Cestrum dumetorum* Schlecht. | aceite de comer C? | |
| | higo (g) C (a) | *Ficus* sp. | cebo blanco (wax) C | |

TABLE 1 (Continued)

| Illness | Herbs | Botanical Identification | Patent Remedies | Other Remedies |
| --- | --- | --- | --- | --- |
| Dolor de |  |  | aceite almendras H |  |
| cabeza (cont'd.) |  |  | pomada de manzana H |  |
| Note: leaves of single species applied with one or more grease |  |  |  |  |
| Dolor de | pájaro bobo (w) C (a) | *Ipomoea pauciflora* Mart. & Gal. | manteca H/C/T | Doctor, Dentist |
| muele | mostaza (w) C (a) | *Nicotiana glauca* Grah. | aceite de comer C |  |
| (toothache) | grilla verde (w) C (a) | *Ricinus communis* L. | aceite rosado H |  |
|  | pirú (w) H (a) | *Schinus molle* L. | pimienta H |  |
|  | confiti (g) H (a) |  | clavo H |  |
|  | ruda (g) ? (a) | *Ruta graveolens* L. | camphor ? |  |
|  |  |  | Other pharmacy pills |  |
| Dolor de | yerba de aigre (w) ? * | *Eupatorium espinosarum* A. Gray |  | Doctor's remedies |
| oído | albahaca (g) ? * | *Ocimum micranthum* Willd. |  |  |
| (earache) | ruda (g) ? * | *Ruta graveolens* L. |  |  |
|  | pata de cabron (w) ? ** | *Solanum rostratum* Dunal |  |  |
|  | mirtoh (g) ? * | *Salvia* sp. |  |  |
| *Leaves (**flowers) placed in ear |  |  |  |  |

NOTE: Spanish labels are used to simplify and to facilitate cross-cultural comparison.

and religious ceremony (e.g., blessing an herbal tea). Apart from these herbal "simples" are herbs used as the major components of ritual cures for all illnesses of evil origin. For sets of symptoms which the doctor does not recognize as "illness" such as the generalized aches, weariness and lack of appetite which characterize magical fright-soul loss, the cure is always a ritual, repeated three times, involving numerous plant elements—copal incense, green bamboo canes, various flowers, and *yerba de espanto*. Similarly, for symptoms which the doctor cannot cure, such as persistent crying and vomiting of babies, wounds which will not heal, various kinds of kidney infections and cancers which do not respond to modern medical treatment, the rediagnosis is again "of evil origin" (witchcraft, evil eye, evil air or sorcery) and the cure herbal—usually a cleaning, sometimes an herbal brew. Diabetes and rheumatism, again illnesses which the doctor can dose but not remove, are also treated from time to time with herbal remedies, as victims search for relief, with or without ritual.

Ritual cures achieve additional therapeutic power and social meanings by using natural olfactory, visual, tactile and auditory symbols as well as cultural symbols which arouse the patient, the assembled social relations, and the curer to therapeutic action. Cleansing or brushing the victim of evil eye with the fragrant leaves of *albahaca*, *ruda*, or *pirú* expresses an ongoing cultural belief that witchcraft has the power to harm, but these particular herbs have the power to dispel evil. By the ritual cure, in which alcohol may also be used to massage and cleanse the victim, the patient is urged toward recovery. The victim of soul loss, after being given the opportunity to confess or bring to mind possible frightening experiences which might have jolted the soul from the body, becomes the focus of concern in a complex ritual. Before her (women are more often subject to fright and soul loss), copal incense is burnt, in order to contact the realm of the supernatural. Next, the earth is struck with green bamboo canes or sanctified palms to raise her soul from the earth, and then she is struck with the same materials to reattach the soul to her body. Spirits (mescal or *aguardiente*) are spat on her back, breast and arms to "shock" the soul back into the body, and the curer also calls to the four corners of the house (symbolizing the four corners of the universe) to regain the victim's soul. The victim later drinks a little bit of the water of the copal, and drinks *yerba de espanto*, the "cooling" foaming green infusion which counteracts the "hot" condition of soul loss. In addition, thirteen classes of flowers and thirteen green bamboo canes are set near the victim's head at bedtime, and the cure is repeated three times.

Such curing ceremonies focus the power and order of nature on the social unit and the patient to provide a cure. In addition to the potential of natural symbols in the plant elements, cultural symbols, such as the persistent sign of the cross, the sacred ritual incense which is used in *all* ceremonies involving saints and the supernatural, the magic numbers thirteen and three, which characterize the numbers of elements and repetitions in other cures, join to restore the victim to physical health and "healthy" participation in the social unit. Just the fact that the social unit has recognized the illness and insisted on a cure is a psychological boost to the victim. It demonstrates their concern. For both patient and social unit, the cure's orderly repetition of basic cultural symbols–the four cornered universe (also the Catholic cross), freshness and coolness in contrast to the "heat" of illness, as well as their implicit acceptance of fright-soul loss as an explanation of symptoms establish their common symbols as well as common concerns. In these ritual cures, herbs are intrinsic to the ongoing cultural religious-cosmological symbolism.

## Preference and Belief

Financial considerations aside, choice of herbal cures indicates one's acceptance of the natural and cultural order which links health and healing to the flora and a particular conceptual view of illness. From the medical perspective, recent analyses of folk healing (e.g., Frank 1973) have stressed the importance of treating the whole person in relieving physical and psychological symptoms of stress and distress, a process which at the same time revitalizes the basic cultural order. From a symbolic perspective, Turner (1967) and Vogt (1976) among others, have described in detail the process of ritual curing in which the individual is brought (even reborn) into society, in the process of being relieved of his physical and/or psychosomatic symptoms. Though the symbolic details of curing are less elaborate in Mitla curing (contrast the Tzotzil fright-soul loss curing ceremony described by Vogt [1976]), the elements of the cure in either case illustrate how the entire social group unites to cure an illness of psychosocial as well as of physical origins, and in the process, makes optimum use of the natural symbols contained in herbs. Less dramatically, even simples have their psychological and symbolic components. The herbal cure continues a tradition of known effective cures, in which the individual may also benefit from the psychological reinforcement of prayers, crosses, or other ritual gestures. Though the very religious also make the sign of the cross and bless the pharmaceuticals they serve in lieu of herbal teas, the chemical cures do not partake of the simple faith that all herbs are natural remedies given by God.

To choose herbs is a recourse to tradition and the healing powers of nature. Some cure only with herbs out of religious conviction: they believe that God's remedies and traditional wisdom should restore them to health.[2] Others state simply that they experience herbs to be more effective. While many of the herbs have been proven to have chemically active substances which produce their claimed therapeutic effects (cf. Ortiz de Montellano 1975; Lozoya L. 1976), the psychological benefits of the cure, which put the individual in a frame of mind to heal himself should not be underestimated (Frank 1973; Ysunza O. 1976). Even the simplest herbal remedy, such as *yerba buena* or *yerba maestra*, involves diagnosis and treatment within the traditional hot-cold system, and therapy from a corpus of natural remedies, usually repeated the magical number three times, to increase the "effectiveness of symbols" (Lévi-Strauss 1963:x) with which the individual is cured. Though *yerba maestra*'s value as an appetite restorant and anthelmintic has been pharmacologically demonstrated as well (Martinez 1944; 129-30; Ysunza O. 1976), this does not negate cultural symbols surrounding its use.

## Financial Considerations

Finally there are financial considerations which enter into some decisions to use herbs rather than their alternatives. For digestive disorders, eye infections, earaches, body aches and pains, the numerous remedies of the countryside and garden are less expensive than patent and prescription alternatives. If one gathers or grows them oneself, herbs are the cheapest route to health if one believes they will work. Nevertheless, prescription pharmaceuticals (sold without prescriptions) for eye infections such as Terramycin ointment and Sulfathiazol for skin infections are the standard remedies for these disorders because they are perceived to be more effective. For female disorders (excessive menstrual flow, infertility) even the traditional specialist for these complaints urges her patient to go to the doctor if she is not quickly successful in treating them. She even saves the cartons from prescription pills in order to prescribe effective physician's remedies to other patients who come to her with similar symptoms and who have insufficient financial means or do not wish to consult the doctor. Inflammation, dysentery, and fevers are also almost exclusively treated by the doctor, whose remedies are perceived to be more powerful and effective.

Also, it should be noted that herbal remedies are not always less expensive than other medical alternatives. Remedies for witchcraft, evil eye, and fright may outprice the cost of a doctor's consultation and prescription.

[2] One woman who expressed stubborn faith in 1972 was brought to the local doctor for her own ailments in 1975 and brought all of her (six) children there one by one for various physical complaints after her herbal remedies did not work.

Though the ordinary cure for fright is cheap (a bargain even after the price went up in 1975 from three cures for nine pesos to three cures for fifteen pesos) in contrast to the doctors' fees (always at least sixty to one hundred pesos), the major curing ceremonies for illnesses of evil origin cost hundreds of pesos in expense and result in time lost to other activities. Choice of herbal remedies thus involves the combined considerations of the nature of illness symptoms, knowledge and availability of herbs, and the costs of the various alternatives for curing a set of symptoms labeled by a particular illness category.

## HERBAL MEDICINE AND CULTURE CHANGE

Both biosystematic and medicinal plant classifications are systems which change content but not structure as new flora or information are introduced, or as known plants or information disappear. The principles of taxonomic nomenclature allow for description, identification, and naming of new taxa, while the medicinal plant classification is constantly modified by the introduction of new flora believed to be effective for curing various ailments and the reanalysis of existing plants in the search for new remedies. This was shown by the rapid absorption of European plants into native gardens and pharmacopeia. By 1580, Mitleños already grew spearmint, among other Spanish herbs (del Paso y Troncoso 1905) while the medicinal system began the process of syncretism between native Mitla Zapotec and Spanish illness categories and treatments (Parsons 1936:516-17).[3]

Today, new herbs are still being introduced, either through the traditional channels of trade and interregional visits or the more unconventional sources of books and herbal doctors. The medicinal plant classification has demonstrated its flexibility, as it adds modern illness labels like diabetes, rheumatism, and amoebas to its curing dimensions. As traditional sets of symptoms acquire new names, herbs good for treating symptoms of aroused bile and *muina*, which include anger, irritability, and thirst, become "good for diabetes." Herb poultices for *latido* and herbs traditionally used to treat dysentery become "good for amoebas." Many different poultices are tried for aches and pains which were traditionally labeled either *aire* or *pasmo* but which now are also called *rumatismo*. A process of translation goes on whereby traditional herbs need not drop out when symptoms are relabeled,

[3]Common Spanish introduced herbs include *yerba buena, oregano, manzanilla dulce, manzanilla amarga, romero,* and *ruda.* Techniques include cupping and various forms of "cleansing" which according to Foster (1953) and Madsen (1965) were Spanish introductions. It is interesting to note that most of the elements of cures for witchcraft, evil eye, and sorcery in Mitla are of Hispanic origin (cf. Messer 1975:333-74).

but instead acquire new medicinal virtues. Conversely, the old symptoms under both the traditional and the new names can also be treated by modern medicines good for the modern illnesses (labels).

Similarly, herbs are maintained alongside modern medicine even where modern pharmaceuticals are used as the principal curing ingredients. People take pills with "hot" spearmint tea to cure a set of symptoms analyzed in the traditional humoral system to be "cold" or a "cooling" tea, like lemon, to accompany the medicines for their hot ailments. Or, they use herbs to modify what they perceive to be the "heating" or "cooling" effects of modern medicines. In this manner, they retain their faith in the hot-cold system and in the efficacy of herbs while accepting modern medicines. People also favor continuing use of herbs because the illnesses which they treat, in spite of modern medicine's deprecations, still exist; the psychological symbolic aspects of herbal healing are still beneficial, and people perceive that the herbs really work. In addition, the psychological climate of Mexico favors ongoing use of herbs. Research groups such as I.M.E.P.L.A.M. (Lozoya L. 1976) are indicative of a country which encourages ongoing úse of (and experimentation with) its natural (herbal) resources.

In spite of the inherent adaptability of the botanical classification and medicinal systems, however, three factors currently are affecting traditional herbal medicine in Mitla.

First, the movement away from agriculture and other nature-oriented pursuits is reducing knowledge of local medicinal flora within the population. In the traditional economy, both men and women had intimate contact with nature. From the age of five or six, young boys traversed the fields and hills with their fathers, pursuing subsistence agricultural activities, and at the same time having the opportunity to learn the different edible, medicinal, or just interesting plants in the environment. Today, with young men more involved in indoor crafts such as weaving or sewing, only outings on a day of rest offer a chance to walk over the hills and fields, and then with less of an eye toward useful plants. Women similarly have and will have increasingly more limited opportunities to learn and collect wild herbs. Their economic roles have enlarged since the inception of large-scale tourist trade goods production and sales; women, who used to spend some time in the fields with their menfolk, now work at home as well. Less time is spent in the search for firewood, which traditionally forced women into the hills in groups. The amount of firewood is decreasing while its cost is increasing, and families, to save time and labor as well as money, are switching to gas or petroleum stoves. Children now receive less instruction about the whereabouts and uses of medicinal plants in their natural contexts. This bodes ill for knowledge among the younger generation. In contrast to cultural settings in which economic changes eradicated plants used for medicine (e.g., the creation of plantations which used herbicides in Puerto Rico, which wiped out local herbal medicine [Steward et al. 1964:483]), plants in Mitla should remain in their proper place in the hills and field borders, but people will no longer recognize or use them.

Second, the large-scale introduction of modern scientific medicaments and disease concepts is undermining the use as well as the underlying medicinal categories of traditional herbs. The declining use of herbs for common ailments is evident in other areas of Mexico. In an area of Nahuatl Indian culture:

> The impact of modern medicine is producing rapid change in the system of home treatment. A generation ago it was necessary for families to grow their own medicinal herbs, but only a few herb gardens remain . . . today. The middle and older generations still use herb medicines, but members of the younger generation often use modern medicines. Patent medicines have been widely accepted as alternatives or replacements for Hippocratic remedies. A stomach ache may be treated with Alka Seltzer, Milk of Magnesia, Epsom Salts, camomile or oregano tea. Sulfathiazol is used for infections and aspirin is taken for a wide variety of ailments. Patent medicines are thus used with the folk theories of disease and disease treatment. [Madsen 1965:126]

Where modern medicine undermines traditional illness beliefs, not only herbal medicine, but also religion, traditional authority, and much of the religious-social structure may disintegrate (cf. Erasmus 1967:106). As a corollary, traditional curing personnel and with them their herbal arts may devolve as modern medicine dominates curing. In Veracruz, anthropologists in one town observed that since the death of most of the distinguished curers (who preferred to describe themselves as herb curers), patent medicines were prescribed more often than herbal ones in home curing (Kelly et al. 1956:80).

Though in Mitla many gardens are still filled with herbs, it has already been mentioned that patent remedies are replacing plants for digestive, eye, female, and infectious complaints. Beliefs in illnesses of evil origin remain functional, though under attack by modern medicine. People have so far been able to compartmentalize the multiple causes of many disorders, and use traditional curers to recover lost souls and dispel evil, while using modern doctors and pharmaceuticals to remove physical infections. Traditional curers still give advice and provide cures for illness of evil origin. It is, in fact, mainly in the area of evil illnesses that herbal cures continue. People now, as in Parsons time (Parsons 1936), rally the family and patient to traditional cultural symbols, including herbs, which are believed to promote health and healing.[4] It is more in the area of mundane cures that herbs

---

[4]It should be emphasized that in contrast to many other Indian communities in Mexico (e.g., Tzotzil, Tzeltal), Mitla has never had a strong tradition of shaman-curer

are dropping out. Also, herbal use surrounding childbirth is lost with modern midwifery. Finally, there are currently few curers who know traditional arts like divination and pulsing, and one can state that the incidence of diagnosis and treatment of illnesses of evil origin has decreased with modern medicine, so that even *susto* and witchcraft cures, with their attendant curers and herbs, persist in attenuated form.

Counteracting the movement away from herbs, though not necessarily fostering the use of local ones, are herbals, mass communication, and modern herbal pharmaceutical research. In Mexico, they continue processes evident since the arrival of the Spanish.[5] In their quest for political-economic-scientific hegemony and spiritual power, the Crown and Church respectively investigated the distribution and medicinal uses of indigenous herbs. In more recent years, pharmacologists, acting out of either profit motives or intrinsic scientific and cultural interests (cf. Martín del Campo 1976) try to evaluate the active principles of herbal medicines and figure out ways to dispense them in low cost, effective forms for developing countries (cf. Lozoya L. 1976). They have in several cases succeeded in isolating the chemical compounds which effect the remedies (or at least the expected reactions) to the illnesses they are believed to treat (Ortiz de Montellano 1975). Yet, with few exceptions, interest is focused on healing with biochemical compounds, not on the role of herbs in culture, which treat the bio-psycho-social individual and his social network (cf. Ysunza Ogazon 1976, for a notable exception). There is little interest in ritual or the kinds of psychological healing which traditional herbs effect. Though such studies may produce an increase in medicinal use of herbs, total cultural medicinal qualities which treat the whole person by drawing on all of the symbolic and empirical relations between plants and people are attenuated when traditional medicinal herbs become drugs.

Similarly, popular herbal culture is enlarging the distribution of herbs and spreading herbal knowledge from one area to another. Those who read can change their sources of authority from the elders or experienced herbalists

---

socio-political religious control, nor a community ethos which uniformly interprets illness as a sign of moral wrongdoing.

[5]Among the first acts of both secular and clerical authorities interested in transporting the plants of the New World to Europe were inventories of local herbs and their uses. The crown commissioned studies, e.g., the inventories of the *Relaciones geográficos* of 1580 (del Paso y Troncoso 1905), and Hernandez' investigations of Mexican flora, 1577. Clerical interests were scientific and spiritual. By understanding the indigenous herbal medicines, they hoped to accumulate a body of effective medicinal knowledge to better treat local ills and at the same time to understand and undermine the local religious-cosmological beliefs which included herbal healers and medicines.

in their own culture to "experts" who have written books. Herbal and pharmacological interests encourage the continued or renewed use of herbs in Mexican culture but not for local cultural or religious reasons. They further a national cultural attitude in Mexico (as in other developing countries) that herbal medicine is good because it is pharmacologically effective and that nations should appreciate and develop indigenous resources such as herbal medicine in order to serve their people. However, in this process, herbs are removed from their local cultural and psychotherapeutic curing contexts.[6]

## PERSPECTIVES

Modern medicine can substantially change not only medicine but world view. So far, in Mitla, people still entertain traditional illness beliefs, the ranks of healers are still being filled and herbs being sought. The cultural system which links plants to illness and cosmological beliefs about the organization of the world and human health within it still exists, and people choose traditional medicine, which is still largely herbal, for those illnesses "which the doctor cannot cure." As long as these symptoms and illnesses persist, non-modern medical practitioners and techniques will be found to cure them.

However, the roles of other local herbs in Mitla culture are changing. In the past, people knew:

Every herb in the countryside is a remedy.
One has only to know. . . .

In the present, the "knowing" and the deep belief that local herbs are part of the natural environment which, with the plan of God, heals, is less pervasive. People still seek herbs as natural remedies, but in the capacity of pharmaceuticals, rather than embodiments of local tradition. They still have recourse to local healers, but they also make use of non-local and spiritualist healers, who use herbs and treat people for illnesses which the doctor cannot heal. With the exception of the very specialized healer who combined local and spiritualist traditions cited above, these practices also change the role of herbs in local culture.

In conclusion, the future of Mitla's indigenous herbal tradition will change as Mitleños look outward, both toward pharmacological herbal medicine and interregional folk medicine, but should not totally disappear. What will

---

[6]The discussion applies only to studies of non-hallucinogenic drugs. Analyses of hallucinogens, even if biochemical, are characterized by psychotherapeutic as well as biochemical aspects (see volumes edited by Viesca Treviño 1976 and Lozoya L. 1976 for the most recent work on the pharmacological and psychotherapeutic aspects of both hallucinogenic and non-hallucinogenic plants).

alter are the ecological, religious, intellectual, and economic considerations governing the decisions to use local herbs. All will continue to affect the "entire range of relations" between human and plant populations in Mitla, not just the medical pharmacological relations between man and herb.

## APPENDIX: COMMON AND LATIN SCIENTIFIC NAMES

| | |
|---|---|
| Albahaca | *Ocimum micranthum* Willd. |
| Cacahuatón | *Calea hypoleuca* Rob. & Greenm. |
| Cuanasana | *Pluchea odorata* (L.) Cass. |
| Epazote | *Chenopodium ambrosioides* L. |
| Espinosilla | *Loeselia mexicana* (Lam.) Brand. |
| Gordolobo | *Gnaphalium* sp. |
| Huesito | *Stylosanthes* sp. |
| Lengua de vaca | *Buddleia sessiliflora* H.B.K. |
| Manzanilla dulce | *Matricaria chamomilla* L. |
| Manzanilla amarga | *Chrysanthemum parthenium* (L.) Bernh. |
| Orégano | *Origanum vulgare* L. |
| Pericón | *Tagetes lucida* Cav. |
| Pirú | *Schinus molle* L. |
| Pitiona | *Lippia alba* (Mill.) N.E. Brown |
| Poleo | *Satureja mexicana* (Benth.) Briq. |
| Romero | *Rosmarinus officinalis* L. |
| Ruda | *Ruta graveolens* L. |
| Salvia | *Lippia graveolens* H.B.K. |
| Suzí | *Jatropha dioica* Sessé |
| Sauco | *Sambucus mexicana* DC. |
| Toloache | *Datura stramonium* |
| Yerba buena | *Mentha* sp. |
| Yerba de espanto | *Loeselia mexicana* (Lam.) Brand. |
| Yerba del aigre | *Eupatorium espinosarum* A. Gray |

## REFERENCES

Erasmus, Charles J.
1967 Culture Change in Northwest Mexico. In: Contemporary Change in Traditional Societies, Vol. 3, Mexican and Peruvian Communities, edited by J. H. Steward, pp. 1-131. Urbana: University of Illinois.

Foster, George M.
1953 Relationships between Spanish and Spanish-American Folk Medicine. Journal of American Folklore 66:201-18.

Frank, Jerome D.
1973 Persuasion and Healing. Baltimore: Johns Hopkins Press.

Harvey, T. E. C., and F. B. Armitage
1961 Some Herbal Remedies and Observations on the Nyanga of Matabeleland. The Central African Journal of Medicine 7:193-207.
Jones, Volney H.
1941 The Nature and Status of Ethnobotany. Chronica Botanica 6:219-21.
1965 The Bark of the Bittersweet Vine as an Emergency Food Among the Indians of the Western Great Lakes. The Michigan Archaeologist 11:170-80.
Kelly, Isabel
1961 Mexican Spiritualism. In: Alfred L. Kroeber: A Memorial. Kroeber Anthropological Society Papers 25:191-206.
Kelly, Isabel, H. García Manzanedo, and Catalina Gárate de García
1956 Santiago Tuxtla Veracruz Culture and Health. Mimeographed.
Kleinman, A.
1973 Medicine's Symbolic Reality. On a Central Problem in the Philosophy of Medicine. Inquiry 16:206-13.
Lévi-Strauss, Claude
1963 Structural Anthropology. Chicago: University of Chicago Press.
Lozoya L., X.
1976 El Instituto Mexicano Para el Estudio de las Plantas Medicinales A.C. (I.M.E.P.L.A.M.). In: Estado Actual del Conocimiento en Plantas Medicinales Mexicanas. Mexico: I.M.E.P.L.A.M.
Madsen, Claudia
1965 A Study of Change in Mexican Folk Medicine. Middle American Research Institution Publication 25.
Martín del Campo, R.
1976 Consideraciones acerca de las Plantas Medicinales Mexicanas y su Posible Proyección Mundial. In: Estado Actual del Conocimiento en Plantas Medicinales Mexicanos, edited by X. Lozoya L. Mexico: I.M.E.P.L.A.M.
Martínez, Maximino
1944 Las Plantas Medicinales de México. 3rd ed. México: Ediciones Botas.
Messer, Ellen
1975 Zapotec Plant Knowledge: Classification, Uses, and Communication About Plants in Mitla, Oaxaca, Mexico. Ph.D dissertation, University of Michigan. Ann Arbor: University Microfilms.
Ortiz de Montellano, Bernard R.
1975 Empirical Aztec Medicine. Science 188:215-20.
Parsons, Elsie Clews
1936 Mitla, Town of the Souls. Chicago: University of Chicago Press.
Paso y Troncoso, Francisco del
1905 Relación de Tlacolula y Mitla. Papeles de Nueva Espana. Segunda Serie, Geográfico y Estadística Tomo 4:144-54. Madrid: Est. Tipográfica "Sucesores de Rivadeneyra."
Rubel, Arthur J.
1964 The Epidemiology of a Folk Illness: Susto in Hispanic America. Ethnology 3:268-83.
Steward, Julian, et al.
1956 The People of Puerto Rico. Urbana: University of Illinois Press.
Swain, Tony (ed.)
1972 Plants in the Development of Modern Medicine. Cambridge: Harvard University Press.
Turner, Victor
1967 The Forest of Symbols. Aspects of Ndembu Ritual. New York: Cornell University Press.

Viesca Treviño, C. (ed.)
1976 Estudios Sobre Etnobotánica y Antropología Médica 1. México: I.M.E.-P.L.A.M.

Vogt, Evon Z.
1976 Tortillas for the Gods: A Symbolic Analysis of Zinacanteco Rituals. Cambridge: Harvard University Press.

Ysunza Ogazón, A.
1976 Estudio Bio-Antropológico del Tratamiento del "Susto." In: Estudios sobre Etnobotánica y Antropología Médica 1., edited by C. Viesca Treviño. Mexico: I.M.E.P.L.A.M.

# PART III
# PRINCIPLES OF RESOURCE UTILIZATION

**INVOIC**

# MUSEUM OF ANTHROPOL

**UNIVERSITY MUSEUMS BUILDING • AN**

**PHONE 313-7**

**SOLD TO**

Andrew Amann

5800 N. Pauline St

Chicago, IL

| DATE | DATE SHIPPED | SHIPPED VIA | YOUR ORDER NO. |
|---|---|---|---|
| 6/24 | 6/25 | LIB | |

| QUANTITY | DESCRIPTION |
|---|---|
| 1 | AP 67 |
| | |
| | |
| | |
| | |
| | |
| | |
| | |
| | |

Form 10188

## INTRODUCTION

Names of plants and their taxonomic relationships are based upon perceptible bio-physical attributes interpreted by a particular culture. How and under what conditions the plants are procured and employed are dictated not by their classification but by prescribed rules. These cultural dictums guide human behavior and, as a result, determine the structure and composition of the local biological communities.

Plant populations undergo continuous disruption. Natural forces such as geological processes or climatic factors affect their presence and productivity. Biological organisms also have a role by eating and otherwise altering the plants. The greatest potential for transforming any plant population directly or indirectly, however, is at the hands of humans.

How human populations manage their plant resources is an important ethnobotanical question. Endowed with even the most rudimentary tools and fire, a human population can easily destroy sizable areas of vegetation. Why they do not results from constraints imposed by necessity, socio-spatial relations, and most significantly, conventions.

Necessity alone is not an adequate explanation. People eat what is available with little regard for the future; if they do not, animals or other people will. The quantity available determines whether a particular plant will provide a snack, a few meals, a seasonal staple, or a storable surplus. Nevertheless, whatever the amount, cultural rules rarely prescribe exactly how much to take but instead guide by very general principles the manner in which a plant should be collected: "take only the part used and leave the rest," "never collect the first plant you desire," etc. These apply to all plant species or broad groupings, such as trees as a life form in Berlin's sense.

Socio-spatial or political relations also may protect floral communities. Some resources may be periodically unavailable because of social tensions or warfare. These regions may become refuges for rare plants. Distance itself may limit how much can be obtained if it must be carried or if other groups' territories must be crossed. Finally, the supply of some plants may be adequate for a small human population but not for a large aggregation. Under such a condition, a plant population may be maintained by protecting a given home range for the exclusive exploitation of a small local population.

Conventional behavior develops by following general rules of conduct, which are reinforced in each situation where they are applicable. Each interaction between a group of humans and a particular plant population or community will be defined by a taxonomy of plants and determined by cultural rules about what to leave and what to remove. Children are taught these by older members of the community until the proper practice becomes automatic. Overall, when direct contact with plants is involved, the actual number of rules and taxonomies are quite limited. On the other hand, prayers to evoke a spiritual response to make an herbal medicine efficacious or to prepare a food dish or ritual potion may be quite complex. This knowledge, which is taught more formally after childhood, is also important for assuring the future availability of plant resources. Even though they may be collected properly, if the post-collection preparation is incorrect, the injured plants, many cultures teach, will leave an area or lose effectiveness. Consequently, resource management can be viewed from a Western perspective of conservation as well as from an ethno-ecological sense that accounts for relative availability in a particular culture's terms.

The tropical areas of the world provide a complex ecological setting for human intervention. Although many plant species are rare and easily exterminated from a local area, this is not a frequent occurrence under the guidance of traditional rules of behavior. Kunstadter illustrates how the Lua' modify their environment and by means of their farming cycle actually create a diversified landscape that does not degrade the environment on a long-term basis. A great number of plants are used but not at the expense of future availability.

Carneiro demonstrates the importance of the tropical forest in the lives of the Kuikuru in South America. Virtually every tree within a territory is named and classified. Even in the course of using the trees, the primacy of the forest in myth and present reality is maintained. Traditionally, there was never doubt that the forest would continue to exist.

Agriculture certainly changes tropical plant communities, but as Dole points out, the processing of manioc in South America requires extensive plant lore for collecting, making, and using objects of material culture. The correspondence of material culture with the preparation of different plant forms, in this case bitter manioc for starch and flour and sweet manioc for fermented beverages, has been ignored by ethnobotanists and consequently the dynamic relationship between forest clearance and the need for forest products has been overlooked. A detailed ethnobotanical study can discover the complementary relationships between food and raw materials and also denote the implications of these interactions for resource management.

The feedback in resource use and resource management may or may not be recognized by members of a particular culture. Their actions may have unanticipated results in the form of the expansion of one plant population over another, the preparation of a habitat suitable for colonization by still other plants, and even the creation of new plant forms. These changes, which result from human activity, may lead to the creation of anthropogenic plant communities simply because the customary rules were followed.

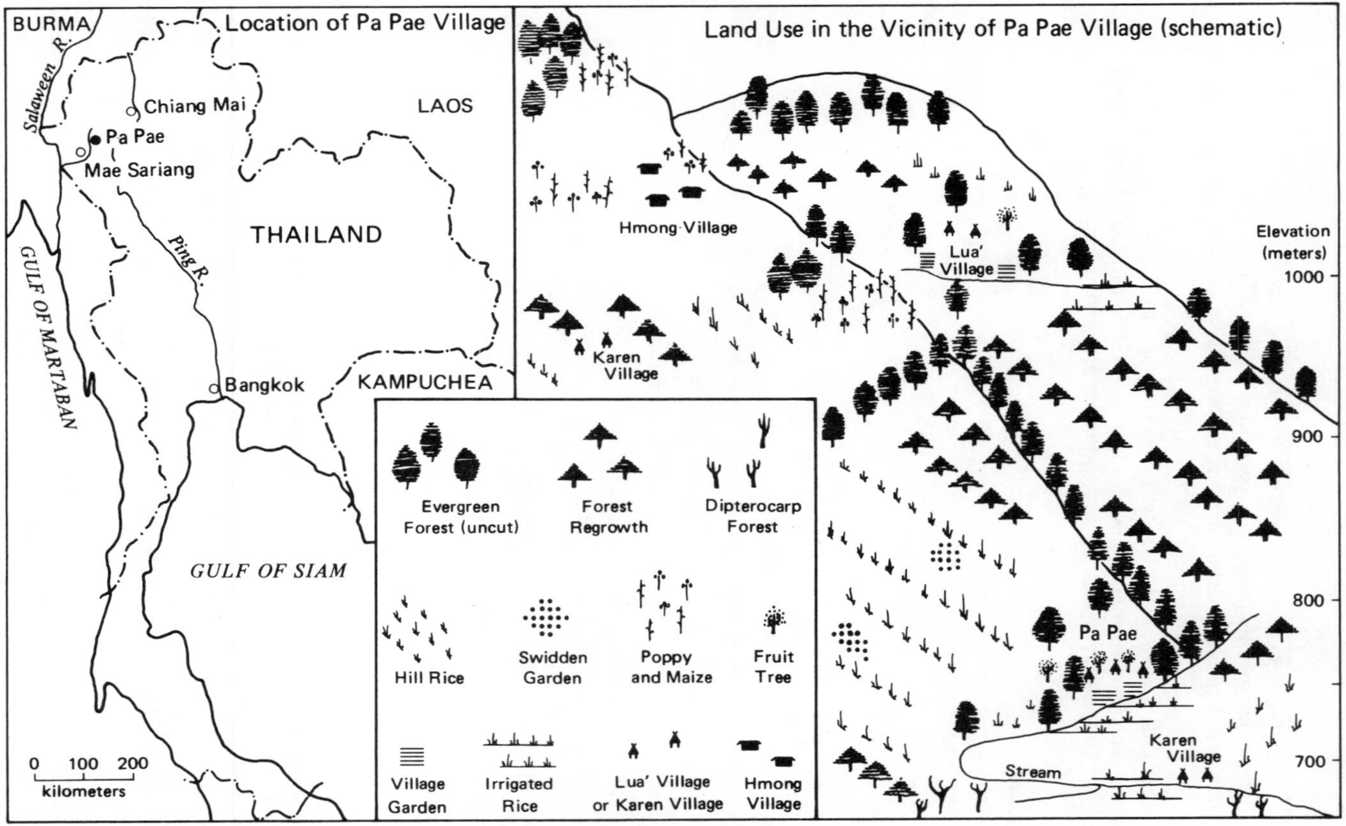
BURMA
Location of Pa Pae Village
Salaween R.
Chiang Mai
LAOS
Pa Pae
Mae Sariang
THAILAND
Ping R.
GULF OF MARTABAN
Bangkok
KAMPUCHEA
GULF OF SIAM
0 100 200
kilometers
Land Use in the Vicinity of Pa Pae Village (schematic)
Hmong Village
Lua' Village
Karen Village
Elevation (meters)
1000
900
800
700
Pa Pae
Karen Village
Stream
Evergreen Forest (uncut)
Forest Regrowth
Dipterocarp Forest
Hill Rice
Swidden Garden
Poppy and Maize
Fruit Tree
Village Garden
Irrigated Rice
Lua' Village or Karen Village
Hmong Village

Fig. 1. Location of Pa Pae village; land use in the vicinity.

# ECOLOGICAL MODIFICATION AND ADAPTATION: AN ETHNOBOTANICAL VIEW OF LUA' SWIDDENERS IN NORTHWESTERN THAILAND

*Peter Kunstadter*
East-West Population Institute

This chapter has two purposes: first to describe the extent of botanical and ecological knowledge possessed by the Lua' swiddeners of the village of Pa Pae in northwestern Thailand, and second to indicate the ecological and economic importance of fallow swiddens and other cultivated areas. In particular I want to suggest the importance to Lua' village economy of the creation and maintenance of a more varied environment than would exist without human intervention. This includes a complex artificial transitional zone established through the process of swiddening.

Discussions of the ecological effects of swiddening have often concentrated on the effects of cutting and burning the forest and usually have been concerned either with the use of cultivated swiddens per se, with the sometimes deleterious effects of forest burning and clearing, or with the ecological implications of density and movement of the populations engaged in slash and burn farming. Other aspects of the cultivation process (e.g., weeding) and economic system (e.g., collection from the fallow swiddens) are often ignored. Data provided here suggest that swiddeners may recognize and use many features of the environment, which they systematically modify, in addition to the more conspicuous features of fire and staple crop farming that are usually noted. Ethnographic data suggest that social organizational as well as technological, environmental and demographic features of the land use system are important in understanding the ecological process of swiddening.

## METHODS[1]

The author spent approximately three years in and around the village of Pa Pae in 1963-64 and 1966-69, doing general ethnographic work and focusing on the ecology of upland cultivation. Plant specimens were collected in 1966-68 by village research assistants, by the author, by Dr. Sanga Sabhasri and his colleagues from the Faculty of Forestry, Kasetsart University, Bangkok, and by Dr. Tem Smitinand, then curator of the Royal Forest Herbarium, Bangkok. Plants were collected systematically in different seasons during the course of biomass studies on different stages of the fallow swiddens, and in old and uncut forests, as well as in the course of research on cultivation practices in swiddens, irrigated fields and gardens. We completed these collections by asking village field assistants for other plants that we may have missed in their environment.

Each specimen was discussed with one or more villagers, and the following information was noted: Lua' name, cultivated or uncultivated, place where specimen was collected, other locations where found, uses, and methods of use. Botanical identifications of most specimens were made in the field by Dr. Smitinand and Dr. Sabhasri. Difficult or unusual specimens were pressed and returned to the Royal Forest Herbarium for identification.

Information on two or more specimens with similar botanical and Lua' identifications was combined, and these were considered to represent a single "variety." When botanical identification varied for two or more specimens with the same Lua' name, or when Lua' name or description varied significantly for two or more specimens with the same botanical identification, these were considered to be distinct "varieties." Information on uses, locations, Lua' name and botanical name was coded and punched for computer processing, and the figures presented here are based on tabulation of those data.[2] Because villagers commonly assume they will make use of almost any variety meeting minimal requirements for such purposes as fuel and temporary swidden fences, a special attempt was made to elicit this information for each specimen. Nevertheless, the number and proportion reported for these uses is probably understated.

---

[1]Research reported here was conducted under the sponsorship of the National Research Council of Thailand, and was supported primarily by a grant from the National Science Foundation. More detailed results of these studies will be found in Kunstadter (1966; 1971; 1974; 1975; 1978), Kunstadter, Sabhasri and Smitinand (1978); Sabhasri (1978), and Zinke, Sabhasri, and Kunstadter (1978).

[2]Because of Dr. Smitinand's interest in orchids, a special effort was made to gather all varieties which could be found. This may slightly inflate the proportion of orchids in the collection and the number of varieties reported as used for "decoration."

There was less than universal agreement among villagers concerning names and uses, and occasionally multiple names were given for the same species. This may slightly inflate the number of varieties reported. No single villager with whom we worked was aware of all the possible uses or knew all the proper names of all the varieties we collected.

In making an initial classification of plants and their uses (Kunstadter, Sabhasri, and Smitinand 1978), we tabulated 967 "varieties," as the term is defined above. Tabulations of numbers of varieties, their locations and uses in this chapter are based on this definition (see Table 1). If a more stringent definition is used, which eliminates probable duplications of species, there are probably about 793 botanically distinct types (565 distinguished as species or probable species, 176 identified only by genus, 21 identified only by family, 18 with distinct Lua' names but unidentified botanically, and 13 with neither botanical nor Lua' name identifications).

With the assistance of Dr. Paul Zinke, School of Forestry and Conservation, University of California, Berkeley, and with the aid of village research assitants, using ground level and aerial photographs and 1:50,000 maps, we surveyed and outlined the areas used for swiddens and irrigated fields. We reconstructed the swidden rotation pattern in this way, as far back as 1944 (Zinke, Sabhasri, and Kunstadter, 1978). Studies of soil fertility and erosion associated with the swidden cycle were made in collaboration with Dr. Sabhasri and Dr. Zinke (*ibid.*; Sabhasri 1978).

In what follows we discuss the different environmental zones as if they were distinct, but the transition between them is sometimes gradual. For example, the distinction between swidden and irrigated field cultivation practices may be blurred. Lua' farmers make small terraces in swiddens to control erosion, and swidden rice is sometimes transplanted from where it has been planted too thickly (sometimes deliberately sown thusly) to where it is growing too sparsely. Likewise, nurseries of rice seedlings for transplanting into irrigated fields may be sown in a swidden clearing. We also contrast cultivated and uncultivated plants, but for some species there are both cultivated and uncultivated varieties of the same species which are used by the villagers. The closely related cultivated and uncultivated species of other plant types suggest this was a common practice in the history of plant domestication. Although we have classified plants by usage, we often note that different parts of the same plant may be used for related or radically different purposes. These circumstances should be understood in interpreting our tabulations of plant varieties, uses and locations.

## THE SETTING

The Lua' are one of a number of Mon-Khmer speaking groups in mainland Southeast Asia whose lives are centered around swidden cultivation of rice.

TABLE 1
DISTRIBUTION OF VARIETIES, LOCATIONS, AND USES OF PLANTS FROM PA PAE

| Uses | Cultivated Varieties | | | | | | | | | | Uncultivated Varieties | | | | | | | | | | | | |
|---|---|---|---|---|---|---|---|---|---|---|---|---|---|---|---|---|---|---|---|---|---|---|---|
| | Swidden | Irrigated field | Village garden | Upland garden | In and around village | Stream bed | Field, garden margin | Purchased | Fish pond | Unspecified | Cultivated swidden | Fallow swidden | Uncut forest | Old forest | Stream bed | Field margin | Along trail | Hillsides | Dry Dipterocarp forest | In and around village | Burned areas | Unknown, unspecified | Total different varieties per use |
| Food | 70 | 5 | 48 | | 19 | | 6 | 3 | 1 | 6 | 4 | 110 | 48 | 45 | 22 | 5 | 9 | 6 | 1 | 27 | | 23 | 295 |
| Animal food | 3 | 2 | | | 1 | | | | | | 1 | 27 | 8 | 4 | 5 | | 1 | 2 | | 2 | | 7 | 44 |
| Medicine | 13 | 3 | 13 | | 12 | 1 | 2 | 11 | | | 2 | 42 | 19 | 11 | 6 | 1 | 1 | 4 | 2 | 23 | | 19 | 123 |
| Weaving, dying | 2 | | 1 | 1 | 6 | 3 | | | | | | 14 | 7 | 7 | 3 | | 1 | | | 6 | | 5 | 33 |
| Construction | 2 | | 3 | | 6 | | | | | | 3 | 45 | 23 | 12 | 3 | 1 | 1 | 2 | 2 | 5 | 1 | 11 | 79 |
| Decoration | 16 | 5 | 6 | 1 | 14 | 1 | 3 | | | | 1 | 16 | 11 | 10 | 2 | 1 | 1 | 2 | | 21 | 1 | 5 | 75 |
| Fuel | | | | | | | | | | | | 22 | 6 | 5 | | | | | | | | 7 | 27 |
| Poison | | | | | | | | | | | | 4 | 2 | 4 | | | | | | 1 | | 2 | 8 |
| Fencing | | | | | | | 1 | | | | | 24 | 8 | | | | | | | | | 7 | 29 |
| Insect repellant | | | | | | | | | | | | 3 | | | | | | | | | | 2 | 5 |
| Miscellaneous | | | | | 3 | | | | | | 2 | 15 | 5 | 4 | 1 | | | | | 2 | | 1 | 23 |
| Not used | | | | | 3 | | | | | | 10 | 157 | 59 | 23 | 7 | 3 | 2 | 6 | 6 | 29 | 1 | 75 | 280 |
| Use unknown | | | | | | | | | | | 5 | 122 | 19 | 10 | 1 | | | 7 | | 4 | 1 | 13 | 162 |
| Total number of different varieties per environmental zone | 84 | 10 | 60 | 1 | 43 | 5 | 7 | 13 | 1 | 6 | 21 | 482 | 164 | 101 | 39 | 9 | 13 | 23 | 9 | 84 | 3 | 147 | 967 |

Source: Modified from Kunstadter, Sabhasri, and Smitinand (1978).

Note: Because one variety may be listed in more than one location and may have more than one use, the total number of varieties is less than the sum of entries in rows and columns.

Permanently settled Lua' villages, ranging in size from about 100 to well over 1000 persons, are found today in the hills of northwestern Thailand. A number of Lua' villages which formerly existed in the lowlands have been assimilated, and their population is virtually indistinguishable from their Northern Thai neighbors.

Pa Pae, with resident population of 231 in December of 1968, is located at an elevation of about 720 m in the small valley of the Mae Amlan, a tributary of the Mae Sariang, which in turn drains into the Yuam, which joins the Salween in its journey to the Gulf of Martaban. The 51 houses of the village are clustered on either side of the stream. The village is for the most part surrounded by uncut, unswiddened groves of trees, preserved by the villagers because of the coolness they provide. Irrigated fields have been built by the Pa Pae villagers since the late 1920s along the major watercourses, and their terraces line the lower levels of the nearby hills, the tops of which are used for swiddening on a regular cycle of one year of cultivation followed by nine years of fallow.

Fig. 2. Pa Pae village, April 1967. Bamboo and fruit trees are planted around houses left center. Forest is preserved around village beyond fallow swidden, right foreground, planted 1966. Low hills behind village were swiddened in 1963. Ash scars dot more distant slope, right center, burned in March 1967.

The villagers deliberately create or maintain a series of environmental conditions, the botanical features of which they know and systematically exploit. These include: *swiddens*, which are cut, burned and cultivated on a regular rotation cycle; *irrigated fields*, cleared, leveled, plowed and harrowed, flooded during the cultivation of wet rice and baked to almost concrete dryness in the hot season; *village gardens*, lining the streams in the dry season, carefully fenced, leveled and watered to produce vegetables; the *village* itself, the area between houses beaten by the repeated passage of many feet, swept clean, frequently grazed over by domestic animals, with fruit trees and other useful plants carefully fenced off until they are large enough to survive; *fish ponds*, dug near water courses in the late 1960s at the urging of Border Patrol teachers, and stocked with *Tilapia* fish and *Ipomea reptans*, a water-growing vegetable; *streambeds* along the cool, moist banks of which several species are planted or gathered; *old and uncut forests*, preserved by village rules, in which gathering is permitted, but cutting of trees prohibited.

The human modifications of these zones, along with their natural characteristics combine to create a range of soil, moisture, sunlight and temperature conditions representing a far greater range than would be present without human interference. The forest, the village and the swiddens were all originally covered with the same type of Moist or Dry Evergreen forest. Within a few meters of the village, the minimum temperature of the densely shaded forest (5.6°C) is about 2.3°C warmer, and the maximum (34.1°C) about 3.6°C cooler than in the village or the swidden, and minimum relative humidity (27%) is about 12% higher than in the village and the swidden. The soil and the air under the uncut forest remain moist throughout the year, compared to the aridity of the burned swidden where forest cover has been removed and where soil surface temperature reaches 75° C in the sunny days before the monsoon rains begin.

The swidden fields range in elevation from about 650 to over 1000 m, with slopes up to 70%. The original vegetation in the area, probably transitional between Moist Evergreen and Dry Evergreen types, was characterized by large trees, 100 cm or more in diameter at breast height, reaching a height of 30 m or more, and casting dense shade from their completely covered canopy. These trees still remain in the uncut, unburned sections of the forest preserved by the Lua'. Swidden regrowth favors fire tolerant species, and trees normally reach a height of 5 to 10 m or so in the nine years of fallow before being cut and burned for another round of cultivation. The monsoon rains drop about 1.4 m of rain annually, almost all between the months of May and November. Maximum temperatures and lowest humidities (over 37°C and 15% humidity) are attained in March and April, and minimum temperatures in December and January. During the rainy season temperatures hover between 20°C and 24°C, with relative humidity near 100% throughout the cloud covered days.

## LAND USE AND POPULATION MANAGEMENT

Total village lands (including fallow and cultivated swiddens and irrigated fields, plus forests reserved around the village, on ridge tops and along water courses) included about 1553 ha, of which 688 were used for swiddens, and 18 ha for irrigated fields. About 70 ha were cut and burned for swiddens each year. Thus under 50% of the village's land was actively used for farming and less than 6% was under active cultivation in any one year. The uncultivated land was, however, in active use for grazing, collecting and hunting, as well as being used passively for the protection of watersheds, for seed sources to reforest cleared areas, and for regrowth essential for restoring soil fertility to the swiddens.

Use of such a small proportion of the total territory for active cultivation may appear profligate, especially by the standards of northern Thailand valleys, where population density on agriculturally used land reaches 800 per $km^2$, as compared with about 33 per $km^2$ within Pa Pae's territory. Traditional Lua' customs have combined to maintain a relatively low population density as well as a low proportion of land under active cultivation. The Lua' system maintains environmental diversity with relatively low human population density. This contrasts with the lowlands, where farmers attempt to create environmental uniformity, controlling the water supply and distribution with irrigation and diking, eliminating effects of slope and exposure by terracing, homogenizing the soil by plowing and harrowing, and reducing species diversity by concentrating on a few crops.

Lua' control of population density has been deliberate, with regulation of formation and size of households, limitation of household control over land resources, and with traditions affecting the number of people living in the village. Lua' farmers know they are able to clear and plant a much larger area than they normally farm, but individual households limit the land they clear and plant to the amount they can weed with their own or reciprocally exchanged labor. The basic constraint in their subsistence swiddening is labor for the tedious job of weeding. Individual household decisions in this regard are backed by the authority of village leaders to reallocate land on the basis of need (according to household size and composition), but this authority is invoked only rarely. Household size, in turn, was traditionally limited not by birth control, but by customs prescribing the order in which siblings are supposed to marry and by limiting the number of married couples supposed to live in a single household. This sort of limitation seems appropriate in a subsistence economic system which has had no opportunity for productive long term investment, and very limited opportunity for storage of wealth except in prestige goods. These customs contrast with those of cash cropping swiddeners, such as the Hmong (Meo), whose average household size is more than double that of the Lua'.

Customs regulating village population size, some of which are clear and direct, evidently have been successful in the past. Migration into the village, except for marriage, is not allowed without the permission of the village religious leader. Migration out of the village is forced on some occasions. Violations of incest taboo for example, are punished by expulsion of the offenders from the village. Out-migration has been voluntary in most cases: individuals or households unable to make an adequate living, or beset by chronic or repeated illnesses or other disasters, move to Karen villages where they "become Karen," or to the lowlands, where they "become Thai," thus dropping out of the Pa Pae population. The net effect, until recently, seems to have been to restrict village population growth and to allow the land-extensive system to persist, even in the face of a relatively high rate of natural increase. As we shall see later, recent changes, including population pressure from other groups, changes in land tenure customs associated with irrigated agriculture, and religious changes are now combining to make the perpetuation of the traditional land use system impossible.

## SWIDDEN TECHNIQUES AND CUSTOMS

Swidden techniques involve only a few simple tools: a long-bladed knife, an iron-tipped planting stick, a short-handled weeding tool with an L-shaped blade, a sickle with a curved, serrated blade, plus a few mats, ropes, winnowing fans, baskets, and fire-making flint and steel, matches, or lighters. The effective use of these tools to provide sustained subsistence yields and to avoid major environmental degradation implies a profound and conscious knowledge on the part of the Lua' farmers. The swiddening system also involves a series of customs and traditions, which, as suggested above, are important in the successful operation of the system. Village religious leaders are responsible for the communal agricultural rituals, the allocation of communally held land for swidden sites, the settlement of disputes over swidden lands between village households, the time of swidden clearing, burning and planting, and the summoning of household representatives to honor their community responsibilities in such tasks as fighting fires in the fallow swiddens. Labor mobilization and cooperation in those tasks requiring a large number of people on a single household's swidden are organized on the basis of kinship obligations and bonds of friendship woven into a network of reciprocal labor exchange. Such cooperation is important primarily in planting and harvesting, both of which must be done quickly.

The annual cycle of swidden work begins in January with the inspection by village elders of the coherent block or blocks of swidden sites prescribed by the round of cultivation and fallow. The state of vegetation regrowth is noted. Size and density of vegetation are the primary indicators to the Lua' of satisfactory restoration of soil fertility during the fallow period. Since an accidental fire in the years shortly before the fallow swidden is scheduled for reclearing will reduce the yield when that site is recultivated, they may choose to modify the normal field rotation if they find evidence of recent fire. The soil may also be tasted ("tasteless" soil is believed to have low fertility), and a few species (e.g., *Aeginita indica*) are recognized as indicators of low potential crop yield. Secondary succession dominated by bamboos or grasses has not been a problem under the Lua' cultivation system, and these types of vegetation, which are useful in their own right, are not regarded as indicating poor soil.

If the area appears satisfactory, households normally return to the site of their swidden ten years previously. If their household composition or size has changed markedly, village religious leaders may reallocate the land as appropriate to changes in the needs of the households and their abilities to farm. After each household takes an omen to assess its chances for a bountiful harvest in its intended site, they cut the brush and smaller trees, leaving stumps about 75 cm tall, from which most of these trees will regrow. Some of the larger trees are felled, but many are merely severely trimmed so as to reduce the shade they will cast on the growing crops. Most of these survive the swidden fire, and regrow quickly during the fallow period. Leaving trees with deep, established root systems speeds forest regeneration in the fallow swiddens since it allows prompt recycling of soil nutrients that are deposited on the soil surface in the ash of the swidden fire but that quickly leach through the shallow root layers of the crop plants. How this process works is not known to the Lua' farmers, but they know the advantages of quick forest regeneration. They are aware of the effects of competition between forest and cultivated vegetation, and could easily kill the trees in their cultivated swiddens if they wanted to do so. Vegetation on ridge tops and along water courses is left uncut with the understanding this will reduce the dangers of erosion.

Cutting is usually completed early in February, and the slash is left to dry until the end of March. Then, accompanied by appropriate ceremonies, it is burned on one of the hottest, driest days of the year, at a time when there is little or no wind, in order to reduce the danger that the fire will spread beyond the slashed swiddens. To give further protection, the swiddeners have cleared and swept a firebreak all around the tops and sides of the swidden block, and they stand watch to try to control any undesired spread of the flames. The fire reaches a temperature of 600°C at the soil surface, but the heat penetrates only a few centimeters below the surface. Most leaves and smaller branches are quickly consumed, but many of the larger logs are only charred. As soon as the fire dies down, the villagers go to their fields to plant maize, some tuber or root crops, and a few flowers, and to scatter cotton seeds. Within the next week or two they clean up the unburned slash. Some logs are piled up and burned. Longer, straighter logs are saved for firewood, used for fencing or to mark household field boundaries, or are placed along the contours of the slopes to retard soil slippage. Brush and branches are also piled into the crevices of the hillsides to retard erosion and to act as frameworks upon which viney plants are grown in the well-watered spots. Fences are built around the tops of the swiddens to keep out grazing animals.

Fig. 3. Planting Pa Pae swidden, April 1967. Charred logs are set along contours to slow water flow and motion. Scars on slopes are ash from piles of wood and brush reburned after swidden fire.

When the chief priest of the village has begun planting his swidden, about the 10th of April, the other households may also begin planting their fields. The swidden fire has desposited ash on the soil surface, and the heat has left the top few centimeters dry and friable. Young men jab their iron-tipped planting sticks into the earth, making a shallow depression, into which older men, women, boys and girls throw a few rice seeds. Tubers and root crops are planted at the bottom of the fields, a row of sorghum is planted to mark the field sides, maize may be planted in various places, and a garden containing peppers, onions, herbs, mustard greens, chili peppers and beans, are planted close to the site of the temporary swidden shelter which is erected in each field.

By mid-May, before the monsoon rains have become steady and heavy, the seeds have begun to sprout, along with myriad weeds, and the annual war with the weeds begins. At this early stage of plant growth, the villagers, including young children, must be able to distinguish between the seedlings of several scores of cultivated varieties they have planted and those of about 150 varieties of weeds that grow in the swiddens. Working with the short-handled tool with an L-shaped blade, they remove all the weeds and chop away at excessive regrowth from tree stumps but do not disturb or loosen the soil for more than about 2 cm below the surface. As the swidden crops grow, villagers take advantage of some of the forest regrowth in cultivated and fallow swiddens by harvesting bamboo shoots (*Bambusa tulda, Dendrocallamus strictus*, etc.), ferns (*Pteris* spp., *Lygodium polystachyum, Tectoria* sp.), mushrooms, wild figs, and many other plants. Uncultivated bamboo shoots, ferns and mushrooms are very important foods during the period before any of the swidden garden crops are ready.

Fig. 4. Pa Pae swidden in June. The rice has grown to about 25 cm and coppice growth has begun from stumps (left). Note trees left uncut on ridge tops and in ravine.

Weeding continues until the beginning of the rice harvest. Each household attempts to complete three rounds of weeding before this time, in order to assure good growth of the rice. Swiddens begin to yield in June and July, and the earliest ripening species include mustard greens (*Brassica* sp.), maize, and many varieties of beans (*Dolichos* spp., *Phaseolus* sp., *Psophocarpus tetragonolobus*). These are important supplements to the diet before the rice

Fig. 5. Weeding with an L-shaped tool, a Pa Pae woman disturbs only the top few centimeters of the soil. Maize is inter-planted with rice. At woman's feet is a bamboo shoot she has collected while weeding.

Fig. 6. Swidden harvest starts in October 1967 with the lower, wetter slopes. Terraces of a small irrigated field are visible in left center. Forest on right is preserved for 1968 swidden.

ripens late in September and in October. Root and tuber crops, squashes, and melons ripen about the time of the rice harvest and are a welcome change from the previously available vegetables. After the rice harvest is completed late in November, the swiddening season is formally closed with a community-wide ceremony to summon the souls of villagers and rice back to the village. Cotton, which ripens late in December or January, is the only major crop remaining to be harvested. Sesame, sorghum, and a few herbs may not ripen until after the rice harvest. Chili peppers may continue to bear for another year or so and may be collected by anyone passing through the tangled vegetation of the fallow swidden.

## KNOWLEDGE USED BY LUA' SWIDDENERS

This brief description allows us to infer some aspects of the knowledge involved in the Lua' swidden system. These are summarized under the following topics: fire control, soil quality, soil erosion, forest regrowth, weed control, pest control, botany, forest typology, and watershed.

Lua' farmers know they must use fire to reduce the bulk of the slash and to deposit ash on the soil surface prior to planting the swiddens. They are aware that they must control fire in order to prevent premature forest destruction and ash deposition. They accomplish these goals through using an appropriate length of swidden-fallow cycle by accepting community obligations to put out fires burning at the wrong time or place, and by preventing the spread of swidden fires to the wrong places with carefully cleared firebreaks.

Soil quality is recognized in terms of the appearance of vegetation, the appearance of the soil, and a few indicator plants. Forest regrowth and fire control are recognized as essential to restoring soil fertility after a single year of cropping in the traditional swidden system. Although Lua' farmers know the techniques of permanent-field wet-rice farming and the use of plowing to restore soil fertility in irrigated fields, they do not plow or fertilize swiddens.

They use animal manure in village gardens and graze water buffalos in fallow irrigated fields and swiddens, but they realize they are unable to gather and transport enough manure to cover their fields and know they are unable to afford or transport chemical fertilizers from the lowlands. Thus, they rely on forest regrowth and fire to restore fertility to the surface of upland fields, and they plow to return leached soil nutrients to the surface of their wet rice fields.

They have seen Hmong (Meo) farmers, who, since the early 1960s, have moved within a few hours' walk of Pa Pae. They are aware of the contrast of Lua' swiddening practices with those of their Hmong neighbors, who plant the same swidden to maize and opium for several years in a row, and who remove all the stumps from their fields and cultivate deeply and cleanly with hoes. They have seen that this system leads to replacement of forest with grassland, and they are aware of the deleterious effects of this form of swiddening on stream water flow. Their choice of swiddening techniques and land management practices thus is conscious and deliberate, not a result of lack of information on alternatives.

The Lua' farmers are aware that cutting and burning expose the soil to the danger of erosion, and take deliberate steps to control erosion, including (1) not cutting vegetation on the steepest slopes, on ridge tops, or along water courses; (2) establishing contour lines on slopes with logs to catch slipping soil and slow the speed of runoff; (3) putting log and brush frameworks across crevices on hillsides and across the bottoms of narrow valleys, thereby reducing the speed of water flow; (4) making minimal disturbance of the soil surface during planting, weeding and harvesting; (5) encouraging forest regrowth.

The forest regrowth, as already indicated, is considered by Lua' villagers to be important for restoring swidden fertility and controlling erosion, and it is also recognized as providing a source of essential or useful plants not found elsewhere. Lua' farmers encourage regrowth by allowing nine years of fallow before recultivation. They deliberately maintain seed sources above the swiddens by preserving ridge top trees, from which seeds will be distributed by gravity. They also ensure quick forest regeneration by allowing coppice growth from established trees which are trimmed but not rooted out.

Weed control is recognized as essential for good crop growth, but weed growth is constrained by techniques already mentioned that prevent soil erosion and promote forest regeneration. Thus hoes and plows are not used in swiddens even though they are used by Lua' villagers in their irrigated fields.

Animal pests (insects, rodents, birds, pigs, barking deer, and bears) are recognized by Lua' villagers as threats to swidden production. The techniques they use to deal with these (traps, deadfalls, snares, noisemakers, guns), along with prayers and sacrifices, are of limited success in controlling their depredations. Since these pests (including the insects) also supply occasional animal protein to the Lua' diet, they are not an unmixed detriment, and no all-out effort is made to exterminate them.

Lua' knowledge of practical botany is indicated by their recognition and ability to name, tell the growth habits and uses, and give many other characteristics of hundreds of varieties of plants found in different parts of their environment. Their sensitivity to plant characteristics is suggested by their multiple use of single species and by their recognition of new uses in recently introduced species. They apply their knowledge when varieties they are seeking are in the company of hundreds of other species in the forest, or when the plants they want to eliminate are only seedlings in the presence of scores of cultivated varieties in their swiddens. They are also practical plant breeders, since they select seeds in order to reproduce the desired characteristics of their highly varied rice crops, which possess recognizable differences in ripening time, grain size and shape, grain color, hull color, taste, etc.

Their understanding of forest typologies is suggested by their recognition of different types of forests which they preserve and, to some extent, create. They know that certain usable species are characteristically found in fallow swiddens at a certain stage of regrowth, while others are found in old forests, preserved by community custom, where trees may not be cut, but from which forest products may be gathered. By community custom, enforced by fines levied against offenders, they also prohibit swiddening in certain areas from which they allow usable species of large size to be cut.

They know the relationship between forest disturbance and the flow of surface water, having observed in recent years the effects on stream flooding of the extensive clearing and cultivation of upper slopes by Hmong farmers, and they recognize in this the purpose of their traditional prohibition on cutting on ridges and along watercourses. It is not certain that they recognize other effects of their swiddening customs on water flow. Moving the sites of swidden blocks from year to year is done in such a way that they are almost never made adjacent to one another from one year to the next, thereby restricting to one part of the watershed the area from which runoff will be rapid.[3] Cutting the fields early in the dry season not only allows the slash to dry before burning but also preserves soil moisture that would otherwise be lost by evapotranspiration of the standing vegetation during the dry season. This subsurface moisture is especially important during the early part of the growing season, before the monsoon rains saturate the soil.

These examples are evidence that Lua' swidden farmers are conscious of the ecological effects of their traditional farming system, including its non-technological aspects, and that they recognize many of its benefits beyond the immediate gains of harvest in a single year. They manage, use and benefit from a much more extensive area than is implied by the amount of land they actually cultivate in a given year.

We turn now to a specific examination of the use of fallow swiddens and other cultivated and non-cultivated parts of the environment, and their importance in the Lua' village economy.

---

[3]In contrast, Hmong practices create a large grass covered area emanating from the village.

## USE OF PLANTS FROM CULTIVATED AND UNCULTIVATED LOCALITIES

The relative importance of different local environments, especially the swiddens and fallow swiddens, as sources of plants used by Pa Pae villagers for food, medicine, animal food, weaving and dyeing, and fuel, is suggested in Table 2. Because 295 varieties are used for food and 119 for medicine, we can consider only food plants and will limit description of medicinal plants to those used to treat postpartum women (since this is a major focus of Lua' medicine), and to those used to treat diarrhea (a leading cause of illness and death at Pa Pae). The majority of cultivated food plant varieties (66 out of 99) are grown in swiddens, while most of the uncultivated varieties (103 out of 182) are found in fallow swiddens. A large proportion of medicinal plants is grown or found in fallow swiddens, and a similar pattern occurs with regard to other uses (Table 2).

TABLE 2*
PLANT VARIETIES FOR DIFFERENT USES FROM SWIDDEN AND ELSEWHERE

| | Food | Medicine | Animal Food | Weaving, Dyeing | Fuel |
|---|---|---|---|---|---|
| Cultivated: Swiddens | 66 | 12 | 3 | 1 | – |
| Cultivated: Elsewhere | 33 | 19 | 3 | 4 | – |
| Uncultivated: Swiddens | 103 | 40 | 27 | 12 | 22 |
| Uncultivated: Elsewhere | 79 | 42 | 11 | 12 | 5 |
| Both Cultivated and Collected: Swiddens | 7 | 3 | – | 2 | – |
| Both Cultivated and Collected: Elsewhere | 7 | 3 | – | 2 | – |
| Total from Swiddens | 176 (60%) | 55 (46%) | 30 (68%) | 15 (45%) | 22 (81%) |
| Total from Elsewhere | 119 (40%) | 64 (54%) | 14 (32%) | 18 (55%) | 5 (19%) |
| Total | 295 | 119 | 44 | 33 | 27 |

*Source: Table 1. Plants from "swiddens" include all those grown and cultivated in swiddens, whether or not they are found in other locations. Plants from "elsewhere" include all those not from swiddens. Varieties "cultivated and collected" include those which are sometimes planted and sometimes taken from where they grow spontaneously.

To some extent, the normal subsistence importance of these varieties is exaggerated by the above numbers and proportions. The primary staple at Pa Pae is rice; it forms the basis of every meal except in times of famine and is normally consuned at the rate of about 700 g per person per day. About half the rice harvest comes from the swiddens and half from irrigated fields. Rice is normally supplemented with maize, beans, taro, cassava, mustard greens, and a number of wild leaves and ferns, all of which are particularly important in the rainy season before rice harvest, when staple stocks run low. Most of these come from cultivated swiddens. Food plants from fallow swiddens and forest are especially important seasonally, as green leafy vegetables for a marginally anaemic population when gardens and swiddens are not producing and as bulk when the rice supply is declining. They are a regular and pleasant source of variety in the diet and an emergency source of calories in times of crop failure.

The fallow swiddens are not a uniform environment. They range from the condition in which they are left after harvest, covered with low-growing herbaceous plants and coppicing tree stumps, to a forest ready for renewed swiddening. The succession in plant sizes and species is important for the pattern of use of the fallow swiddens. In the second year of fallow there is a mean of 56 species (two test plots). The greatest species variety in the fallow swiddens–a mean of 77 species in four test plots–is found in the middle years of regrowth. In later years, as the trees grow taller, the low-growing grassy and bushy species are crowded out, resulting in a decline to a mean of 60 species ( in four plots) in the ninth year after cutting (Sabhasri 1978:162, 164, 165). Many of the useful species, especially herbs and shrubs, are characteristically found in the earlier regrowth stages, but are suppressed in the later years of the fallow period. An important example is *Imperata cylindrica* grass, which is used for roofing thatch. It is gathered in the six to eight year regrowth, but by the time fallow fields are ready for recutting and burning, *Imperata* is almost gone. Given the ten year cultivation/fallow cycle, *Imperata* is naturally suppressed and does not become a problem in Lua' swidden succession but is instead a species of great use in the local economy. This contrasts with the situation reported under different cultivation systems in Thailand and elsewhere in Southeast Asia, where the spread of *Imperata* comes to dominate the treeless secondary growth. The Hmong (Meo) for example, generally practice clean weeding and deep cultivation, removing all stumps from their swiddens, repeatedly burning over the same field which is kept under cultivation for a number of years and cutting swiddens clear up to the ridge tops. Under Hmong cultivation, there is no coppice growth from stumps and there are no nearby seed sources for trees. Regrowth is predominantly grassy, and reforestation takes place very slowly (Keen 1978: 213, 215; Kunstadter and Chapman 1978:8-10; Kunstadter et al. 1978: photo 124).

Height and size succession are also important to Pa Pae villagers in terms of fuel, almost all of which comes from uncultivated species in the fallow swiddens. Woody regrowth is fast. When fallow fields are cut, after nine years of regrowth, trees are dominant, with stems rather than branches, leaves, grasses, bamboos or vines making up the bulk of the biomass. New growth stems that reach 10-15 cm in diameter and 5-10 m in height are adequate for firewood, fencing, and many construction uses. Thus we may speak of a succession of uses of the fallow swiddens as composition and sizes change.

The net effect of these successions is to even out the seasonal availability of useful varieties. This is particularly important in the Lua' economy, given the lack of effective preservation or storage techniques for such things as green leafy vegetables and the extreme seasonality of their cultivated varieties. The distribution of different food plant varieties from different environmental zones throughout the year, and in different stages of the swidden cultivation/fallow cycle is suggested in Figures 7 and 8.

## FOOD PLANTS

Different types of food plants come from different zones in the Lua' environment. As might be expected, almost all of the grains (both in variety and in volume) come from the swiddens and irrigated fields. The most important grain plant is rice, of which two basic types, non-glutinous (*Oryza sativa*) and glutinous (*O. glutinosa*), are grown in both swidden and irrigated fields. Important Lua'-recognized varieties of both types which are grown in both kinds of fields are designated by maturation (early, which matures in September; medium; late, which ripens the end of October or early November), grain color (white, red, black), fragrance (fragrant and ordinary), length of grain (short, long), etc. Other grain plants are grown exclusively in swiddens. These include a Labiatae species, white and black varieties of sesame (*Sesamum indicum*), millet (*Setaria italica*), white and black varieties of tall-growing sorghum (*Sorghum vulgare*), and corn (*Zea mays*), which ripens a month or two before the rice and is an important item in the rainy season diet. *Eleusine coracana*, grown in the swiddens of the nearby Lua' village of La'up, is occasionally brought to Pa Pae for use in brewing rice liquor. Millet and sorghum are also used in brewing. One variety of Job's Tears (*Coix lacryma-jobi*) is used in brewing liquor and is planted in La'up swiddens. Two other named varieties are planted in Pa Pae gardens and along stream beds but are used for decorative beads, not as food. The seeds of a few other plants, such as *Scleropyrum wallichianum*, are collected from uncut forest and eaten.

At least seven species of starchy tubers and roots are grown in swiddens and/or gardens. These include *Amorphophallus* sp., *Capparis* sp., *Colocasia* (?) sp., four distinct named varieties of *Dioscorea alata* (yams), *Ipomoea batatas* (sweet potato), *Kaempferia* sp., and two varieties of *Manihot*

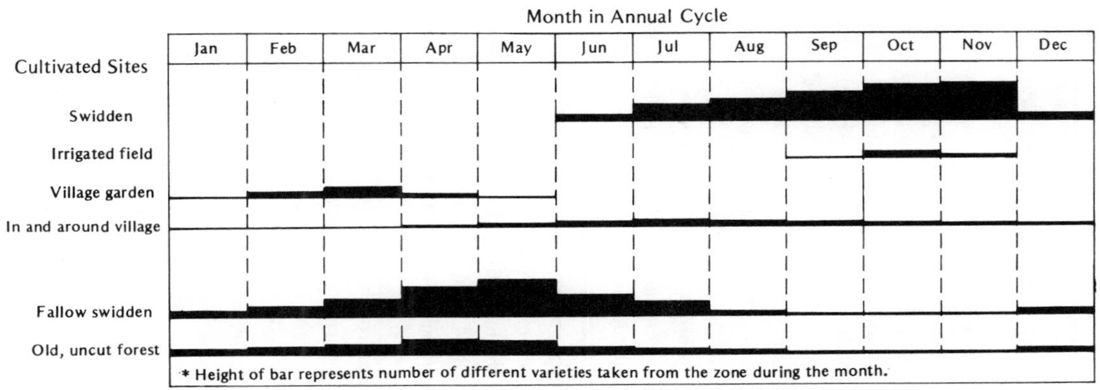


Fig. 7. Seasonal fluctuation in numbers of plant varieties used for food from some different zones in the Lua' environment.*

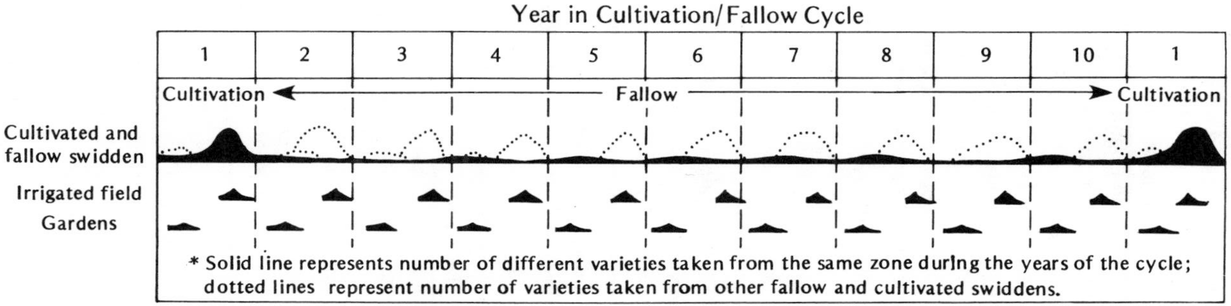


Fig. 8. Relation of swidden cultivation/fallow cycle to numbers of plant varieties from some different environmental zones.*

*esculenta* (cassava). Three species are collected, including *Asparagus filicinus* from the woods near the village, a wild variety of *Dioscorea alata*, collected in fallow swiddens, and *D. pentaphylla*, also from fallow swiddens. Three species of onions and garlic (*Allium ascalonicum, A. porrum, A. sativum*) are grown both in swiddens and village gardens. The leaves and bulbs of these species are common ingredients of most cooked dishes.

Gardens and swiddens plus the uncut forest also provide most of the species commonly used as spices. These include *Alpinia galanga*, a ginger-like root, planted in gardens and collected in fallow swiddens and uncut forest; *Anisochilus siamensis*, an aromatic herb cultivated in swiddens and collected around the village; *Capsicum frutescens*, chili peppers, planted in gardens and grown from seed at home before transplanting into swiddens; the bark of the roots of *Cinnamomum subavenium* and *C. tamala*, from uncut forest, is used in seasoning; *Coriandrum sativum*, coriander, planted in swiddens; *Curcuma longa*, turmeric planted in gardens and collected from the forest, with mild and strong-flavored varieties; *Cymbopogon citratus*, lemon grass, planted in swiddens and gardens; *Foeniculum vulgare*, fennel, planted in swiddens and gardens; *Heracleum burmanicum*, whose leaves and flowers flavor curries, planted in swiddens; *Mentha arvensis*, mint, grown in swiddens and gardens, and *M. arvensis rufus* planted in gardens; the fragrant flower of *Mesona* sp., grown in swiddens; *Ocimum basilicum*, basil planted in gardens; the flower of *Oenanthe* sp., grown in swiddens and gardens; the root of a *Piper* sp., collected from the forest and used like ginger or *Alpinia*; and *Zingiber officinale* (ginger), planted in gardens and occasionally gathered from the forest.

Four of the five species of beans grown by Pa Pae villagers are planted both in village gardens and in swiddens. They are an important supplement to the diet when they ripen in the gardens toward the end of the dry season, and especially when they become available from swiddens in the middle of the rainy season. Species include *Dolichos lablab* (two varieties), *Psophocarpus tetragonolobus*, and six different named varieties of *Vigna sinensis*, four of which are planted exclusively in swiddens. One of these, which does not grow well at Pa Pae, is successfully cultivated in La'up swiddens. A fast ripening variety of soya beans (*Glycine max*) is planted in swiddens and along the dikes of irrigated fields.

The importance of swiddens, irrigated fields, and village gardens in providing the basic Lua' staples is apparent from the preceding paragraphs. It is with regard to fruits, berries, and leafy plants that other environmental zones are particularly important. Uncultivated zones are especially significant as sources for fruits and berries, with 94 collected varieties as compared with 35 cultivated varieties. Uncultivated zones also predominate as the source for edible leaves, shoots, and flowers. There are 25 cultivated varieties, compared with 110 collected, plus an additional 13 ferns and 4 grasses, all of which are collected. Thus most varieties of green leafy foods and of fruits are collected rather than deliberately planted.

Both the fallow swiddens and the old, uncut forests are important environments for collecting: 39 fruit and berry varieties are taken from the fallow swidden, and 31 from the old and uncut forests; 30 plant varieties with edible leaves, shoots or flowers, plus 5 ferns and 1 vine come from the fallow swiddens; another 35 leafy varieties, 3 ferns and 1 vine come from the old and uncut forests. The fallow swiddens and forests are also a source for collection of grassy plants such as bamboo shoots.

As compared with swiddens and irrigated fields, village gardens and other less formally cultivated spots around the village are especially important as sources of fruits and berries (22 of the 35 cultivated varieties come from these sites) and locations for planting leafy varieties (17 out of 25 cultivated varieties).

Taken together, this description of locations from which food plants are derived suggests that Lua' villagers get different kinds of foods from different environments and at different times of the year. Their deliberately varied environment helps to decrease fluctuations both in quantity of food and in number of varieties which would result if they were restricted to a less varied environment. As population and agricultural intensification increase, the systematic variations in the Lua' environment can be expected to decline. This will heighten their vulnerability to crop fluctuations and may lead to nutritional deficits associated with an increasingly monotonous diet.

## MEDICINAL PLANTS

It is more than a metaphor to refer to the forests and fields of the Lua' environment as their drug store. From the plants in their environment the Lua' derive well over a hundred medicinal species used internally, as baths, as poultices, etc., for a wide variety of ailments to which they are heir, including their most common illnesses, such as diarrhea, coughs, fevers, conditions associated with childbirth and menstruation, aches, pains, etc. We have space here to consider in detail only those related to diarrhea (a common complaint and leading cause of death, especially among children), and those conditions associated with childbirth and menstruation.

The frequency of diarrhea and other stomach complaints in the lives of Pa Pae villagers is reflected in the numbers of plant varieties they employ for treatment of these conditions. The boiled bark of three different Lua'-named varieties of *Engelhardia spicata*, one from the fallow swidden, one from the uncut forest, and one from an unknown location is boiled for diarrhea or stomachache medicine. Another variety, known and designated by the villagers by its red color, is not used for medicine. The bark of *Ficus hispida*, collected in fallow swiddens, is said by some to be used for diarrhea medicine, but this use is denied by others. The fruit is also eaten. Roots of *Flacourtia cataphracta*, growing wild near the village, are boiled for diarrhea medicine, and the fruits are eaten. Another *Flacourtia* species, *F. rukam*, not distinguished in name by Pa Pae villagers from *F. cataphracta*, is collected from fallow swiddens. It is used similarly and is also a medicine for stomachache and mucous stools. The *Ochna integerrima* plant or, some say just root, collected in fallow swiddens, is boiled for diarrhea medicine. Opium (the dried sap of *Papaver somniferum*) has long been used as a medicine for diarrhea, stomachache and other pains, such as those associated with tatooing. It is not grown at Pa Pae but is purchased from Hmong (Meo) growers or peddlers from nearby villages. The young leaves of guava (*Psidium guajava*), planted in the village or growing wild in the forest, are boiled for diarrhea medicine. The fruits are also eaten. The sweet shoots of *Rubus ellipticus*, found near the village, are boiled and pounded with salt as medicine for childhood diarrhea, or the medicine may be given to the nursing mother of a sick child. The fruit is also eaten. The root of *Rubus moluccanus*, collected from beside streams, is boiled and drunk for diarrhea medicine. Berries of this plant are eaten, and a medicine made by pounding and squeezing small branches is used for treating infected umbilicus. *Spondias pinnata*, the hog plum, is planted in the village for its fruit, which is eaten, and for the bark, which is boiled for diarrhea medicine. The boiled bark of a *Sapindaceae* variety (location unknown) is given to children for diarrhea. The anti-diarrheal properties of *Holarrhena antidysenterica*, which is collected for fuel in the fallow swiddens are widely known among Northern Thais, but this use was denied by Pa Pae village informants.

*Alpinia* sp., the common *kha* root spice in Thai and Lua' cooking, is ground up for stomachache medicine. It may be collected where it grows along cool-flowing streams and is also planted in village gardens. The shoots of *Cassia tora*, collected from around the village, are boiled, and the liquid is drunk for stomachaches; shoots are also eaten for food. This species is now much less common than formerly in the village area. Teak wood (*Tectona grandis*), collected from the unswiddened forest, is boiled for stomachache medicine. The finely chopped root of a plant known in Thai and Lua' as *mənəi* (location unspecified) is boiled and mixed with opium to relieve bad childhood stomachaches. This root is also given to postpartum women.

The largest group of medicinal varieties from the Pa Pae collection was used as components of "hot medicine" (Lua': *jiə kowic*; Northern Thai: *ja hən*). Many kinds of hot medicine, with different formulations, are known among Lua' and Northern Thais. The main use of these medicines is for women who have recently delivered a child and who want to "clean the uterus" so it will be ready for the next pregnancy. The medicine may also be used deliberately as an abortifacient, but its effectiveness as such has not been demonstrated. Most of the plants or plant parts used in hot medicine are aromatic spices or herbs, which are also used for other purposes.

A mixture of this type for use as hot medicine, purchased by Pa Pae villagers from the lowland market, was found to contain the following: *Amomum krehvanh*, cumin seed (*Cuminum cyminum*), cloves (*Eugenia aromatica*), *Foeniculum vulgare* (fennel, also grown in village gardens and swiddens), *Illicium* sp., *Myristica fragrans*, *Piper chaba*, and *Piper nigrum* (black pepper).

Varieties used in hot medicine which were grown or collected from the village environment included the following: *Adhatoda vasica*, the roots of which are sliced, pounded and dried, deliberately planted, or collected where it grows wild around the village; *Allium ascalonicum* (garlic, used also as food), planted in gardens and swiddens; *Alpinia galanga* (*kha* root, a common ingredient in Thai and Lua' cooking), planted in village gardens and collected in fallow swiddens and old forests; *Angiopteris evecta* ( a fern, also used for food), collected from around the village; *Asparagus filicinus* roots, collected from fallow swiddens; pounded *Brassica* sp. seeds (Lua' variety of mustard greens which are the major cultivated vegetable), planted in gardens and swidden; sliced, pounded, dried root of *Carica papaya* (papaya, also used as food and for a cooling application to heads of postpartum women), planted in village; roots of both red and yellow varieties of *Celosia argentea* (flowers of which are used ceremonially), planted in swiddens and growing wild near the village; bark from roots of *Cinnamomum tamala*, collected from old forests north of the village; liquid from boiled *Croton laevigatus*, collected in fallow swiddens; *Heracleum burmanicum* (used as an herb with chicken and pork), collected from around the village; *Hydrocotyle javanica*, from field margins (also eaten as food); *Kaempferia* sp. (also a food), planted in gardens and upland fields; boiled roots of *Moghania stricta* (also used by men to give them strength), collected from unspecified location; pounded, sliced and dried root of a basil (*Ocimum gratissimum*), collected from around the village; *Phaulopsis* sp., prepared in the same way, collected from around the village; *Piper sarmentosum* (also eaten), found around the village; *Piper* sp. (a different pepper vine, also eaten), collected around the village, dried and pounded to a powder; *Plumbago indica*, collected near another Lua' village to the north of Pa Pae; sliced, pounded, dried root of *Plumbago zeylanica*, collected in the vicinity of Pa Pae; *Sambucus javanica*, from around the village; sliced, pounded and dried root of *Solanum incanum*, from the village vicinity; several Lua' named varieties of *Spilanthes acmella* (leaves and flowers are eaten, root is boiled and the liquid sucked for toothache medicine), the uncultivated variety, collected around the village, is stronger than that grown in swiddens primarily for food; root of *Strobilanthes* sp. (also an ornamental), grown in the village; *Tinospora* sp., found around the village and in the uncut forest near the village; *Zingiber officinale* (ginger, commonly used as a spice), planted in gardens and sometimes collected in the forest; and a botanically unidentified vine known in Lua' as *nɨəʔ səkiək*, (also put in rice liquor to give a red color), found around the village.

Many of these species are known in Lua' by names that are either the same as Northern Thai or very similar to the Northern Thai terms, suggesting that knowledge of their use may be from the Northern Thai, unlike others that have names which are clearly of Lua' origin.

Several other kinds of medication are used for postpartal women, and for the relief of the "vapors" associated with menstruation. The cut sections of papaya fruit (*Carica papaya*), are placed on the head of a postpartal woman suffering from fever; pounded, powdered roots of *Celosia argentea*, already mentioned as "hot medicine," along with several other ingredients, are used for menstrual problems; women who have given birth inhale the fumes of *Sambucus javanica* leaves gathered from around the village which are spread on hot rocks where the woman is lying by the fire in traditional postpartum confinement (these leaves are also toasted and the juice then squeezed on congested joints). The finely chopped root of a botanically unidentified plant known as *mənəi* in both Northern Thai and Lua' is boiled and the water given to postpartum mothers to drink.

The importance of postpartum ailments to the Lua' is suggested by the large number of plant varieties used in their treatment. Lua' diagnosis and treatment of these ailments and even the names of some of the medicinal plants, are similar to those among Northern Thais. The fact that Pa Pae villagers both purchase "hot medicine" plants and sell them in Northern Thai markets suggests that knowledge of this type of disease and its treatment has gone in both directions. Clearly medicine and medicinal plants have been a major focus of inter-ethnic contacts for the Lua'.

"Hot medicine" and tonics consist mainly of spicy plants, while traditional medicine used to treat fevers are mainly perceived as "cool." Thus there is a rough parallel between these plant uses and the hot-cold conception of disease causes and cures which is so widespread in Asia, even though this is not a very explicit theme in Lua' medicine (see Kunstadter 1975).

The fact that Lua' villagers make medicinal use of even introduced weeds (e.g., *Eupatorium odoratum* leaves, chewed and packed in wounds) suggests that they actively familiarize themselves with new plants in their environment, as well as simply using their traditionally transmitted knowledge. The depth of Lua' knowledge of the characteristics of the plants in their environment is suggested by their use of different parts or different preparations of the same plant for different purposes, the range of which is only suggested by this abbreviated outline of Lua' medicinal plants and their uses.

## ANIMAL FOOD PLANTS

Forty-four plant varieties were reported to be used for animal food, of which 6 were cultivated and 38 were uncultivated. Of the uncultivated varieties, 16 were reported only from fallow swiddens, and 11 were found both in fallow swiddens and in other locations. The cultivated varieties included non-glutinous and glutinous rice from both swidden and irrigated fields. All of these varieties that are fed to animals are used primarily for human food. Bananas and banana stalks (*Musa sapientum*), used for animal food, are grown around the village primarily for human consumption. Important uncultivated varieties of animal food include *Colocasia antiquorum*, a wild-growing taro, the leaves of which are collected from stream beds and fed to pigs. Many kinds of grasses (*Panicum montanum*, *Saccharum procerum*, etc.) occuring primarily in the fallow swiddens, are eaten by domestic animals. The farmers know that their animals also graze on shoots, leaves, and fruits of many other plants in the fallow swiddens.

## PLANTS USED FOR WEAVING AND DYEING

Thirty-three varieties were reported to be used for weaving or dyeing, the most important of which, cotton (*Gossypium* sp.), is planted in upland swiddens along with rice, as well as in separate upland swidden gardens. Some cotton thread for weaving, as well as some dyestuffs, are purchased in the lowland market, but all women's clothing and most men's clothing and blankets are woven at home, mostly of homespun, home-dyed cotton. The importance of the fallow swiddens for weaving and dyeing is suggested by the fact that eight of the plant varieties were reported to be found only in fallow swiddens, while another six were found both in fallow swiddens and elsewhere.

The rotted leaves of four important dye plants, *Marsdenia jekinsii*, *M.* cf. *tinctoria*, *Phlogacanthus curviflorus*, and *Strobilanthes flaccidifolius*, are used for producing the common dark blue or black dye for women's skirts. These varieties are planted in cool, wet places near the village, and *S.* ? *flaccidifolius* may also be collected from an uncut forest north of Pa Pae. *Curcuma longa* (turmeric), grown in swiddens and village gardens and collected from fallow swiddens and uncut forests, is used as a food flavoring, a medicine for sores, as "gold" in offerings to spirits, and as a yellow dye. *Morinda tomentosa* is planted around the village, and the root is used for another yellow dye. *M. coreia* may be planted around the village or collected in the nearby uncut forest and used similarly; *Lithocarpus dealbatus* is collected from the fallow swiddens and used for a red dye. The bark or berries of *Mallotus philippinensis*, collected from an old forest north of Pa Pae, can be pounded and boiled for a red dye. The leaves and roots are also boiled and sucked for sore gums. *Perilepta siamensis*, collected in fallow swiddens, is used for making a black dye, as is a botanically unidentified variety, known in Lua' as *məkham.* The bark of *Ternstroemia gymnanthera* is finely chopped and boiled for use as a red or black dye, or to intensify red dye. It is found in fallow swiddens or in the disturbed vegetation near the village. Another red dye is made from the fruit of *Tithonia diversifolia*, found in the forest north of the village. The bark of an uncultivated variety of *Lagenaria leucantha* is used for yellow or brown dye, or, when mixed with mud, as a black dye. The bark of *Oroxylon indicum*, found in the uncut forest near the village, is mixed with lime for a light-colored dye; its bark is used as medicine for tiger bites on buffaloes, and its fruits and shoots may be eaten.

The dye-resistant leaves of *Curculigo latifolia* (from uncut forest near the village) and *C. recurvata* (location unspecified) are used to bind thread which is dyed for weaving *ikat* designs. The stems of *Saccharum procerum*, a large grass collected in fallow swiddens, are used by Lua' women as cylinders on which they make rolls of ginned and fluffed cotton prior to spinning. This plant is also used as food for domestic and wild animals. The bristly dried infructescences of two or more *Pandanus* varieties found in uncut forests are used to comb or card cotton thread to prepare it for weaving. Leaves of the wild banana *Musa acuminata* are collected from the forest, and the powdery substance from their undersides is used to lubricate cotton thread to keep it from catching during weaving. These leaves may also be used for wrapping cooked rice to carry it to the fields.

Loom parts are made from several species of plants, including *Craibiodendron stellatum*, collected in old fallow swiddens. The wood from *Dalbergia dongnaiensis* and another *Dalbergia* sp., also found in fallow swiddens, is used to make the beater or sword used in weaving with a backstrap loom. *Dalbergia cultrata* (location unspecified) is used for the same purpose. *Bambusa tulda*, a bamboo that is planted around the village and that grows wild in fallow and cultivated swiddens and in other parts of the forest, is used for construction of loom parts, for weaving baskets, or for other construction. The shoots are also eaten.

## FUEL

The importance of fallow swiddens as a source of fuel in the Lua' economic system is indicated both by the customs regarding forest use (dead wood can be collected but trees cannot be cut in certain nearby "reserved" forest areas) and by the preponderance of varieties reported to come from the fallow swiddens. Of the 27 varieties said to be used for fuel, 22 came from fallow swiddens, and 14 of these were reported only from fallow swiddens. Women customarily bring firewood back to the village each time they return to the village from the upland fields, unless they happen to be carrying some other swidden product.

The need for fuel for cooking and for heating houses through the cold winter nights is obvious. Familiar with the burning qualities of various woods, the villagers collect them for specific purposes, such as making charcoal for blacksmith fires, or for making long-burning, sparkless cooking fires.

## RECENT CHANGES IN THE TRADITIONAL VILLAGE ENVIRONMENT

One of the major changes in recent years has been the increase of population to the point where there is little room for new settlement, and traditional patterns of Lua' land use and land tenure are being quickly eroded. Little is known of the details of Lua' settlement in this area before the nineteenth century. There were evidently more Lua' villages than exist today, and they extended much further both north and south than today. In the mid-nineteenth century, small numbers of Karens moved into the area, which was apparently once exclusively Lua'. They were allowed by Lua' villagers to settle in the less desirable spots at higher elevations. Until about 1900, when the central Thai government took control of the area and prohibited the practice, these Karens paid their Lua' landlords rent in the form of one tenth of their rice crop.

Karen village histories suggest that from a few migrants their population size doubled each generation, and they now vastly outnumber the Lua' in these hills. They occupy most of the areas between Lua' villages, and have often swiddened in areas traditionally reserved for other purposes by Lua' villagers. Today Karen swidden areas adjoin those of Pa Pae in most years' swidden locations, and little by little the Karens have acquired land on the edges of Pa Pae territory. The Karen economy and technology is very similar to that of the Lua', but Karens are somewhat less conservative of land resources, paying less attention to the control of fires and erosion, and being less rigid in regular rotation of swidden sites. Although Karen villagers apparently started out following roughly the same swidden rotation pattern as their Lua' neighbors, population pressure now seems to be forcing them to shorten their rotations, to cultivate more marginal areas, and to push into Lua' lands wherever they can. Pa Pae village religious leaders, who were once in a position to control the burning of the swiddens, must now negotiate with the leaders of Karen villages to coordinate swidden burning, and Pa Pae villagers must now protect their territory against fires which spread from Karen fields.

Wet rice technology was introduced in Pa Pae about 50 years ago. Because no field rotation is required, it is much more productive per unit area than is swiddening, and this has helped to relieve some of the population pressure. Lowland Northern Thai concepts of wet rice land tenure were also introduced, leading to the individualization of irrigated land holdings, and the ability to exchange land for money, which was never done traditionally with swiddens. Through foreclosure of mortgages on irrigated fields, Karens have gained control of many wet rice fields within Pa Pae territory, and this land is no longer subject to Pa Pae control. Religious change within the village—the conversion of some Pa Pae residents to Christianity—further diminishes the authority of village religious leaders. It may affect their ability to control land use patterns and customs within the village, as it has already affected their ability to mobilize villagers for communal agricultural rituals. These circumstances have combined to weaken the ability of village leaders to control land use and to reallocate land among villagers, so that even swidden land (although it is supposed to be a community resource) is now occasionally rented or sold by an individual at Pa Pae to an outsider.

At the same time as pressure is increasing from the outside, the population of Lua' villages appears to be growing rapidly, due to natural increase. Religious conversion, by eliminating the cost of traditional animistic sacrifices for the converts, allows villagers who are economically unsuccessful to remain in the village. Before the option of conversion existed, people who were unable to keep up with village or household religious expenses (for animal sacrifices) left to "become Karen" or "become Thai," thereby decreasing village population size and the drain on village resources. In this sense, religious conversion is associated indirectly with a decline in the standard of living and in turn with the gradual deterioration of the village environment.

As already mentioned, Hmong groups have begun moving into the immediate vicinity of Pa Pae since the early 1960s, further inhibiting the ability of Pa Pae villagers to control land use in their area and putting additional strain on local forest and watershed resources. The ridge tops and watercourses above Pa Pae are no longer protected, and stream flow in the river is much more subject to floods than in the past. This makes the task of irrigation more difficult.

Other pressures on upland resources external to the village are related to demands in the lowlands for resources found in the hills. In other Lua' villages, this has meant loss of swidden lands to pine plantations, roads and other governmental schemes, and in more distant areas, to mines. There has been little or no compensation to the economy of these villagers to accompany these changes. Upland technological assistance in this area has led to the introduction of a few new crop plants (e.g., tomatoes) and improved varieties of other crops (e.g., hybrid corn), but these are relatively unimportant. The major advances of modern tropical agriculture—use of fertilizers, pesticides, improved strains of high yielding rice—are not available to these hill villagers for economic and environmental reasons.

The net effect of these changes has been to reduce the amount of land available to Pa Pae villagers, to reduce the control over the land which can be exercised by village leaders, and to reduce the extent and variety of different environmental zones available to the villagers. At the same time, there has been no basic improvement in the productivity of their fields, except for the gradual expansion of irrigation, and these uplanders are becoming more dependent on the lowlands for wage labor to supplement their agricultural income. At a time in their history when gathering supplementary foods and other products from the forest may be becoming more necessary, their opportunities for doing so are decreasing. As more area is being converted to irrigated fields, as old, uncut forest areas are diminishing, and, to the extent that swidden rotations become shorter as the variety of their botanical resources decrease, the importance of these supplementary resources will become more apparent.

Land use practices of Hmong and other cash cropping swiddeners in northern Thailand suggest that cash cropping both decreases environmental diversity and also increases the amount of land used per person. Lua' at Pa Pae in 1967-68 farmed about 0.38 ha per capita per year. Walker (1976: 176) found that the opium- and chili-growing Lahu villagers he studied used about 0.46 ha per person per year. Hmong at Metho, a village a few kilometers east of Pa Pae, used more than this (Geddes 1976). A major increase in need for land was noted by Scholz (1969) when an Akha village he studied began growing corn as a cash crop. Thus, to the extent that Lua' villagers change from subsistence to cash oriented farming, we can expect that the amount of land used per capita will increase. The only ways they can find this land will be to cut their reserved forests or to reduce the length of their swidden/fallow cycle. Either option can be expected to decrease the species variation in their environment. This in turn will probably increase their dependency on the market for subsistence.

## SUMMARY AND CONCLUSIONS

Pa Pae villagers have knowledge of large numbers of plant varieties, their characteristics and their locations. Many of the plants have several different uses (dyestuffs, food, medicine), and different parts of the same plant are often used in different ways. The environment that is created and maintained for essential food plants by the deliberate activities of Pa Pae villagers is more varied than would be found within their territory if they were to give up swiddening, lengthen their swidden cycle, convert their entire area to swiddens or convert it all to wet rice fields.

Our description of the Lua' swidden system suggests some modification of Geertz' generalizations (1971:16 ff.). In irrigated farming, he says correctly that the variability of land forms, water supply and vegetation species are rigidly controlled. By contrast, "swidden agriculture . . . maintains the general structure of the natural ecosystem rather than creating and sustaining one organized along novel lines and displaying novel dynamics." In fact, the Lua' systematically change the forest structure by clearing and admitting sunlight necessary for the growth of grassy crop plants, in contrast with the suppression of such plants by shade in the forest. They concentrate species diversity in small cultivated areas, in contrast with the wide dispersal of different species in the forest. They localize nutrient recycling in time and space by slash and burn to make nutrients available for shallow rooted crops, and they maintain a continual successional sequence, thereby increasing the variety of non-cultivated species (cf. Connell 1978). A closed canopy is properly seen as a feature of Lua' swiddening during the fallow period, but not while the fields are being cultivated.

The villagers capitalize on different zones of their territory and more or less deliberately create and maintain the different environmental conditions in these zones within which different plants can thrive. To the extent that Pa Pae villagers remain in a subsistence economy, these different environments are essential to the maintenance of their traditional way of life. For example, plants used in weaving and dyeing include both cultivated and uncultivated varieties which come from almost all parts of the Pa Pae environment. If uncut forests or streambeds are extensively modified, the environment from which Lua' dye plants come will be lost. The villagers will either be forced to give up the colors once provided by these plants or purchase similar colors in the market.

Population pressure and the declining ability of Pa Pae villagers and their leaders to control what has traditionally been their land base are resulting in a narrowing of the range of environmental types available to the villagers. Effects on the basic subsistence system will include a decrease in the ability of Lua' villagers to prevent fires in their fallow swiddens and may force them into a shorter swidden cycle, which will decrease soil fertility and lower their crop yield. The effects will probably be felt first through the destruction of the uncut forest habitats of certain plants used for food, medicine, dyes, or other purposes. The loss of these forests will also deprive the Lua' villagers of their traditional margin of safety for times of crop failure, or forced dispersal of the village as a result of epidemics, when they traditionally gathered their subsistence in the forest. This will mean further dependency of Pa Pae villagers on the more modernized market economy and society of the Thai lowlands.

If the villagers' increased needs for cash are met through cash cropping, the amount of land used per capita can be expected to increase, and the environmental diversity can be expected to decline. This seems to be a general pattern of change affecting the part of the world in which the Lua' live.

## REFERENCES


Connell, Joseph H.
1978 Diversity in Tropical Rain Forests and Coral Reefs, Science 199:1302-10.

Geddes, William R.
1976 Migrants of the Mountains, Oxford: Clarendon Press.

Geertz, Clifford
1971 Agricultural Involution: The Process of Ecological Change in Indonesia. Berkeley, University of California Press for the Association of Asian Studies.

Keen, F. G. B.
1978 Ecological Relationships in a Hmong (Meo) Economy. In: Farmers in the Forest, edited by Peter Kunstadter, E. C. Chapman, and Sanga Sabhasri, pp. 210-21. Honolulu: The University Press of Hawaii.

Kunstadter, Peter
1966 Residential and Social Organization of the Lawa of Northern Thailand. Southwestern Journal of Anthropology 22:61-84.

1971 Natality, Mortality and Migration in Upland and Lowland Populations in Northwestern Thailand. In: Culture and Population: A Collection of Current Studies, edited by S. Polgar, pp. 46-60. Cambridge, Mass.: Schenkman Publishing Company and Carolina Population Center.

1974 Usage et Tenure des Terres Chez les Lua' (Thailande). Etudes Rurales 1974 (53, 54, 55, 56):449-66. Paris.

1975 Do Cultural Differences Make Any Difference? Choice Points in Medical Systems Available in Northwestern Thailand. In: Medicine in Chinese Cultures, edited by Arthur Kleinman, Peter Kunstadter, E. R. Alexander, and James L. Gale, pp. 351-83. Washington, D.C.: Department of Health, Education and Welfare.

1978 Subsistence Agricultural Economies of Lua' and Karen Hill Farmers, Mae Sariang District, Northwestern Thailand. In: Farmers in the Forest, edited by Peter Kunstadter, E. C. Chapman, and Sanga Sabhasri, pp. 74-133. Honolulu: The University Press of Hawaii.

Kunstadter, Peter, and E. C. Chapman
1978 Problems of Shifting Cultivation and Economic Development in Northern Thailand. In: Farmers in the Forest, edited by Peter Kunstadter, E. C. Chapman, and Sanga Sabhasri, pp. 3-23. Honolulu: The University Press of Hawaii.

Kunstadter, Peter, E. C. Chapman, and Sanga Sabhasri (eds.)
1978 Farmers in the Forest. Honolulu: The University Press of Hawaii.

Kunstadter, Peter, Sanga Sabhasri, and Tem Smitinand
1978 Flora of a Forest Fallow Farming Environment in Northwestern Thailand. Journal of the National Research Council of Thailand 10:1-45.

Sabhasri, Sanga
1978 Effects of Forest Fallow Cultivation on Forest Production and Soil. In: Farmers in the Forest, edited by Peter Kunstadter, E. C. Chapman, and Sanga Sabhasri, pp. 160-84. Honolulu: The University Press of Hawaii.
Scholz, Friedhelm
1969 Zum Feldbau des Akha-Dorfes Alum, Thailand. Yearbook of the South Asia Institute, Heidelberg University, 1968/69, 3:88-99.
Walker, Anthony R.
1976 The Swidden Economy of a Lahu Nyi (Red Lahu) Village Community in North Thailand. Folk 18:145-88.
Zinke, Paul J., Sanga Sabhasri, and Peter Kunstadter
1978 Soil Fertility Aspects of the Lua' Forest Fallow System of Shifting Cultivation. In: Farmers in the Forest, edited by Peter Kunstadter, E. C. Chapman, and Sanga Sabhasri, pp. 134-59. Honolulu: The University Press of Hawaii.

# THE KNOWLEDGE AND USE OF RAIN FOREST TREES BY THE KUIKURU INDIANS OF CENTRAL BRAZIL

*Robert L. Carneiro*
American Museum of Natural History

The Amazonian rain forest, perhaps the most complex biotic community in the world, has for millennia served as the habitat of many hundreds of Indian tribes. This vast forest affords such a wide range of resources that once a tribe adapted to it, it would lack for very little. Material for food, clothing, shelter, tools, weapons, utensils, ornaments, medicines, etc., are found there in plenitude. In this paper I would like to focus on the major component of the forest—the trees themselves—and to describe the knowledge and use of these trees by the Kuikuru Indians of central Brazil.

How long the Kuikuru have lived in their present-day habitat near the Kuluene River in the Upper Xingú basin is difficult to say. It has certainly been for centuries, and of course their ancestors lived somewhere in Amazonia for thousands of years before them. Through the accumulated observation and experience of many generations, the Kuikuru have gained a very detailed and precise knowledge of their rain forest environment, and this becomes evident as soon as one begins to probe this knowledge.

The Kuikuru utilize the forest in many ways. For one thing, they fell it to provide land for planting their gardens. This may seem only a negative use, but it is not really so. After all, it is the forest that creates soil conditions favorable for planting. The very existence of forest on a tract of land permits a Kuikuru to say, "Here is well-drained, easily-worked soil where I can plant my manioc and be sure of obtaining a reasonably good crop." But beyond that, burning the trunks and limbs of felled trees releases large amounts of mineral nutrients which contribute substantially to crop yields. Moreover, the charred logs in a burned garden provide firewood which, over the ensuing months, is cut up and brought back to the village. Even if nothing more were done with the trees of the forest, this would be enough to rank them as a major asset.

So much has been written about tropical rain forests in recent years that it seems unnecessary for me to attempt to describe it here. One aspect of the rain forest that does require comment, though, is its extraordinary diversity of tree species. Thus, while the entire state of Michigan has only 89 different species of native trees (Otis 1926), a single acre of rain forest in Amazonia often has over 100 species. In a tract of forest in British Guiana 1.5 ha (3.75 a) in size, P. W. Richards (1952:230) found 91 species of trees 4 in or more in diameter, and Ghillean T. Prance, et al. (1976:11-12) recorded in 1 ha (2.47 a) of forest near Manaus no less than 235 species with a diameter of 2 in or larger.

The forests of the Upper Xingú are not as luxuriant as those of the wetter parts of Amazonia. While the yearly rainfall in the Upper Xingú, some 75 in, is enough to produce a respectable forest, the marked seasonal skewing of this rainfall, with no rain falling at all during June, July, and August, acts to hold down the height of the forest and to limit the number of species it contains. Thus, a number of trees well known in the more humid parts of Amazonia do not occur in the Upper Xingú. These include mahogany (*Swietenia macrophylla*), Spanish cedar (*Cedrela odorata*), lignum vitae (*Guaiacum officinale*), Brazil nut (*Bertholletia excelsa*), rosewood (*Dalbergia nigra*), Pará rubber (*Hevea brasiliensis*), and cacao (*Theobroma cacao*).

## KUIKURU FOREST CLASSIFICATION

The forest cover of the Upper Xingú is not the same everywhere, of course. The Kuikuru, in fact, classify the stands of trees in their habitat into several types. First and foremost there is *itsuni*, primary rain forest, the vegetation that covered the entire region in the days before human habitation, and which still covers much of it. For the most part it is *itsuni* that the Kuikuru clear to plant their manioc.[1]

When a garden is abandoned, the plot is invaded by a type of vegetation the Kuikuru call *tafuga*. This is essentially secondary forest, but it may contain a good many primary forest species which grow either from coppice

[1] The fact that *itsuni* is primary forest does not mean that it is necessarily *virgin* forest. There is archaeological evidence that the Upper Xingú was fairly densely populated for centuries, and consequently much forest must have been cut over for agricultural purposes. After walking through the *itsuni* near the present-day Kuikuru village as well as examining aerial photographs of the forested region to the south, I think that the *itsuni* near the village may well be a forest that has regenerated over the last 200 or 300 years after the area was cleared and later abandoned.

shoots or from seeds derived from the surrounding *itsuni*. In evolving back to *itsuni*, *tafuga* is replaced by one of at least two types of transitional forest. The next stage in forest succession is usually either *agïpe* or *agafagïpe*, so named for the species of tree that is most prominent in it: *agï* in *agïpe*, and *agafagï* in *agafagïpe*, the suffix '*-pe*' meaning "thing."

The Kuikuru recognize yet another type of forest called *egepe*, whose principal distinguishing feature, as far as they are concerned, is less the kinds of trees growing in it than the color of the soil underlying it.[2] While *itsuni* grows in red earth, *egepe* grows in black earth. The Kuikuru say *egepe* soil is more fertile than *itsuni* soil and invariably plant their maize there. Manioc tubers will grow much thicker in *egepe*, they say, but since there is not much of it around and since manioc does tolerably well in *itsuni* anyway, that is where most of their manioc gardens are planted.

Finally, there is *indagïpe*, strips of forest growing along rivers and around the shores of lakes. *Indagïpe* is thus equivalent to our "gallery forest."

Because cleared and abandoned land is repeatedly burned, preventing, or at least retarding, reforestation, there is a good deal of savanna in Kuikuru territory. As long as this savanna is predominantly grassland, it is called *oti*. Nevertheless, certain trees commonly grow there, especially cashew (*Anacardium occidentale*), the sandpaper tree (*Curatella americana*) and mangaba (*Hancornia speciosa*).

## KUIKURU FAMILIARITY WITH TREE SPECIES

In addition to recognizing forest types, the Kuikuru identify individual trees. I was much interested in learning how many species of trees occurred in their habitat, and whether they would have names for each of them. Having just made a brief study of Yąnomamö tree recognition without finding a single tree they could not name, I was curious to see if the Kuikuru would do as well.

For this purpose, and also to observe exactly how the Kuikuru cleared the forest for planting, I selected and roped off 1/6 of an acre of *itsuni* about 2 mi south of the Kuikuru village.[3] Then I went through this tract

---

[2] Nevertheless, my best informant, Jakalu, once contrasted "*egepe*" and "*itsuni*" by listing trees characteristic of each. Typical of *egepe*, he said, were ***jukuku, uagi, ukefu***, and *ñoṇiñoṇi*, while typical of *itsuni* are ***tifa, fala, tsuitsî, tafaku, ikuengî, kugisoki, tufaga, ñagı̈ifo***, and *epiṇgipe*.

[3] The reason 1/6 of an acre was chosen as the size of the study tract was to provide comparability. When I made my first forest inventory in Luquillo National Forest in Puerto Rico, I found that in the time available I could complete only 1/6 of an acre instead of the full acre I had intended. Accordingly, when I later studied two Yąnomamö villages, I decided to use sample forest tracts of 1/6 of an acre so I could compare my

numbering every tree growing in it that measured at least an inch in diameter at breast height. All told, there were 172 such trees in the tract.

Over the next two days I went through the tract again, this time with an informant, and asked him the name of every tree. Actually, I used two informants. One of them, Mafukakumá, identified trees numbered 1 through 52, and the other, Aṇafïtá, identified trees numbered 53 through 172. They were able to identify every tree without difficulty. They did so by looking at the trunk, and if they could not tell what tree it was from this, they looked up at the crown and scrutinized the leaves and branches, and any flowers or fruits the tree might have. Occasionally they slashed the bark and examined what lay beneath, since the color, odor, or taste of the wood or sap is often diagnostic of the species.

The only problem Aṇafïtá had in naming trees was sociological rather than botanical. Four of the trees I asked him to identify, *fetétepïgï*, *tefugukueṇgï*, *kajïtifagï*, and *kajïïgikeki*, all of which he knew perfectly well, he nonetheless could not name since they contained the names of certain of his in-laws, which by Kuikuru custom he could not utter. He had to prompt some boys who were accompanying us to say the names for him.[4]

When I returned to the village and tabulated the results of this inventory I found that the 172 trees in the tract represented 45 different species.[5]

For comparative purposes I present here the results of comparable inventories I made in three other sectors of tropical forest. At the El Verde station in Luquillo National Forest in Northeastern Puerto Rico I had identified for me 31 species of trees in a 1/6-acre tract. In a tract of the same size near the village of Hasuböwateri on the Upper Orinoco, Yaṇomamö informants identified 44 different species of trees, while near the village of Tayariteri

---

results with those obtained in Puerto Rico. A sixth of an acre may seem small, but since in the four tracts of this size that I studied the number of trees contained ranged between 125 and 172, it seems to provide a reasonably large sample of trees.

[4] Anafïtá's linguistic inhibitions did not end there. While we were identifying trees, a noise was heard in the foliage nearby. It was a monkey, but Aṇafïtá could not tell me so because the word for monkey, *kajï*, had been the childhood name of one of his in-laws. And when a little later a boy was stung by a caterpillar called *ïmpe*, Aṇafïtá likewise could not tell me since '*ïmpe*' was also the name of an in-law.

[5] Perhaps I should say "kinds" of trees instead of species since I cannot be sure that there is a one-to-one correspondence between the tree names given to me by the Kuikuru and botanical species. However, Richards (1952:232) writes: "Among the Arawak Indians of British Guiana . . . several hundred vernacular names of trees are in common use and, for the most part, each name represents a single taxonomic species." And my impression is that Nicholas Guppy (1958:7, 49, 65, 83, 284n), a tropical botanist who made a study of rain forest trees of the British Guiana-Brazil border, felt that the tree names he got from his Indian informants were almost invariably equivalent to botanical species.

further down the Orinoco, the Yąnomamö there identified no fewer than 56 different tree species in a 1/6-acre tract.[6]

Some weeks after my inventory of the *itsuni* tract, I decided to try a second experiment in tree identification. I selected another area of *itsuni* about half a mile from the first one and cordoned off a plot exactly 10 ft on a side in a part of the forest floor where only seedlings were growing. From this plot I pulled out every living plant, no matter how small, totaling 153. I wrapped these plants in a large plastic sheet and brought them back to the village.

Sitting in the men's house I unwrapped the sheet, presented the plants one by one to the assembled men, and asked if they could name them. I expected this to be harder than identifying trees since in seedlings the characteristics that will distinguish them in later life are often undeveloped. Nevertheless, the men were able to identify every seedling. This was not always done immediately, and sometimes, over the more unusual or less distinctive specimens, there was some discussion and even disagreement before an identification was fixed upon. But without fail, a consensus was ultimately reached on every specimen. And I was convinced that it was a genuine consensus and not one arrived at simply to provide me with a name to write down.

The 153 plant specimens I had pulled out of the 10-foot square represented 29 species of trees, 6 species of shrubs, 7 species of vines, and 1 species of grass. Of the 29 tree species 14 had also been present in the 1/6-acre tract, while 15 had not.

The Kuikuru had proved themselves excellent identifiers of trees. But I had one more test for them. How well would they do at naming species when the only plant material they had to work with was dead leaves? Back I went to the *itsuni*, randomly selected a section of forest floor, and marked off a plot one foot square. From this plot I picked every leaf or major leaf fragment and carefully put them in a plastic bag.

The top leaves were relatively newly fallen, and though already mostly dry, they were still largely intact. At lower levels, though, the leaves were

---

[6]By using an index of diversity we can compare the degree of floristic variety of these four forest tracts. The simplest such index is the cumulative number of species per 100 individuals, which gives us the following results: El Verde 24.0, Kuikuru 26.2, Hasuböwateri 35.2, and Tayariteri 38.1. While it is difficult to draw valid conclusions from a comparison of index values for widely separated regions of rain forest, different index scores from tracts within the same region might prove very significant. Since a fully climax rain forest has the highest index of diversity of all, a tract of forest with an index value substantially lower than the highest index value recorded in an area would very likely be younger, regenerated forest. This information might reasonably lead an ethnologist or archaeologist to conclude that a certain sector of forest, which to casual observation appeared virgin, had once been cleared, settled, and abandoned.

progressively more deteriorated, either through having been eaten by insects or worms, or attacked by fungi and other lower forms of life. Finally I reached a level where the leaves were badly decomposed and their remains matted together in a gummy mass. Here I stopped. Altogether in this 1-foot-square I had picked up 190 leaves.[7]

Returning to the men's house with my plastic bag, I drew out the leaves one by one. Despite the care with which I had tried to handle them, 13 of the leaves had fallen apart, so I showed the Kuikuru only 177 leaves. Patiently, the men looked at each leaf in turn. Many of them they named immediately. Some, however, either because they did not differ much from other species or because they were in poor condition, caused more of a problem. Occasionally, men who were elsewhere in the village at the time but who were thought to be particularly good at plant identification were called in to give their opinion. After due deliberation, the Kuikuru once again identified every specimen presented to them.

The 177 leaves I showed them turned out to represent 37 species of trees, 7 species of vines, and 1 species of shrub. Of the 37 tree species, 22 had previously been identified in either the 1/6-acre tract or the 10-foot-square plot, while 15 were "new" species. Altogether, then, the three test areas had yielded a total of 75 different species of trees.

I might mention at this point a couple of minor problems that had to be recognized before an accurate list of species could be made. First of all, there was the problem of synonymy: a few trees had more than one name. For example, in inventorying the forest tract I had gotten the name *enufe-kueṇgï* from Mafukakumá for several trees, and later, from Aṇafïtá, I obtained the name *epiṇgipe* for several more trees. These turned out to be different names for the same tree.

I had anticipated this problem but not the following one. While inventorying the forest tract I had been given names for what I thought were four separate species—*isitisu*, *tafaku*, *ataka*, and *tafatago*—and had recorded them as such. Only later did I learn that *isitisu* is the name given to the *tafaku* tree when it is young, and that *ataka* is likewise the name of a young specimen of *tafatago*.

Since my field work centered primarily on other things, I made no effort to obtain an exhaustive list of all the kinds of trees growing in the Kuikuru habitat. But I often asked people the names of trees we passed on trips through the forest, as well as the name of the wood from which various

[7]Counting only those leaves in reasonably good condition, we can make the following calculation. Assuming that 90% of the forest floor in Amazonia is leaf-covered, there are 7,448,760 dead leaves per acre, and 4,767,200,000 per square mile. Assuming (conservatively) that there are one million square miles of rain forest in Amazonia, this means that at any given time there will be on the forest floor nearly five quadrillion dead leaves.

artifacts were made. In preparing this paper I brought together all the scattered bits of information about trees and their uses obtained on my field trip of 1975 as well as those gathered during my initial field work in 1953-1954.

After tabulating and cross-checking these data, I found that I had recorded a total of 187 different species of trees.[8] I have listed these species in alphabetical order by native name in Table 1, with those few for which I have a positive or tentative botanical identification being listed again in Table 2. This is a substantial number of tree species, yet there are undoubtedly many more species growing in Kuikuru territory. Several types of evidence point to this conclusion.

First let me cite the linguistic evidence. There is a suffix, *-kueṇgï*, in Kuikuru which means "another kind of." A number of tree names have this suffix incorporated in them indicating that they closely resemble another tree. Thus there is a tree called *falakueṇgï* which is named after one called *fala*, a tree called *ipejikueṇgï* named after one called *ipeji*, and so forth. All told, there are 19 such pairs of trees on the list. Now, there are also several tree names ending in *-kueṇgï* (or its variant, *-gueṇgï*, *-fueṇgï*, or *-tsueṇgï*) for which no companion tree is listed. Thus there is an *iñatsikueṇgï* but no *iñatsi*, a *tamatamakueṇgï* but no *tamatama*, and so on. There are 13 such names altogether.[9] We may infer that such companion trees exist in the Kuikuru forest but that their names simply never came up. This would bring the total of tree species to 200.

The most persuasive evidence of all that there are a good many more than 187 tree species in the Kuikuru habitat is afforded by species counts from other parts of Amazonia. I have already cited the fact that Prance et al. (1976:11-12) recorded 235 tree species in a single hectare of forest near Manaus. It is true that this part of the Amazon rain forest is wetter than that of the Upper Xingú so we cannot expect the same great diversity of species in the latter area. The forest survey most nearly applicable to the Upper Xingú is one carried out by J. A. Ratter, P. W. Richards, and their associates some 80 mi east of the present-day Kuikuru village. In an area of forest that is probably not as diverse as that of the Kuikuru, Ratter et al. (1973:489-92)

---

[8]This number excludes palms. Not many palms grow in the forests of the Upper Xingú. More grow along lakes and rivers. I made no special effort to obtain the names of palms, so the following list, which includes all the names I have, is surely incomplete: *ikiṇi*, which is burití (*Mauritia flexuosa*), *kafugu*, which is macaúba (*Acrocomia sclerocarpa*), *naïga*, *kaṇaga*, *faká*, *iṇukuagï*, *kutúi*, *fegita* and *kïa*. *Heliconia*, called *afepa* by the Kuikuru, the wild relative of the banana plant so common in the wetter forests of Amazonia, is virtually absent from Kuikuru forests.

[9]The tree called *togokigekueṇgï* bears the suffix in question, but since *togokige*, after which it is named, is the cultivated cotton plant, which is a shrub rather than a tree, I have not included *togokigekueṇgï* in this count.

## TABLE 1
## TREE NAMES RECORDED AMONG THE KUIKURU INDIANS

Uses of trees indicated as follows:
A—to make artifacts; B—for painting, annointing, or decorating the body; C—ceremonial, shamanistic, or magical use; D—medicinal use; E—fruits or nuts eaten; F—for firewood; G—grown in gardens; H—for housebuilding; L—for lashings or fibers; M—figures in mythology; O—for ornaments; P—provides poison; R—yields latex or resin for non-ornamental use; S—leaves used for sanding, lining, wiping, etc.; W—to make watercraft, including caulking; X—for making soap; Y—for making salt

afafueṇgï
afaṇite (A,F,H)
afatata (A)
afíatandaṇagï (A,C,F)
afíatandaṇagïkueṇgï
afiño (S)
afusagï (D)
agafagï (H,X)
agafḯtagï (D,Y)
agakagï (H)
agakasi (D,H)
agapi (E)
agatḯgokogo
agiña (A)
agï (C,E,H)
agugáiṇkuṇu (A)
aguka
aka
akaga (E)
akïti
aku
akugíataṇagï (A?)
akuguṇgueṇgï (D,F)
amigi (A)
aṇa (A,B,D,E)
aṇakueṇgï (F)
asatafa (O)
atamai (M)
atati (O)
atḯ (E)
ẹẹti (A)
ẹẹtitsueṇgï (D)
egeífe
egeikajḯ
egïfotsoño (E)
ẹịpogukueṇgï
ekïti
eṇufe (E)
epiṇgipe (F)
etiṇukikueṇgï
etsekétiñatipïgï (H)
etuṇi (O)
fagakatso
fagatïgï
fala (B,F,H,W)
falakueṇgï
fata (D)
fetétepïgï (D,F)
fetiñï
figatïgï
fiṇugi
fitsá
fitsakueṇgï
fïgafegï (E)
fïgéfuti (A)
fïgéfutikueṇgï (D)
fïtaguengï
fọ (B,E)
fomíamitsï
fonoto
fotafueṇgï
fugikï
fulokóigu
gḯkïgi (F)
guifífaṇagï (H)
ifagati (A,F)
ifagatsueṇgï
igeífe (D)
ịịte (P)
ïkú (B,M)
ikueṇgï
impe (B,E,G)
inansi (D)
iṇgugo (D)
iñambe (D)
iñatsikueṇgï (D)
iñúi (D,F,H,P)
ipeji (A)
ipejikueṇgï (F)
ipogofo (F)
isike (M)
itekueṇgï
itsati (B)
itsunítikuniñïgï (D)
itsutu (A,B)
jukuku (M)
kagafïgï (L)
kagate (A,M)
kaifa (A)
kainái (D)
kaïntï (D)
kaïntïkueṇgï (D)
kajḯigikeki (D,E,F)
kajïtifagï
kamajáfïgifutisï
kaṇaṇgatï (A)
kaṇúa (F)
kaṇuakueṇgï
kaṇuguso
kañifekú (D,F,L)
kapula
katsegï (A,H)
katuga (D,E,R)
kejite (A,C)
keunti (E)
kïntï (A)
kofï
kofïkueṇgï (A,D)
kuakugu (D)
kugipisi
kugisoki (A,D)
kụọ́ (A,D,F,H,S)
kutuni (D)
majafi (A)
mefupe (A)
megimegi (E,F)

TABLE 1 (Cont.)

| | | |
|---|---|---|
| miṇakueṇgï (H) | tagïpuiñï | tolougu (A,F) |
| mukugoti (E) | taguka (B) | tsuịtsï (S) |
| netufe (P) | takisi (A) | tufaga |
| ñagíifo | tali (B,M) | tugé (D,F) |
| ñájifo (A) | tamatamakueṇgï | tugokogo |
| ñoṇiñoṇi (F) | taṇoṇo (D) | tumatumakueṇgï |
| okoño (F) | taṇoṇokueṇgï | tuṇufi (A,C,D,E,F) |
| onto (B,G,L) | tapogo (E) | uagi (A,B,E,M,W) |
| pugupugu (C?,F) | tatï (A) | ueṇgïfi (A,C,H,M) |
| pugupugukueṇgï | tefuku (B) | ufaku (H) |
| sandaki (B,D) | tefuguguеṇgï (E,F) | ufitsa |
| ta (A) | teniñíụuï (C) | ugafagu (A) |
| tafaku (A,C,D,F,H,L) | tifa (A,B,D,M,W) | ugafe |
| tafakúiti (L) | tifakueṇgï (R) | ugagati (A,B,D,S) |
| tafatago (A,H) | tifigu (L,M) | ugagu (D) |
| tafisi (S) | tiṇukikueṇgï (D) | ugififaṇagï (C,H) |
| tafitséugu | tiñá (F) | ugigi |
| tafofisiñï (A) | tiñafó (A) | ugiti (A,E,F,H) |
| tafofota (E) | tipitsïgiñï (L) | ugitikueṇgï (E) |
| tafoti (E) | togokigekueṇgï | uguta |
| tafueṇgï | tolóagafïgi (D,Y) | uikïpa (A) |
| tafutafu | tolófïgï | ukefu |
| tagïfïgitiñi (M) | tolófomisï (B,D) | utsueṇgï (F) |

identified some 270 species of trees. This figure may be taken as representing the minimum number of tree species that occur in the Kuikuru forests. Quite likely, the number is considerably higher.

A rough estimate of the number of tree species in the Kuikuru forests may be obtained by applying the concept of a "species/area curve" (Richards 1952:235; Longman and Jeník 1974:68-70; Holdridge et al. 1971:674-78). This is a graph showing how the cumulative number of tree species increases as the area of forest increases. Beginning with the fact that there were 45 species of trees in a 1/6-acre tract of Kuikuru *itsuni*, and considering that there are probably in excess of 300 tree species in their habitat, I have constructed a formula which allows one to estimate the number of tree species to be found in an area of any given size in the same type of forest. The formula is as follows:

$$N = 64^{\frac{a}{a^{0.95}}}$$

where $N$ = the number of tree species measuring an inch or more in diameter

TABLE 2
TREES OF THE UPPER XINGÚ FORESTS
POSITIVELY OR TENTATIVELY IDENTIFIED BOTANICALLY

| Kuikuru Name | Portuguese Name | English Name | Botanical Name |
|---|---|---|---|
| aŋa | genipapo | | *Genipa americana* |
| eipogukueŋgï | | | *Miconia* sp. |
| fïgafegï | cajú | cashew | *Anacardium occidentale* |
| impe | piquı | souari | *Caryocar brasiliensis* |
| kagate | cuieté (?) | calabash tree | *Crescentia cujete* |
| kainái | angico | | *Piptadenia rigida* (?) |
| kañifekú | pimenta do sertão | | *Xylopia aromatica* (?) |
| kapula | piquı | souari | *Caryocar* sp. |
| katuga | mangaba | | *Hancornia speciosa* |
| kugipisi | imbaúba | trumpet tree | *Cecropia* sp. |
| majafi | pau d'arco | | *Tecoma violacea* |
| onto | urucú | anatto | *Bixa orellana* |
| tafaku | pindaıba | lancewood (?) | *Xylopia amazonica* |
| tafakúitï | imbaúba | trumpet tree | *Cecropia* sp. |
| tafatago | | | *Miconia* sp. |
| tafoti | goiabeira | guava | *Psidium guajava* |
| tifa | jacareúba | | *Calophyllum brasiliense* (?) |
| uagi | jatobá | courbaril | *Hymenaea courbaril* |
| ugagati | lixeira | sandpaper tree | *Curatella americana* |
| ukefu | angico | | *Piptadenia peregrina* (?) |

at breast height, and $a$ = the area of forest in acres. This formula, arrived at partly by guess work, and no doubt mathematically unorthodox and inelegant, nonetheless is probably not far from the mark. On the basis of it I have made the computations shown in Table 3. Whatever the actual number of trees in their forests, though, it would surprise me if the Kuikuru did not have a name for every one.

## KUIKURU USE OF TREES

The Kuikuru's familiarity with the trees in their habitat is by no means limited to being able to identify them all. They know a great deal more about them. Thus they can tell you of the *aku* tree that the tapir likes its seeds, of *iñúi* and *egeikajï* that monkeys eat their fruit, of *fiŋugi* that its fruit are relished by the toucan and the jacú. And they can go further and say that one kind of grasshopper called *fakuasa* likes the roots of the *takisi* tree, while another kind called *augïgo* eats the leaves of the *tifa*, which, they might add, are also favored by a leaf-cutter ant known as *kïakekueŋgï*. And so on.

TABLE 3
CONJECTURAL SPECIES/AREA CURVE
FOR TREES IN THE KUIKURU FORESTS

| Area of forest in acres | Estimated number of tree species |
|---|---|
| 1/6 | 45 (actual) |
| 1 | 64 |
| 2 | 74 |
| 5 | 91 |
| 10 | 106 |
| 25 | 132 |
| 50 | 157 |
| 100 | 189 |
| 250 | 240 |
| 500 | 291 |
| 640 (1 sq mi) | 313 |

But besides knowing their names and the animals that feed on them, the Kuikuru also make very extensive use of the trees of the forest. Millennia of close association with them have taught the Kuikuru much about the useful properties of the wood, bark, sap, resin, roots, limbs, leaves, fruits, seeds, etc., of many trees. In Table 1, after the name of each tree, I have indicated with code letters what kinds of use are made of it. This list, however, is far from complete. Except for the 75 species represented in the three test areas, I made no systematic effort to discover all the uses made of each tree. And even with those trees, uses sometimes turned up later that had not been given to me when I first inquired about them.[10] Moreover, the man from whom I learned the medicinal uses of the trees in the forest tract told me that several men in the village knew more about this subject than he did. Thus we can be sure that a substantial number of additional uses exist for these trees. Even so, the list gives a fair idea of the wide use the Kuikuru make of the forest.

The Kuikuru are well aware of the physical properties of the wood of each tree. If in no other way, they learn this partly through having to cut them all down in making clearings for their gardens. They know the hardness, weight, strength, ease of carving, straightness of grain, flexibility,

[10] When I asked an informant what use the Kuikuru made of the *tuṇufi* tree, he said none. But I was able to inform him that twenty years before the Kuikuru had told me that the digging stick was made of *tuṇufi*, thus providing me with the ethnologist's rare and delightful satisfaction of being able to tell a native informant something about his own culture.

resistance to decay, etc., of every tree, and take these properties into account in making artifacts from them.

For certain uses the Kuikuru may avail themselves of any one of several different kinds of wood. Thus, a good many trees are used for house construction, even though there is a tendency to favor certain ones for center posts, others for rafters, still others for wall posts, and the like. Many (but not all) trees make suitable firewood, and the Kuikuru consciously select these rather than picking up the first piece of dry wood they come across. Sometimes they girdle or even fell a tree known to make especially good firewood, like *iñúi* or *kụọ́*, so it will dry out and be ready to use when the time comes.

However, a number of artifacts are customarily made from only one kind of wood. For example, the sharpened stick employed during the ear-piercing ceremony known as *tipoño* (see Basso 1973:65-70) is always made from the wood of the *aṇa* tree (*Genipa americana*). This close connection between tree and use has resulted in the naming of certain artifacts after the tree whose wood, bark, resin, etc., is used for it. Thus *kofï* has given its name to the mortar, *tuṇufi* to the digging stick, *etuṇi* to the female pubic covering, *tafaku* to the bow, *kïntï* to one type of fish trap, *ta* to the carrying ring, and *katuga* to the rubber ball game.

The most specialized use of trees, though, involves medicines. For example, an infusion of the bark of the *kaïntï* tree is used to cure dryness of the eyes. The bark of *itsunítikuniñïgï* is said to reduce enlarged glands in the neck. And if a person is struck by lightning, he passes the leaves of the *taṇoṇo* tree over his body to protect himself from being struck again.

Some trees have very limited use. The only use made of the *teniñíuṇuï* tree, for example, is that its leaves are employed in wrapping cigarettes, while the only use mentioned for the *netufe* tree is that its roots are boiled and given to dogs in order to poison them.

On the other hand, some trees have multiple uses. The sandpaper tree, *Curatella americana*, called *ugagati* by the Kuikuru, gets its English name from the abrasiveness of its leaves. The Kuikuru use these leaves for sanding the teeth of their combs and for smoothing the inside of gourds.[11] *Ugagati* wood, which is hard and variegated in color, is favored for making stools. And various parts of this tree have medicinal uses as well.

The wood of the *kụọ́* tree was formerly used to make a spade for loosening the soil in preparation for planting manioc. *Kụọ́* was also used for the handle of the stone axe. It makes good firewood as well, and its bark produces a low flame over which arrow cane may be safely heated

[11] The leaves of this tree, which occurs throughout the drier parts of the American tropics, were once used by the silversmiths of Taxco to burnish their jewelry.

and straightened. Parts of the tree are used medicinally, and its leaves are favored by mothers to wipe their babies' bottoms.

*Tafaku*, pindaíba (probably *Xylopia amazonica*), is one of the commonest trees in the forest and, to the Kuikuru, one of the most versatile. Its name is the same as that of the bow, suggesting it was once the usual wood for making bows, although this is no longer the case.[12] Stripped of their bark, the straight, hard, corrugated trunks of young *tafaku* trees are favored for making the wall posts of a house. The slats laid across a large pot to rest the manioc strainer are sometimes carved of this wood. *Tafaku* is also the choicest firewood of all, not only because it burns with a hot flame and does not readily go out, but because it splits easily and evenly, yielding long, straight pieces that are ideal for making the large star-shaped fires the Kuikuru build under their cooking pots. Indeed, so valued is *tafaku* as firewood that a suitor places a big load of it against the house of his prospective mother-in-law in an attempt to gain her favor. The outer bark of this tree makes good lashings, and is thus useful for house building, and its roots, bark, and resin are all employed medicinally.

The *tifa* tree (possibly *Calophyllum brasiliense*) is one of the tallest in the forest and has a large, regular, cylindrical bole. Years ago, the Kuikuru hollowed out *tifa* trunks to make drums. When they first learned to make dugout canoes a few years ago, they selected *tifa* as the best wood for the purpose. From the inner bark of the tree an emetic is prepared. When the bark is removed, a sweetish liquid flows that may be drunk. Shamans use *tifa* resin to make a *foti*, a disc which is rubbed on the body of an apprentice shaman as part of his initiation.

The last tree whose uses I will list is *uagi*, the jatobá (*Hymenaea courbaril*). The bark of this tree is an inch thick, tough, and even. Before the Kuikuru learned to make dugouts, they removed large, semi-cylindrical slabs of jatobá bark and fashioned their canoes from them. Inside the seed pod of the jatobá is a dry brownish fluff which, although disdained by adults, is occasionally eated by children. Jatobá, sometimes called the South American copal, produces great quantities of resin. This resin drips off the tree in globules and lies on the ground around its base.[13] When it crystalizes,

---

[12] In modern times the Kuikuru have made their bows of *ipeji* wood rather than *tafaku*. They say that the best bows of all, though, are made of *majafi* wood, which is pau d'arco (*Tecoma violacea*), a tree that does not grow within present-day Kuikuru forests. Nevertheless, they obtain some *majafi* bows in trade from the Kamayurá, in whose habitat the tree does grow, and who are specialists in making bows from it (Oberg 1953:30-31).

[13] This resin does not decay in the ground and is preserved archaeologically. I found small nodules of it two feet deep in a test pit. Mors and Rizzini (1966:43) say of the commercial exploitation of jatobá resin: "The most sought-after variety is half-fossilized and must be dug out of the ground beneath the trees."

*uagi* resin sparkles, and the Kuikuru say the jaguar's eyes, which shine brilliantly at night, are made of it. *Uagi* resin is used both medicinally and cosmetically. Moreover, the *uagi* tree figures in tribal legend, and this brings us to the role that trees play in the supernatural beliefs of the Kuikuru.

## TREES AND KUIKURU BELIEFS ABOUT THE SUPERNATURAL

At one time, it seems, trees were people. One Kuikuru myth tells of a village populated by tree-people whose chief was Jukuku, a tree which bears bright yellow flowers in August. Jukuku's daughter married Atsiji, the bat, and bore him a son named Kuantïn̹ï. Kuantïn̹ï is one of the Kuikuru's principal culture heroes, and in the myth told about him, trees figure rather prominently. In very abbreviated form the myth runs as follows.

One day Kuantïn̹ï was in the forest when he suddenly found himself surrounded by a band of jaguars. The jaguars wanted to kill him, and to save his life he promised their chief his two beautiful daughters. When he returned home, though, he could not bring himself to send his daughters away. So he conceived the idea of carving women out of wood and sending them instead. He carved two female figures out of each of four different trees: *uen̹gïfi*, *fata*, *ikú*, and *epin̹gipe*. The two girls carved of *epin̹gipe* (a tree useful only as firewood) failed to come to life, but the other six did. Of these six girls only the two carved of *uen̹gïfi* survived the journey to the village of the jaguar chief and became his wives. The older of the two, Itsan̹itsegu, became pregnant by the jaguar and, although killed by her jaguar mother-in-law, gave birth posthumously to the Sun and the Moon. Itsan̹itsegu herself ascended into the sky where today she is chief of the village of the dead (Carneiro 1977:6).

Of the more than 200 trees known to the Kuikuru, only four are considered to be spirits (*etseke*). These are *uen̹gïfi*, *tali*, *tifa*, and *uagi*. It is not entirely clear to me if the *uagi* tree is a spirit in its own right, as *uengïfi*, *tali*, and *tifa* clearly are, or if it is simply that *tuguá*, the spirit of the whirlwind, resides in this tree.[14]

The *etseke* or spirit of the *uengïfi* tree is recognized by the Kuikuru as the chief of the forest. *Uen̹gïfi* is also the tree with the greatest social and ceremonial importance. Its trunk is used for the center posts and door posts of a chief's house. Moreover, when a chief dies and is buried, a low fence is built outlining his grave in the plaza, and the logs used for this fence are

[14] Some informants seemed to say that there was another tree, *tin̹on̹iñï*, "a kind of dark jatobá" (possibly the purple-heart tree, *Peltogyne confertiflora*), which was also a spirit. However, others indicated that '*tin̹on̹iñï*' was not a tree but rather another name for the spirit *tuguá*. Because of this uncertainty, I have not listed the name in Table 1.

of *ueṇgïfi*. Years after the death of a chief (or someone of the chiefly line), when a feast of the dead is held to commemorate him, a *ueṇgïfi* trunk is used for the memorial post.

In addition to the *etseke* of individual trees, the forest shelters a number of other spirits which are not themselves trees. One such spirit is *afasa*, always depicted with a gourd mask covering his face, who is said to be owner of the forest. Because so many *etseke* live in the forest, there is a certain element of danger there. The mere seeing of a spirit by anyone but a shaman is said to cause serious illness or even death. Thus, when the day following my leaf-counting expedition I became unaccountably ill, the Kuikuru were convinced that it was due to the baleful influence of the spirits of the forest.

Despite the Kuikuru's belief that dangerous *etseke* reside there, it would be incorrect to say that they fear the forest. They do not hesitate to walk through it, or to fell any tree growing in it, including the four thought to be spirits. Nor is there any propitiation of these spirits when their trees are cut down. The closest thing to apprehension that I learned of with regard to the felling of a tree (other than the physical danger of being hit by it) was the belief that after a man cuts down a *tuṇufi*, he should smear his infant's body with a potion made with its aromatic leaves, since the act of felling this tree is said to make the child ill.

## CONCLUSION

And there, having barely begun it, we must leave the study of the Kuikuru's adaptation to the forest. Only recently has the subject of a tribe's relationship to the trees that surround it begun to be painstakingly explored. It deserves to be pursued much further. The general importance of a rain forest environment to Amazonian Indians has long been recognized and accepted. Now we need to examine this relationship in minute detail in order to reveal the fine grain of the process at work. Not only will the result of such a study be significant in and of itself, it is bound to deepen our knowledge of all other aspects of Amazonian Indian life.

## REFERENCES

Basso, Ellen B.
1973 The Kalapalo Indians of Central Brazil. New York: Holt, Rinehart and Winston, Inc.

Carneiro, Robert L.
1977 The Afterworld of the Kuikuru Indians. In: Colloquia in Anthropology, edited by Ronald K. Wetherington, pp. 3-15. Taos, New Mexico: The Fort Burgwin Research Center Vol. 1.

Guppy, Nicholas
1958 Wai-Wai. New York: E. P. Dutton & Co., Inc.

Holdridge, Leslie R., et al.
1971 Forest Environments in Tropical Life Zones. Oxford: Pergamon Press.

Longman, Kenneth A., and J. Jeník
1974 Tropical Forest and Its Environment. London: Longman.

Mors, Walter B., and Carlos T. Rizzini
1966 Useful Plants of Brazil. San Francisco: Holden-Day Inc.

Oberg, Kalervo
1953 Indian Tribes of Northern Mato Grosso, Brazil. Smithsonian Institution Institute of Social Anthropology, Publication 15.

Otis, Charles Herbert
1926 Michigan Trees, a Handbook of the Native and Most Important Introduced Species. Ann Arbor: University of Michigan Botanical Garden and Arboretum.

Prance, Ghillean T., William A. Rodrigues, and Marlene F. da Silva
1976 Inventário florestal de um hectare de mata de terra firme km 30 da Estrada Manaus-Itacoatiara. Acta Amazônica 6:9-35.

Ratter, J. A., et al.
1973 Observations on the Vegetation of Northeastern Mato Grosso, I. The Woody Vegetation Types of the Xavantina-Cachimbo Expedition Area. Philosophical Transactions of the Royal Society of London, B. Biological Sciences 266:449-92.

Richards, Paul W.
1952 The Tropical Rain Forest; An Ecological Study. Cambridge: Cambridge University Press.

# THE USE OF MANIOC AMONG THE KUIKURU: SOME INTERPRETATIONS

*Gertrude E. Dole*
The American Museum of Natural History

Manioc is a "root" crop that provides staple starch for most of the people in tropical South America. Popularly called *mandioca, cassava* or *yuca* in Latin America, it has been represented as two species, "bitter" manioc, *Manihot utilissima* (called *mandioca* or *yuca brava*) and "sweet" manioc, *Manihot aypi* (*Mandioca mansa, palmata* or *dulce; aipim; macacheira;* or simply *yuca*). According to botanical analyses, however, all manioc cultivars are varieties of a single species, now called *Manihot esculenta* (Ferreira 1942; Rogers 1963; Rogers and Fleming 1973). All varieties or strains of this species contain some cyanogenic glucoside which produces a toxin when the tubers are exposed to air. The toxin is prussic, or hydrocyanic, acid (HCN), a substance that must be removed before the tubers can be eaten by human beings or most other mammals.[1]

However, the botanical evidence for a single species is contrary to native interpretation among manioc users, who regularly distinguish relatively harmless strains from those that contain large enough amounts of poison to require special treatment. Therefore, since the data in this paper are ethnographic, I shall follow the native practice and distinguish harmless ("sweet") strains from the more toxic ("bitter" or poisonous) varieties.

Literature on the use of manioc contains a number of general assumptions that appear to be based on the use of bitter manioc, probably because

[1] Inadequate removal of HCN and failure to neutralize residual poison with sufficient protein in the diet apparently has resulted in general illness among some African groups who have adopted this crop as a staple (Spath n.d.).

that type was first observed by Europeans and has been most thoroughly described. It is sometimes assumed, for example, that wherever manioc is used it is always processed by draining the pulp in a sleeve press (*cibucan* or *tipití*), and further that the purpose of draining it is to get rid of the poison (Oviedo 1851:271; Wagley 1964:66; Reichel-Dolmatoff 1971:12; Rogers n.d.:8). Other common assumptions are that manioc is regularly used in the form of flour (variously called *farinha, fariña, farina, harina,* or *farine*), and that only the poisonous strains are used in making flour.

Although these generalizations may represent the most common practices, accumulated data on the variety of manioc use in tropical South America now warrant a critical examination of the validity of these assumptions. The Carib-speaking Kuikuru and other peoples in the Upper Xingú region of central Brazil cultivate only bitter manioc. However, the techniques used to process the tubers in that region differ considerably from those used elsewhere in Amazonia or the Antilles. I will describe the techniques used by the Kuikuru and contrast them with practices elsewhere. In this way I hope to put the Kuikuru data in perspective and at the same time evaluate some of the currently accepted generalizations about the use of manioc.

## PROCESSING MANIOC AMONG THE KUIKURU

During the dry season Kuikuru women harvest and process great quantities of bitter manioc. Although $CN^-$ occurs throughout the starchy pith of bitter manioc tubers, much of it is concentrated in the shallow dermal layers, including the cortex directly under the skin proper (Schery 1952:428-29; Spath 1973:53). Hence processing begins with removing a generous layer of skin. First the manioc tubers (called *kuigi* by the Kuikuru) are washed, but not soaked for any length of time as is done by many native peoples and rural Brazilians. The washed tubers are scraped with freshwater bivalve shells (*efete*), or currently with sharp metal instruments usually fashioned from cans. In this way not only the outer skin is removed but also the cortex, and with them a large part of the toxin.

After being peeled, the tubers are reduced to a pulp by grating them on a wooden board (*iñagí*) about two feet long, with slightly concave sides, and about six inches wide at the ends. In making the grater a hand-carved board is first grooved in diagonal lines with dogfish (*Raphidon vulpinus*) or piranha (*Serrasalmus* sp.) teeth. Short spines from a prop root of a *paxiuba* palm (*Iriarta exorhiza*) or trimmed spines of a *buriti* palm (*Mauritia flexuosa*) are set in the grooves over a rectangular area of approximately 24 square inches midway between the ends of the board (Fig. 1).

To grate manioc, the board is laid across the very sturdy everted rim of a wide flat-bottomed pot (Fig. 2) obtained in trade from the neighboring

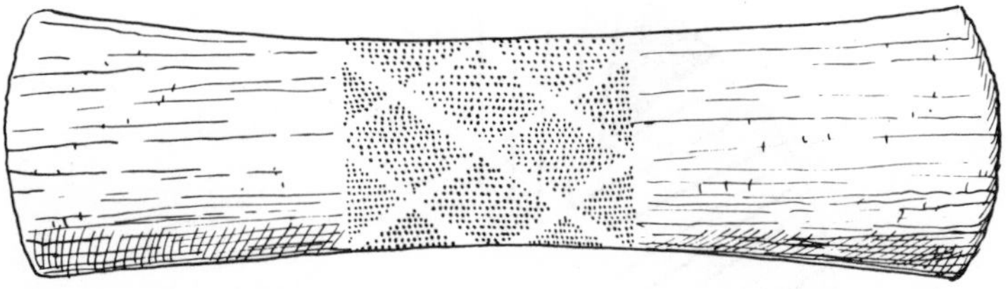

Fig. 1. Manioc grater board studded with dogfish teeth. (The scale of this drawing is ca. 1:6).

Waurá.[2] Seating herself on a mat before such a pot, a Kuikuru woman often braces one end of the grating board against a post and the other against her abdomen. Holding a tuber upright with both hands she rubs it vigorously back and forth on the palm spines to shred it to a watery pulp, which falls or is brushed into the pot below. This grated pulp corresponds to the mush obtained among some other groups by soaking tubers in water for three or four days and then mashing or pounding the semi-fermented mass.

The next step in the process serves three functions: (1) it removes much of the residual poison from the pulp; (2) starch also is separated from the pulp; and (3) the pulp is partially dried. The most widely known device for accomplishing these ends is the *tipití*, or sleeve press. However, the Kuikuru do not have the *tipití* and accomplish the same ends by using a very simple device consisting of a twined "roll-up" mat (Fig. 3). Roll-up mats rather than *tipití* are used for straining manioc by all Upper Xingú peoples. They are distinctive of a restricted region in the center of the South American continent, principally the headwaters of southern tributaries of the Amazon.[3]

---

[2] All pots used by contemporary Kuikuru are obtained from the Waurá, an Arawak-speaking group who are now the only group in the Upper Xingú region who make pots, except that the Arawakan Mehinacu make some for their use only. The Kuikuru, who speak a Carib language, formerly made their own pots in the same general shape as contemporary Waurá pots but have abandoned potting (see Dole 1961-1962). The style of pots made for processing manioc is very well adapted to their function, being not only flat-bottomed but also shallow and very wide, somewhat like washtubs (see Fig. 2).

[3] Twined mats for straining manioc have been reported among the Carajá on the island of Bananal in the Araguaya River (Krause 1911), the Bororo (Frič and Radin 1906) and Paressí-Kabishí (M. Schmidt 1914) in Mato Grosso, the Bauré, Huanyam and Moré in the Guaporé region (Nordenskiöld 1924; Rydén 1942), and the Movima in the Mojos (Denevan 1963).

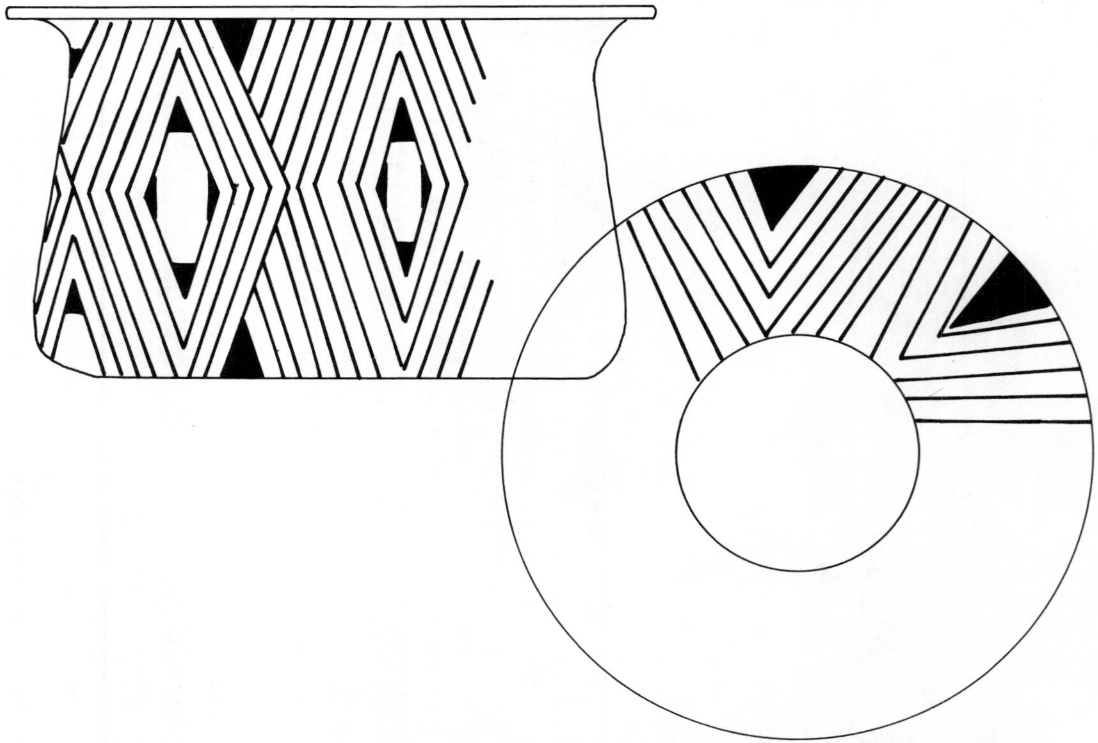

Fig. 2. Schematic drawing representing the shape of cooking pots obtained from the Waurá and designs on the body and bottom of some of those pots. (Scale of this drawing is ca. 1:4.)

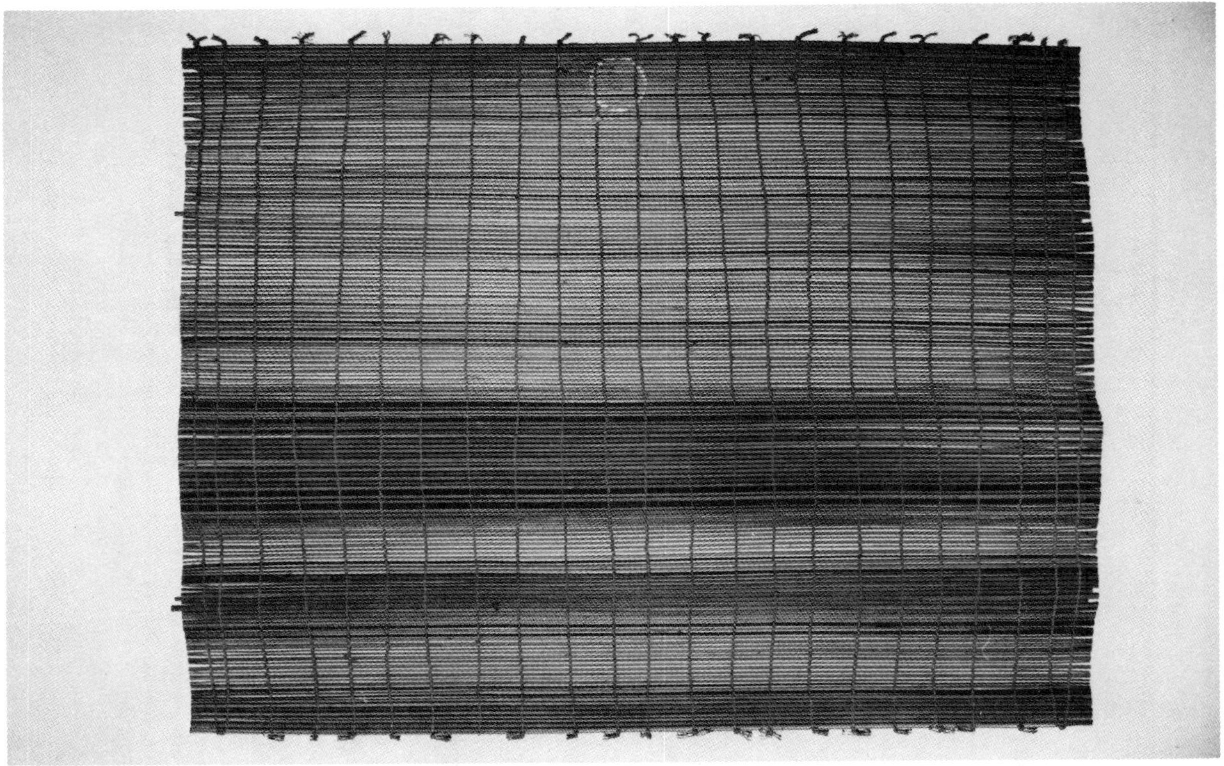

Fig. 3. *Tuafi*, or roll-up mat used in straining manioc. (Scale of this photo is ca. 1:4.)

The Kuikuru straining mat, called *tuafi*[4] is approximately 16 inches wide and about two feet long. It is made of stiff midribs of *buriti* palm leaflets tightly twined together at about one-inch intervals with strong, durable cord made of fibers shredded from leaves of wild pineapple, *kafaga* or *tafaga* (*Ananas sativus*). The *tuafi* is rigid along the axis of the midrib elements but is very flexible along the axis of the twined elements, like a "Venetian blind," thus it can be rolled or folded on itself. The narrow spaces between the midribs allow only liquids or fine solid particles such as flour to pass through.

In many Amazonian societies manioc pulp is pressed without being rinsed. In the Upper Xingú, however, the pulp is well rinsed with water. For this purpose an enormous flat-bottomed pot is used. This type of cauldron, sometimes measuring over two and one-half feet in diameter, is especially prized and ordinarily is not put over a fire. As with common pots, the interior surface is sealed with several coats of rubbed soot from charred tree bark. The exterior is slipped with an emulsion of *urucú* (*Bixa orellana*) and red clay, and later painted red with a mixture of *urucu* (*umïŋi* in Kuikuru) and oil of the *piquí* fruit (*imbene, Caryocar butyrosum*), forming a background on which designs in red, black and white are painted, especially on the bottom (see Lima 1950; Dole 1961-1962). The designs on the bottoms of these cauldrons suggest the special function to which the pots are put, as will be made clear later. They often consist of a circle cut into rounded arcs (Fig. 4).

To rinse the grated manioc wooden bars (*apo*) carved out of *pindaiba* wood (*Xilopia* sp.) are placed across the mouth of a cauldron and a *tuafi* is spread over them (Fig. 5). A mass of wet pulp is placed on the *tuafi* and water is poured over the mass with a dipper made of half a bottle gourd, *kuginta* (*Lagenaria*, Fig. 6a). The mass is pressed with the hands to remove much of the water and manioc juice. The pulp, which is too coarse to pass through the mat, is then spread out, and the mat is neatly folded on itself in three or four folds in accordion fashion. The worker presses the folded mat against the bar supports and also wrings it with her hands until as much

---

The Guayakí in Paraguay use them for sifting palm pith flour (Métraux and Baldus 1946:436); and among the Chorotí and Ashluslay in the Chaco small twined mats are used for straining honey (O'Neal 1949:75). Similar mats were formerly used among the Patamona and Makushi in the Guianas but have fallen into disuse (Roth 1924:310). Among the Desana in the Vaupés region of Colombia a very long roll-up mat was used to cover a canoe full of manioc beer (Koch-Grümberg 1967, Vol. 2:223-24).

[4]The same term *(tuafi)* is used also by the Trumaí, and the Tupian Camayurá *(tuaví)* as well as by other Carib-speaking Xinguanos. A similar term *ătoă* is used by Arawakan Paressí-Kabishí for similar mats (M. Schmidt 1914:206), which suggests the possibility that the utensil and the term may have been adopted by Xinguanos from Arawak-speaking peoples to the west.

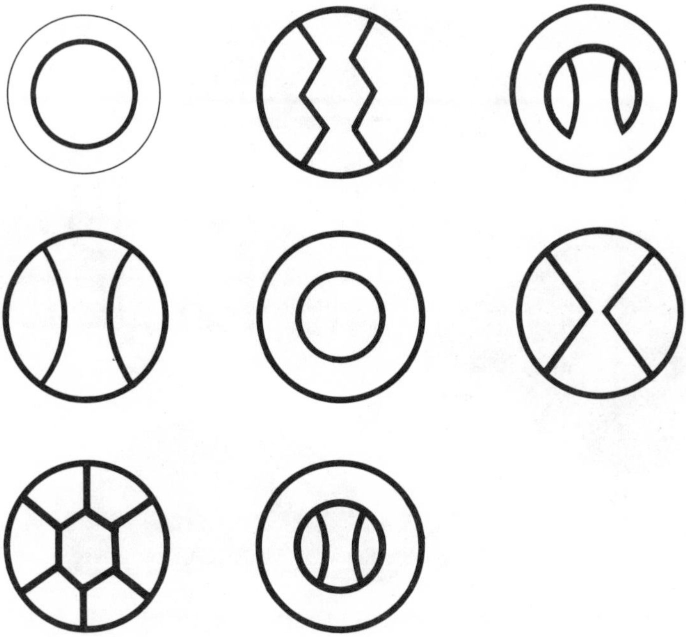

Fig. 4. Designs on the bottoms of some large Kuikuru pots.

Fig. 5. Upper Xingú woman rinsing manioc pulp. Photo: Serviço de Proteção aos Índios.

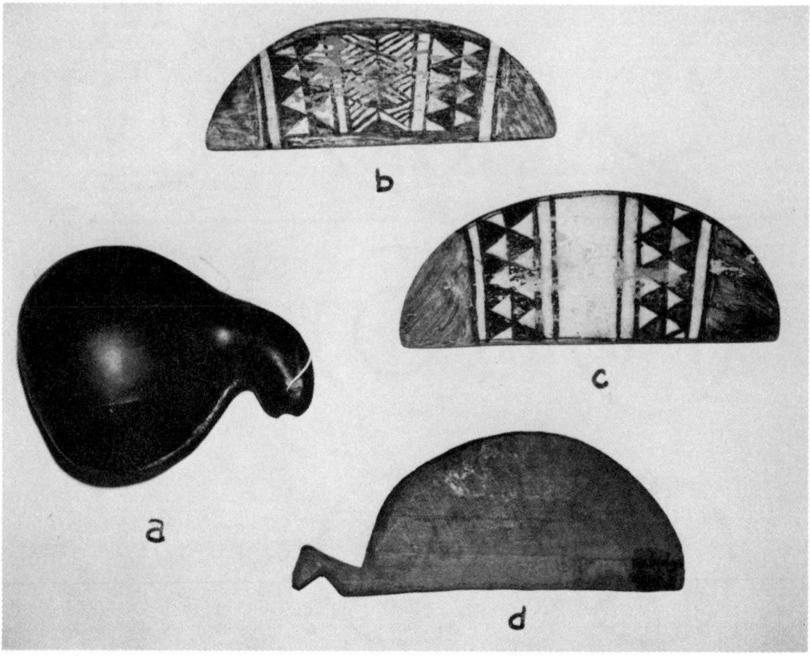


Fig. 6. Utensils used in processing manioc: gourd dipper: a, 6 1/4 in.; spatulas: b, 7 1/2 in.; c, 8 in; d, 8 1/8 in.

liquid as possible has been squeezed out. This step is repeated, using both expressed liquid and more fresh water. The liquid thus pressed out contains in suspension starch that has been freed by breaking up the cell structure in the grating process (see Schery 1952:429).

Contrary to the practice of some other Amazonian groups, all three products–the leached pulp, the liquid, and the starch rinsed out with the liquid–are used as food by the Kuikuru. The pulp is made into flour and used in flatbread, beverages and pastes or sauces mixed with other foods; the liquid is boiled and consumed as a soup; and the fine starch is used in making soup and flatbread.

In preparing these foods a number of alternate steps may be taken. These steps are represented in a flow chart (Fig. 7) and the accompanying pictorial diagram (Fig. 8).

Although freshly pressed pulp (*timbuku*) may be used in small amounts to prepare daily meals of soups or flatbreads, most of it is dried and stored for future use. If a woman wants the pulp to dry rapidly, she spreads it thinly on whatever mats are available. These may be circular trays (*kaŋagitafu*,

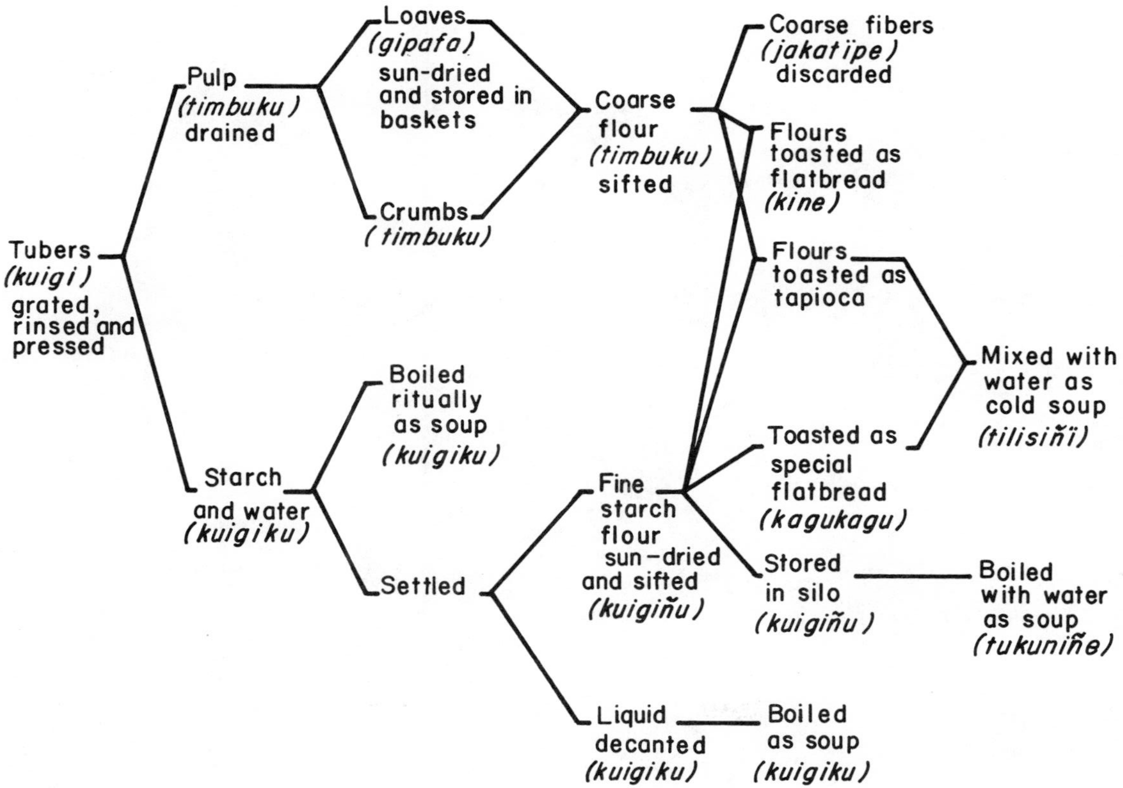


Fig. 7. Alternate steps in the preparation of manioc among the Kuikuru.

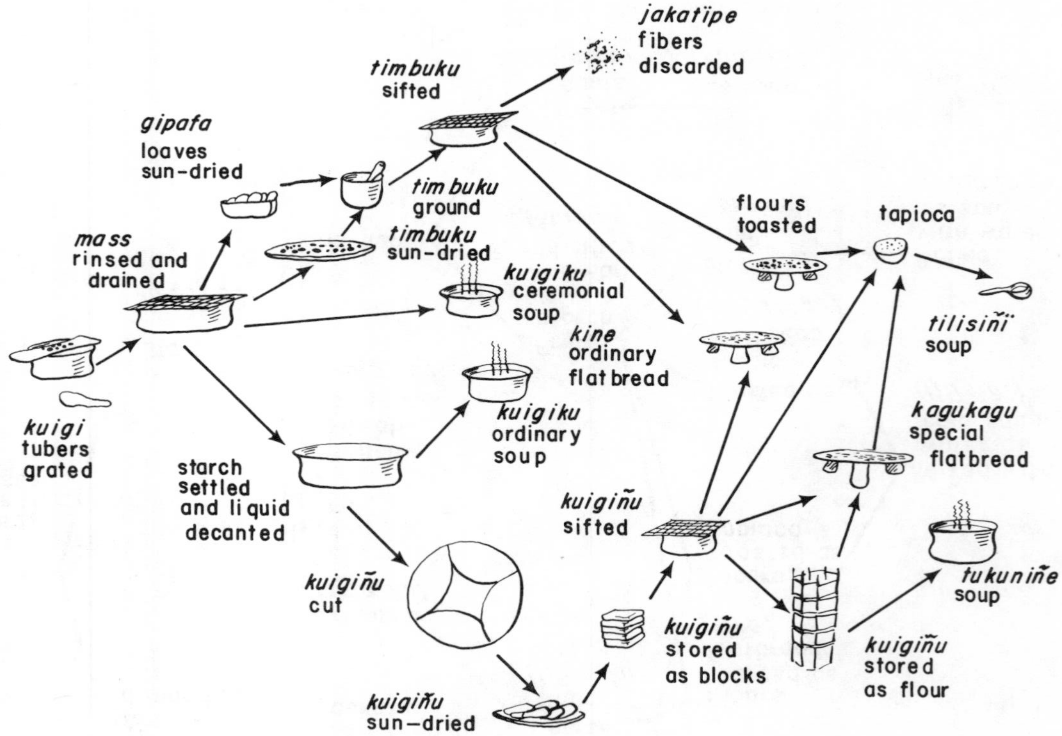


Fig. 8. Pictorial representation of steps in the preparation of manioc among the Kuikuru.

Fig. 9) made of twilled leaflets of a small palmate palm (*kanaga*) similar to *buritı*, or they may be large rectangular mats (*fugutaŋa*, Fig. 10) constructed of long *kaŋaga* palm petioles. If the pulp is to be stored, however, it is pressed into loaves (*gipafa*). When all the juice has been squeezed out, the folded *tuafi* is deftly opened up and the three or four strips of rather dry pulp that were pressed in the folds are piled on one another, making one long bar. Either end of this bar of pulp is then folded inward or broken off and piled onto the middle segment to form a solid ovoid ball about four to five inches by seven or eight inches. These loaves are sun-dried on a tall platform erected for the purpose outside the house, on a mat, or in a woman's rectangular carrying basket (*tatofoŋo*, Fig. 5) made of twilled palm leaflets.

It may be noted that only in a pronounced dry season such as occurs in the Upper Xingú region could thick solid loaves of moist meal be thoroughly dried outdoors. Here virtually no rain falls during three months (June, July and August), and very little during May and September, when dry, large numbers of *gipafa* loaves are stored in men's very large leaf-lined carrying baskets (*tsaŋgo*) inside the house to be used as needed through the rainy season. Since the pulp has been thoroughly rinsed, it contains little sugar and can therefore be stored for months without absorbing moisture from the air.

The liquid that is pressed out of the grated pulp may be boiled immediately and consumed as a festive meal. When heated, the starch granules are broken down and become partially soluble, thickening the liquid to make a glutinous translucent soup (*kuigiku*, "manioc juice") that resembles thin tapioca pudding. When this festive beverage is made, a man is designated as a formal taster, thus publicly assuming responsibility on behalf of the host for protecting the guests from harmful effects of the toxin. He sips the soup after it has been boiled for some time and always pronounces it to be unsafe to drink because of residual poison. It is then boiled some more, and again pronounced unsafe. Only after being boiled and tested a third time does the official taster declare it safe to drink. When the HCN has finally been volatilized by boiling, the *kuigiku* is taken to the plaza in large half-gourd bowls and passed among all the guests, each one sipping from the bowl and passing it along to another.

Ritual testing of the beverages demonstrates in dramatic manner native awareness of the danger of ingesting the juice of bitter manioc before the toxin is removed. As further evidence of this awareness, the Kuikuru say that manioc is sometimes used to intentionally kill people. In fact there are numerous accounts of the lethal effect of consuming unprocessed bitter manioc. In the Antilles, for example, raw manioc was reportedly used by many natives who were "driven to despair by the cruelty of the Spaniards"

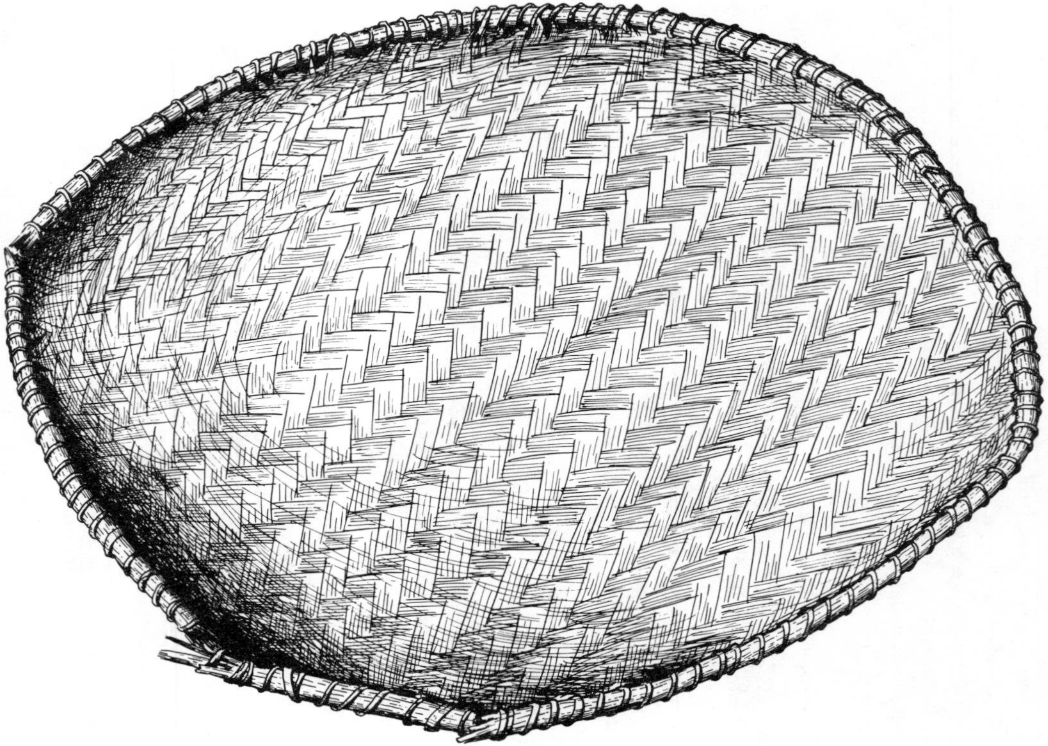

Fig. 9. Basketry tray used for drying manioc. (25 1/2 in. round.)

Fig. 10. Manioc drying on mats and trays.

and wished to die rather than submit further (Roumain 1942:34; see also Oviedo 1851). In the Guianas women committed suicide in the same way because of unhappy "affairs of the heart" (Roth 1924:560).

Even the vapor that escapes during the cooking of bitter manioc is noxious. Among the Urubú, who prepare manioc flour for trade, it is felt that the process of toasting it is dangerous, and they say that "the person who roasts it will get such a knot in his anus that he will never be able to defecate" (Huxley 1956:155). And the Mehinaku of the Upper Xingú region believe that processing manioc outdoors during the dry season causes the days to be short by releasing quantities of poisonous fumes that "give the sun a headache and he speeds across the sky to get home quickly" (Gregor 1977:39).

However, in spite of the poison in bitter manioc juice, ritual testing occurs only on special occasions among the Kuikuru. Ordinarily the rinsing water is left undisturbed for several hours in the special large flat pots to let the starch settle. The liquid is then decanted or dipped off and the sediment is pressed in the bottom of the pot to release the last bit of juice, which is poured off. The decanted liquid is boiled and consumed as a soup (*kuigiku*) without any ritual testing.

Although draining does not get rid of the poison, it does remove from the pulp a substance that is regarded as very important by the Kuikuru, namely, starch. As they settle, the grains of starch (*kuigiñu*) cling to one another to form a firm hard layer, much like moistened cornstarch or the sediment in raw potato juice. The layer of starch may be one and one-half to two and one-half inches thick, depending on the quality and quantity of tubers that are leached. Being solid, this layer can be cut and removed in blocks. A woman marks off and cuts four rounded arcs with a shell scraper or wooden spatula (*kutigu*, Fig. 6b-d). The pattern thus produced in the starch resembles designs painted on the bottom of some of the large pots mentioned above (Fig. 4). The five cakes of *kuigiñu* are lifted with a spatula and removed to be dried in the sun and later used as flour. If rain threatens before the blocks of *kuigiñu* are completely dry, they are stacked inside the house. In the semi-darkness of the house a bright orange fungus coating sometimes develops on the damp raw *kuigiñu*. If this occurs, the fungus is scraped off and discarded before using the starch.

When thoroughly dry, cakes of *kuigiñu* are crumbled and rubbed through a twined sifting mat (*anagi*) that is usually a little larger and coarser than the straining *tuafi*. Sifting mats are made of somewhat larger palm midribs and are twined together more loosely. For this purpose cord made of the cortical fiber of *buriti* palm leaves (*tate*) or of native cotton (*togolkige*) is used rather than the more durable fiber of pineapple.

The starch flour is a fine powder. It is stored inside the house in leaf-lined wicker silos (*kuigiñúofefijï*) about 18 inches or more in diameter and five or more feet high. This fine flour is used for beverages and bread. Foremost in the daily fare is a thick soup (*tukuniñe*) made by lightly cooking starch flour in water. One woman made hers according the the following recipe. About one and one-half quarts of *kuigiñu* was combined with four or five quarts of water, and mixed with her hand. This mixture was then poured into about three gallons of boiling water and stirred as it thickened with one of the wooden supports (*apo*) used in the straining and sifting processes. The resulting gruel was further diluted and at the same time cooled by dipping about three-fourths of a cup of it into about one and one-half pints of cool water and stirring it with a gourd dipper that also served as a drinking cup (Fig. 11). The Kuikuru never drink plain water but instead drink soup made of manioc (or maize or *piquí* in season) many times during the day.

Another daily staple is flatbread (*kine*), usually made of coarse and fine flours combined. Coarse flour is made by pulverizing *timbuku* (pulp) in an upright hollow-log mortar (*kofï*) set in the dirt floor of a house. Two women usually cooperate to pound the pulp with very rapid alternating strokes of two-foot wooden pestles (*kutofo*) until it is pulverized. The resulting flour is rubbed through a sifting mat. Whatever cellulose fibers (*jakatïpe*) cannot be pushed through the mat are discarded.

Coarse flour made in this way is combined with fine flour to make flatbread. First, a small amount of water is mixed with *kuigiñu*, which is then

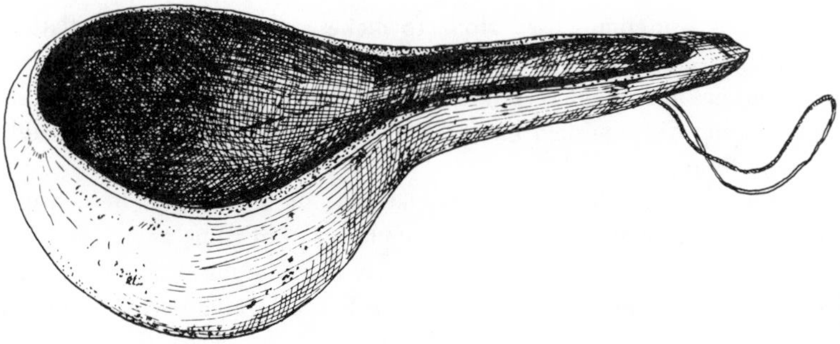

Fig. 11. Gourd dipper or cup. (6 in. long.)

kneaded and sifted. The slightly moist *kuigiñu* is mixed well with the coarse flour, the proportions varying according to individual taste, the available supply of each type of flour, and the purpose for which the bread is made. For ordinary bread a large proportion of *timbuku* is used by most women because manioc yields more of this coarse flour than the fine type. If toasted only lightly, a higher proportion of *kuigiñu* makes the bread softer and gives it a better flavor.

Three or four pints of mixed coarse and fine flour is spread over the surface of a heated ceramic griddle (*alato*), which among the contemporary Kuikuru usually consists of the bottom of a broken cooking pot. The rim of the griddle is raised only slightly. The Kuikuru have neither enormous griddles with high rims nor "ovens" (ceramic collars around the firebed), such as are characteristic of the northwest Amazon region. Instead of ovens, the Kuikuru use a set of three ceramic potrests to support a griddle.

A little water is sprinkled on the flour, and it is pressed down evenly with the hands and a spatula to form a layer of one-quarter inch or less. The bread is toasted on one side for about a minute and a half. Then a previously toasted bread is placed over it, toasted side up, for about two minutes and removed. After about a minute more the freshly toasted bread is lifted from the griddle with the spatula. Two of these flatbreads are folded together, untoasted surfaces facing inward. This bread is not dried in the sun to be preserved. It is softer and thinner than the cassava bread made by peoples in the Guiana and Orinoco regions.

The Kuikuru do not store their bread, but eat it within a day or two. It is far more palatable immediately after being toasted than a day or two later. Flatbread is torn into pieces and eaten as open sandwiches with such foods as fish or *piquí* paste, toasted grasshoppers, large *saíba* (leaf-cutting) ants, and chili peppers (*fomi*, *Capsicum* sp.), at three more-or-less regularly spaced meals.

*Kuigiñu* is sometimes used alone to make a choice, very thin flatbread (*kagukagu*).[5] Being pure starch, *kuigiñu* is more highly valued than *timbuku*, which contains many finely broken fibers. Hence on special occasions such as entertaining distinguished guests, the starch flour is used without adding any coarse flour.

Flour is also used to make a special type of beverage to be consumed on trips away from the village. For this purpose both fine and coarse flours are mixed together and a little water is added. The mixture is heated on a griddle briefly, but not long enough to harden the starch particles or change its color. Starch granules are altered by heat to form pearly grains known as tapioca, a word derived from a Tupian term, *typyáca* (cf. Camayurá *tibuák*, "starch flour," Oberg 1953:22). As already mentioned, these grains are partially soluble and when added to water they quickly thicken it to a jelly-like consistency. If allowed to stand for several hours the tapioca swells and the mixture becomes quite firm. When traveling, a handful of this moist tapioca is stirred into a gourd cup of water to make a cold soup (*tilisiñï*). A variant of this cold soup may also be made by soaking *kagukagu*, thin bread made of *kuigiñu* alone. Prepared in these ways manioc is a true "convenience food" which people carry with them when they leave the settlement for more than a few hours.

Manioc flour is used also to make a thick paste with boiled fish or to make a sauce with very hot peppers that are cultivated by the Kuikuru. These "dishes" are often made by men and are spread on pieces of flatbread. They are reminiscent of the pepperpot of the northern regions of Amazonia and the Antilles but differ in significant ways: they are made in small amounts as delicacies and are eaten immediately rather than being preserved by reheating day after day; nor are they used to flavor or preserve meat.

## COMPARISON AND DISCUSSION

From the foregoing description it is clear that the preparation of manioc among the Kuikuru differs in several respects from usage among many other

[5] From an ethnolinguistic point of view it is interesting that an especially fine bread made of starch flour alone was called *xauxau* (or *jaujau*) by the Arawakan Taino in the Antilles (Roumain 1942:33).

Amazonian peoples. The differences include:

not soaking tubers to soften or ferment them
grating rather than mashing or pounding
rinsing the pulp
use of a straining mat rather than the *tipití*
drying and storing pulp in loaves
use of starch as flour
use of rinsing liquid for soup
low-rimmed griddles
ceramic potrests rather than ovens
flatbread not sun dried to preserve it
starch-flour flatbread
tapioca soup for travel
lack of pepperpot

Thus some of the traits generally assumed to characterize use of manioc throughout Amazonia do not apply to the Upper Xingú regions: the *tipití* is not used here, nor is farinha the main product, since the peoples in this region use a variety of other products, including tapioca, but prepare very little of their manioc as farinha. Furthermore the assumption that manioc is drained in order to get rid of the poison is not strongly supported by the Kuikuru data, since draining does not of itself dispose of the prussic acid. To be usable, the starch and juice both must still be cooked, as must the pulp also. I will return to this problem later.

The generalization that only bitter manioc is used for flour is not challenged because the Kuikuru use only bitter varieties. To test this assumption, more comparative data are needed. Moreover, the many contrasts between the use of manioc in the Upper Xingú and techniques used elsewhere pose additional questions: Why, for example, do some peoples crush the tubers by grating while others soak and pound them? What is accomplished by draining off the juice? And further, what is the purpose of rinsing the pulp? Why are very poisonous varieties used by some rather than the less toxic ones? And why do some societies ferment manioc and others not?

With respect to the last of these questions first, fermentation of sweet manioc is very widespread. The first written accounts of manioc relate that fermented beverages were made of sweet varieties (see, e.g., Magalhães de Gandavo 1922:169). To understand why sweet varieties were used for this purpose it is necessary to consider the details of preparing sweet manioc. To begin with, some very pragmatic considerations suggest that sweet manioc was originally preferred and preceded the use of bitter varieties as food among some Amazonian peoples. One factor is convenience. Techniques used in processing non-toxic varieties are usually very simple. Because the

HCN in sweet varieties is mostly confined to the dermal layers, peeling them may remove virtually all the toxin. Indeed, the level of toxin in some varieties is so low that the tubers were eaten raw among the Tunebo of eastern Colombia and the Aimoré of eastern Brazil (Ghisletti n.d.:66; Cardim 1925; 198-99).

Another consideration is taste. It is significant that the term "sweet" is commonly used to refer to the relatively non-toxic varieties. The flavor of cooked sweet manioc is often compared to roasted chestnuts, and early European colonists preferred it even to that of fresh wheat bread, which in sixteenth-century Europe was not sweetened with added sugar as it is today (Magalhães de Gandavo 1922:45). By contrast, according to Franz Caspar bitter manioc does not taste good either boiled or roasted without further preparation (personal communication, Dec. 13, 1961). The difference in taste of bitter and sweet manioc is undoubtedly caused by the difference in amounts of both bitter prussic acid and natural sugar. Hence in addition to being less toxic, "sweet" manioc may have been preferred in part because it tasted better, given simple means of preparation.

To return to the question of fermentation, the factors of taste and ease of preparation may have a bearing here, too. The process of fermenting sweet substances is sometimes very simple. Foods are fermented in societies of all levels, very often by accident, and a taste for fermented foods very likely often develops in response to accidental fermentation, especially in warm climates.

Sweet manioc is frequently soaked in water to loosen the skin. When soaked for a few days, the skin can be rubbed off easily. Soaking in still water causes the natural sugars to produce alcohol through "microbial conversion" (see Rogers n.d.:9). Thus varieties that have the most natural sugar ferment most readily. This fact together with the ease of preparation by soaking may explain why sweet varieties are often made into fermented beverages, variously called *cauím*, *cas(s)irí*, *caxirí*, or *masato*. In fact, ethnohistorical data suggest that fermentation of sweet manioc preceded the making of flour among several groups. The Tukuna, for example, formerly only fermented sweet manioc and seem to have learned to make flour from Neobrazilians recently (Nimuendajú 1952:14, 18, 21). Likewise, among the Tucanoan, Desana beer was formerly the chief product of manioc, and flour has been made in quantities only in recent time (Reichel-Dolmatoff 1971: 13-14). In the Peruvian Montaña several Panoan groups had only sweet varieties aboriginally. They did not grate the tubers or make flour until recently but boiled them as a pot vegetable and made a mildly alcoholic masato. Among the Amahuaca, for example, the wife of a Peruvian lumber patron introduced the technique of making flour a few years ago. The Sharanahua make flour for themselves only occasionally, but one young man has decided to make flour and sell it to Peruvians (Siskind 1968; cf. also Kensinger et al. 1975:50, for similar data for the Cashinahua).

Fermentation is not restricted everywhere to sweet manioc. The same device is used to process bitter varieties among diverse and unrelated peoples from the Antilles to headwaters of the Amazon and from Colombia to coastal Brazil. Fermentation has been most highly developed in the northern and western parts of tropical South America, where periodic drinking parties are most widely celebrated. On the Amazon, for instance, in the sixteenth and seventeenth centuries the Tupian Omagua harvested tubers and then reburied them in huge leaf-lined pits during the annual floods. Although the tubers "rotted" under water level, the black mass was dug up later and used, reportedly without having lost any of its value (Fritz 1922:50; de Acuña 1859:66). Essentially the same techniques are still used today by Tukuna in an area formerly occupied by Omagua (Bolian n.d.).

It has sometimes been reported that the purpose of fermenting, like draining, is to extract the poison. Among the Witoto of northwestern Amazonia and the Muñoques in Venezuela, for example, manioc was fermented in part for the expressed purpose of removing poison (Whiffen 1915:131; Alvarado 1945:69; cf. Lecointe 1922:335). In this connection a botanist has recently suggested that in the fermenting process microorganisms may convert the nitrogen of HCN into protein for their own growth and thereby "help to eliminate the poisonous properties" (Rogers n.d.:8-9).

In spite of the evidence for a causal relation between fermentation and detoxification, fermentation alone apparently does not destroy all the poison in every case. The Amanayé, for example, drain their fermented manioc pulp and throw away the liquid to prevent domestic animals from poisoning themselves (Lange 1914:224-26). On the other hand, fermentation is unnecessary for the removal of HCN. The poison is often volatilized by the heat of cooking *before* fermentation begins. Because of the extreme volatility and solubility of this substance it can be eliminated by heat or by thorough rinsing, especially if the cell structure is broken down to free the juice (Lecointe 1922:332-33). Many peoples do in fact render bitter manioc edible by soaking (Schultz 1959:113), boiling, toasting, sun-drying, drying before the fire (Southey 1822:243; Oberg 1953:22), or a combination of these techniques. The Cubeo use ashes to absorb and remove poison from the starch (Goldman 1963:62).

Referring to the Island Carib treatment of bitter manioc one writer has said, "we must marvel that a plant which is poisonous to eat should have been made edible by a process so complex" as that used to detoxify manioc (Taylor 1972:5). Far from being a marvel, it should not be surprising that Amazonian peoples have removed toxin from manioc by these various means, since there are numerous reports of other bitter or poisonous foods having been treated in the same ways. Removal of toxins by leaching or boiling or both is widely practiced by gatherers as well as horticulturists and thus is not a new invention by users of manioc.

Since fermentation may not be sufficient to remove all the HCN, and since other techniques are effective in volatilizing it, one wonders why so many groups regularly ferment bitter manioc. Research may someday yield a satisfactory answer, but in the meantime a few advantages are apparent. First of all, fermentation adds flavor to a very bland food. Moreover, in changing the starch to sugar and the sugar to alcohol, fermentation produces the means of inebriation, the social lubricant.

A related and more practical benefit is the preservation of manioc itself. To provide for times of need, the Nambiquara bury tubers and dig them up "half rotten, weeks or months later" (Lévi-Strauss 1964:265). It is said that those buried by the Omagua on the Amazon could be preserved for as long as two years (de Acuña 1859:66; cf. Staden 1928:138 for a similar report on the use of "rotting" manioc among the early coastal Tupinamba). In fact, fermentation is quite effective in preserving manioc (cf. Ehrenreich 1948: 41). According to Martín Matos Arvelo, the Muñoques fermented manioc to preserve it up to two and even three years (Alvarado 1945:69).

Let us turn now to the question of why bitter manioc is preferred by some societies. As I have indicated, there was a general knowledge of techniques for removing poisons from starchy foods even before the domestication of manioc. This general knowledge made it possible to use bitter manioc, but it can not explain a preference for that type over sweet varieties. To solve this problem it is necessary to compare the use of bitter and sweet manioc.

Where toxic varieties are used for fermented beverages, more elaborate steps are generally taken to process them. After a generous layer of skin is removed, the tubers are always crushed and cooked, which of course destroys natural enzymes that might change starch to sugar. Hence, to encourage fermentation of the cooked manioc, masticated material is often added, introducing salivary enzymes that initiate the transformation of starch into sugar. Alternately, sweet material such as cane syrup or manioc that has been charred to break down the starch may be added. Sometimes natural yeasts and molds also are supplied from leaves or small amounts of fermented foods.

The usual way of crushing bitter manioc tubers is by grating them on a hard surface, a rough stone, a spiny section of *paxiuba* palm root, or most frequently a fabricated grating board. Grating breaks up the fibrous structure and thereby facilitates the elimination of poison by freeing the juice, which contains most of the residual poison after the dermal layers have been removed. Grating manioc, in the words of Yolanda and Robert Murphy, "is the nastiest job of all, for it gives backaches, grated fingers and knuckles, and it is hard work besides" (1974:125). Alternately, some groups use the soaking technique already described for sweet manioc. This method is sometimes preferred, even at the expense of some caloric value (see Murphy 1960:65).

As with fermenting and draining, it has sometimes been assumed that the primary purpose of grating is to remove the poison. It is true that grating helps to release toxin. However, non-poisonous varieties as well as poisonous ones are sometimes grated, as among the Pauserna (Métraux 1942:101). Since this step is not necessary for removing poison from sweet varieties, processing them in this way suggests that there is some other impelling reason, as suggested by Simpson (1940:409), to spend the added time and energy required for grating.

If we look again at the extensive archaeological and ethnographic data on the processing of other types of roots as well as grains, seeds and palm pith, it becomes clear that the ultimate objective of grinding and grating is to obtain flour. (In the region of the Paraná delta, people even ground fish to make a powder similar to pemmican that could be kept for some time [Lothrop 1946:182].)

In this connection, a major problem in tracing the use of manioc in tropical South America has been the relation of griddles to manioc flour. Some prehistorians have inferred the cultivation of manioc from the archaeological presence of griddles (Cruxent and Rouse 1958:244). This inference is difficult to substantiate since griddles are used in tropical America and elsewhere to toast maize as well as flours of seeds and various kinds of stems and roots other than manioc. However, on the basis of their Momil findings in Colombia, the Reichel-Dolmatoffs have proposed a modified interpretation of the presence of griddles. Recognizing that small griddles may have been used for maize tortillas in prehistoric Colombia, they suggest only that the occurrence of the distinctive very large griddles with sharply upturned rims such as are found today in parts of Amazonia indicate the cultivation and preparation of bitter manioc as "flatbreads or simply manioc flour" (1956: 271). This seems to me to be an important suggestion, even though archaeological griddles may not mark the introduction of manioc cultivation or even its earliest use as a staple, since flatbreads of manioc have been baked also on more primitive devices such as stone slabs (Roth 1924:289; Métraux 1928:102-04). Moreover, many peoples toast manioc flour in other ways, using earth ovens, ashes, or wide dishes, including some with rounded bottoms. Making "buns," "pies," and "cakes" of various kinds of flours is a very old, widespread practice that antedates either the cultivation of manioc or the manufacture of griddles. Furthermore, perhaps the most widespread and oldest use of flours, including manioc, is in liquid foods such as soups and not toasted bread. Thus manioc flour may well have been used as a staple without griddles.

Ethnographically, as the Reichel-Dolmatoffs point out (1956:271), the distinctive very large griddles—up to one and one-half meters according to Lecointe (1922:333)—with unusually high rims are associated with manioc in Amazonia. It should be noted, however, that high rims may actually be a hindrance in making flatbreads, but since toasting farinha requires that the particles be stirred to prevent overcooking and thus hardening the starch grains too much, high rims serve the important purpose of keeping loose particles on the griddle as the Reichel-Dolmatoffs point out (1956:271). Carrying their suggestion a step farther, it might be fruitful to relate this type of griddle specifically to the production of large quantities of manioc flour (farinha) and not merely to the use of manioc as flatbread or flour.

Flour has very important advantages over other products. It can be prepared in large quantities; it can be stored for extended periods of time (see Schwerin 1970:24); and it can be carried easily on the hunt or on longer journeys. The Tchukahamai, a "Cayapó" group in central Brazil, use farinha as their provision on treks away from the settlement, carrying it in woven palm-leaf pouches (Villas-Boas 1955:84).

Another advantage of flour is that it can be made into flatbreads and carried or stored in this form. Thus Miller wrote of the enormous Piaroa flatbreads,

> The heat causes the particles to adhere, forming a tough, round wafer. . . . This is the bread of the Orinoco and is always carried as the main article of provision by Indians and [other] travelers alike. [1917:276; cf. McGovern 1927:109]

With these advantages, the development of a cultivable source of flour has enormous implications for social development, and specifically for the stability of settlements and their internal social organization, and for mobility of forest-dwelling groups and consequent intergroup relations. When people began to use manioc flour extensively for trade, for journeys, or to store for use in a season when processing the tubers was difficult or impossible, large quantities of this "convenience food" must have been processed during short periods. In these circumstances very large griddles with raised rims would have made the processing much more efficient. Certainly the use of enormous griddles with high rims has spread in historic time as a response to opportunities to trade farinha to European colonists. This fact and the use of large quantities of farinha or flatbreads for trade and travel (McGovern 1927:109; Humboldt 1887:108) point to the possibility that large high-rimmed griddles may have developed with widening trade networks, migration, and raiding in Amazonia.

Similar conditions were reported in the sixteenth century among the Taino, who carried manioc flour on voyages among the islands and to the mainland from what is now Puerto Rico (Oviedo 1851:271). At the same time Tupian peoples on the coast of Brazil used a kind of flour made especially for taking on war expeditions.

> They have two kinds of flour, one called farinha de guerra [war flour] and the other farinha fresca [fresh flour]. War-flour is made of the same root. . . . The war flour is very dry and is made in a way to keep longer and not spoil. [Magalhães de Gandavo 1922:45, 158; Brackets are the translator's.]

Referring to information obtained from Donald Lathrap, Spath has noted that "by 1000 B.C. there are significant manioc-based Barrancoid expansions in the Orinoco basin and the Amazon basin and the . . . Chavín expansion throughout much of Peru" (1973:61).

Since flour is a versatile food and one that affords greater control for delayed use, and since flour has been widely represented in the literature as a staple of choice, I conclude that a major object of the complicated, time-consuming and onerous task of processing bitter (and sometimes sweet) manioc is to obtain flour.[6]

In spite of the great advantages of farinha, this is not the most useful or most highly prized product of manioc, however; a finer flour, the pure starch, is even more prized because of its caloric efficiency, its convenience and its versatility. Since pure starch flour lacks fibers it provides more calories per unit of volume,[7] and lacking any sugar, it is even more durable than farinha. Rinsing the pulp maximizes the yield of starch flour. In the rinsing process sugar leaves the pulp along with the starch, but since sugar is soluble in water it is removed when the liquid is decanted, leaving a product that can

---

[6] A similar conclusion is reached by Spath.

> The more toxic varieties of manioc grown particularly in South America and Africa are subjected to rather elaborate processing to obtain a variety of storable starchy products. . . . Though ninety percent or more of the toxic glucosides are removed by these more elaborate processing techniques, the processing itself seems to be aimed at the production of the storable and transportable flour or cakes. The removal of toxins, which can be accomplished as effectively by much simpler means, nonetheless remains an important by-product. (n.d.:1)

Spath reasons further that "considering the wide range of variation in the cultivated strains of manioc, toxic strains would have been selected against or simply discontinued if they did not possess some redeeming quality," and since "the more toxic strains of manioc are used to produce farinha and casabé because they produce a more durable and higher quality starch . . . these varieties would not have been selected for until storable products became important" (1973:46).

[7] This is not to say that a diet lacking fibrous foods is healthful for human beings.

be thoroughly dried and stored indefinitely without spoiling because in the raw state it does not absorb moisture and therefore does not ferment. The Taino, for example, kept their *jaujau* bread made of starch flour up to three years (Sturtevant 1969:181). As O. F. Cook put it,

> Separated from the sugars and other readily soluble substances which retain or absorb moisture, the starch of the taro, cassava [manioc], arrowroot, canna, and other root crops can be quickly and thoroughly dried and will then keep indefinitely. [1903:489]

In addition to being extremely durable, the starch flour is the most convenient of native foods. A particularly useful property is that when heated lightly the grains "explode" (Jacques 1958:138) or agglutinate in small pellets that are soluble in water (Ferreira 1942:36) without further cooking. This, of course, is tapioca or the Kuikuru toasted *kuigiñu*, which is mixed with water to make a cold soup. In this form it is unsurpassed as a convenience food in preindustrial societies. The raw starch can also be made into a hot soup by boiling with water, or it can be toasted to make bread. It is the starch that holds particles together in the toasting of flatbread, which explains why starch flour is added to farinha by those people who rinse the pulp. On the other hand, in societies where manioc pulp is drained without rinsing, the resulting farinha retains more of the starch and can be used by itself for making flatbread.

Manioc starch is sometimes referred to as "precious" (McGovern 1927:134; Ferreira 1942:36). This is the "finest of the forms of manioc food," according to Barandiarán, "and a favorite of the Makiritare" (1962:5). Isolation of starch has been compared to the development of grains in importance.

> In the absence of cereals this simple expedient might well be deemed an epoch-making discovery, since it rendered possible the accumulation of a permanent, readily transportable, food supply, and thus protected man from the vicissitudes of the season and the chase. [Cook 1903:489]

For the production of flour, tubers are crushed by either mashing or grating. Crushing the tubers not only frees the juice but also exposes more of the starch by breaking up fibers. This mechanical device of exposing a larger surface area apparently facilitates the transformation of the glucoside into glucose and HCN and at the same time makes it possible to both volatilize the HCN and cook the starch more readily. Starch in pure form can be obtained only from the juice of pulp that has been crushed or grated, and preferably rinsed as well. Since rinsing maximizes the yield of starch flour, I think it may be inferred from the advantages provided by starch that a major objective of rinsing the pulp is to obtain as much of this "high quality" durable starch flour as possible.

Bitter varieties of manioc generally have more starch in addition to producing larger tubers, and have a higher HCN content (Lecointe 1947:267, 280; Wagley 1964:66). The reasons for the correlation of HCN with size of tubers and starch content are not entirely clear. Genetic potential is undoubtedly a primary factor, since in the same environment bitter varieties generally can be left in the ground longer without rotting and produce larger tubers than sweet varieties. They also are slower to mature and can tolerate cooler climates and drier soil conditions. These environmental features also clearly influence the amount of both HCN and starch as well as the size of tubers. Manioc that is grown in cooler and drier climates or in more acid soils produces larger tubers and is more toxic than manioc grown in moist fertile soils. Moreover, tubers that are left in the ground to full maturity increase not only in size but also in the amount of starch as well as HCN (Ferreira 1942; C. B. Schmidt 1951; Rogers 1965).

The burdensome process of making flour appears to have been associated with bitter rather than sweet manioc originally, and since native users of bitter manioc value the starch most highly it seems probable that bitter manioc has been developed or adopted for the advantages of its higher starch content.

Bitter manioc appears to be correlated with the use of presses, and there are indications that the correlation was higher aboriginally, before flour making was adopted by peoples who used only sweet varieties. As has already been mentioned, the most widely known type of press is the sleeve press or *tipití*. However, as we have seen, a different type of press is used among the Upper Xingú peoples. (See Dole 1960 for a variety of other devices for expressing the juice.)

Twined roll-up mats similar to the Kuikuru *tuafi* are relatively simple and are used throughout the world for a variety of functions. In Amazonia such mats occur in outlying regions and are sometimes used for purposes other than draining manioc. Hence their use for pressing manioc appears to predate the use of the *tipití*. This conclusion is supported by evidence that the *tipití* was a late development that was widely diffused especially by migrating Tupians shortly before European colonization (Métraux 1928).

An ultimate product of the laborious process of separating starch from pulp is the decanted liquid, which contains small amounts of starch and sugar. This liquid is not used by all groups that make farinha and starch flour because it also contains the largest proportion of HCN left in the tubers after peeling. Among some native peoples, however, it is made safe by boiling and is variously made into soups, as among the Kuikuru, pepperpots in northern Amazonia and the Antilles, or the highly prized sauce and preservative *cassaríp*, or *túmali* of the Island Carib. This sauce is made by evaporating most of the water, leaving a thick syrup or paste. According to science writer Margaret Kreig, the botanist Richard Evans Schultes found this sauce "delicious . . . when spread on meat or fish as a condiment" (1966:48).

It is possible to detoxify the liquid with sunlight alone. When placed in glass bottles and exposed to the sun for two or three weeks, the HCN is volatilized. Treated in this way the thin liquid, called *tucupí*, is used as a meat sauce by Brazilians (Lecointe 1922:335; Galvão 1962:130; Wagley 1964:65).

## CONCLUSION

A comparison of the manner in which manioc is processed reveals a greater variety than is generally recognized.[8] To some extent the variations reflect differences in degree of cultural development, but in part they appear to be related also to the types of manioc used.

The primary purpose of processing manioc is to make it edible. As with cultivation of the plant, the various techniques used in processing the tubers provide means of controlling the properties of the plant. To make edible the carbohydrate of sweet manioc requires no more elaborate means than peeling and cooking or fermenting. If, however, a society wishes to produce not only edible but also storable and transportable carbohydrate, more elaborate techniques are required.

Although the technology of the Kuikuru and other peoples in the Upper Xingú region is relatively simple, it nevertheless provides a high degree of control over the properties of bitter manioc. Without using the *tipití* and without fermenting, Xinguanos are able to detoxify and use the pulp, the starch, and the juice. Moreover, the starch and pulp are prepared in such a way as to preserve them for extended periods. The Kuikuru and other peoples who isolate the starch value this product most highly. Since starch yields more calories per volume than either the juice or pulp, and since it is more durable as well as more versatile and convenient to use, it seems likely that bitter manioc is preferred by some societies for its greater yield of starch and that this part is a primary objective in grating, rinsing, and draining.

Use of the starch by the Kuikuru does not indicate an advanced degree of cultural complexity. Indeed, societies that aboriginally used the *tipití* were considerably more advanced than the Kuikuru. With only the *tuafi* as a draining device, processing manioc for domestic consumption alone occupies most of the Kuikuru women's days, especially during the dry season. It is doubtful that they would have sufficient time or energy with their simple technology to process large quantities of flour for trade or

[8] Schwerin has reviewed the techniques used among 202 societies (n.d.).

sale as do the Tukuna, Mundurucú and others who use the *tipití* and large griddles with "ovens."

Recent ethnological and botanical research has clarified some of the problems posed by the great variety of manioc strains and ways in which they are used. However, many problems remain unsolved and will require much work because of the complex nature of the cultural, botanical and environmental variables involved. To begin with, culture-historical problems of the development of manioc technology are complicated by widespread diffusion of both techniques of preparation and types of manioc used. In order to determine the sequence of development and spread of technology it may be necessary to inquire into some botanical issues. For example, the precise nature of the relation between the amount of toxin and the starch and sugar content of mature tubers is not yet well understood, nor is the extent to which fermentation detoxifies and preserves manioc. Other problems are the relation of growth potential to amount of toxin, and the complex relations among the types of manioc used, the amount of trade or commercial production, and techniques used to prepare manioc. Some of these problems will undoubtedly require investigation from several different approaches.

## REFERENCES


Acuña, Cristoval de
1859 A New Discovery of the Great River of the Amazons. In: Expeditions into the Valley of the Amazons, 1539, 1540, 1639, edited by Clements R. Markham, pp. 41-134. London: Hakluyt Society.

Alvarado, Lisandro
1945 Datos Etnograficos de Venezuela. Caracas: Escuela Tecnica Industrial Talleres de Artes Graficas.

Barandiarán, Daniel de
1962 Actividades Vitales de Subsistencia de los Indios Yekuana o Makiritare. Antropológica 11:1-29. Sociedad de Ciencias Naturales La Salle.

Bolian, Charles E.
n.d. Manioc Cultivation in Periodically Flooded Areas. Paper presented at the annual meeting of the American Anthropological Association, New York, 1971.

Cardim, Fernão
1925 Tratados da Terra e Gente do Brasil. Rio de Janeiro: Ed. J. Leite and Co.

Cook, Orator F.
1903 Food Plants of Ancient America. Smithsonian Institution Annual Report 1903:481-97.

Cruxent, José M., and Irving Rouse
1958 An Archeological Chronology of Venezuela. Pan American Union. Social Science Monographs 6, Vol. 1.

Denevan, William M.
1963 The Aboriginal Settlement of the Llanos de Mojos: A Seasonally Inundated Savanna in Northeastern Bolivia. Ph.D. dissertaion. Ann Arbor: University Microfilms.

Dole, Gertrude E.
1960 Techniques of Preparing Manioc Flour as a Key to Culture History in Tropical America. In: Selected Papers of the Fifth International Congress of Anthropological and Ethnological Sciences, 1956, edited by Anthony F. C. Wallace, pp. 241-48.
1961-1962 A preliminary Consideration of the Prehistory of the Upper Xingu Basin. Revista do Museu Paulista N.S. 13:339-423.
Ehrenreich, Paul
1948 Contribuições para a Etnologia do Brasil. Revista do Museu Paulista N.S. 2:7-135.
Ferreira Filho, João Candido
1942 Cultura da Mandioca. In: Manual da Mandioca, edited by Amadeu A. Bargiellini, pp. 5-74. São Paulo: Chacaras e Quintais.
Frič, Vojtěch, and Paul Radin
1906 A Contribution to the Study of the Bororo Indians. Journal of the Royal Anthropological Institute 36:382-406.
Fritz, Samuel
1922 Journal of the Travels and Labours of Father Samuel Fritz in the River of the Amazons between 1686 and 1723. London: The Hakluyt Society.
Galvão, Eduardo
1962 Elementos Básicos da Horticultura de Subsistência Indígena. Revista do Museu Paulista N.S. 14:120-44.
Ghisletti, Luis V.
n.d. The Tunebos: An Anthropological Linguistic Study. Ms.
Goldman, Irving
1963 The Cubeo. Indians of the Northwest Amazon. Illinois Studies in Anthropology 2.
Gregor, Thomas
1977 Mehinaku. The Drama of Daily Life in a Brazilian Indian Village. Chicago: University of Chicago Press.
Humboldt, Alejandro de
1887 Estado de la Agricultura de Nueva España–Minas Metalicas. Appendix to La Naturaleza 7:94-145.
Huxley, Francis
1956 Affable Savages. An Anthropologist among the Urubu Indians of Brazil. New York: Capricorn Books.
Jaques, Harry E.
1958 How to Know the Economic Plants. Dubuque, Iowa: William C. Brown Company.
Kensinger, Kenneth M., et al.
1975 The Cashinahua of Eastern Peru. The Haffenreffer Museum of Anthropology, Brown University, Studies in Anthropology and Material Culture 1.
Koch-Grünberg, Theodor
1967 Zwei Jahre unter den Indianern. Reisen in Nordwest-Brasilien 1903-1905. 2 vols. in one. Graz, Austria: Akademische Druck- u. Verlagsanstalt.
Krause, Fritz
1911 In den Wildnissen Brasiliens. Bericht und Ergebnisse der Leipsiger Araguaya-Expedition, 1908. Leipsig.
Kreig, Margaret B.
1966 Green Medicine. New York: Bantam Books, Inc.
Lange, Algot
1914 The Lower Amazon. New York: G. P. Putnam's Sons.
Lecointe, Paul
1922 La Culture et la Préparation du Manioc en Amazonie. Revue de Botanique Appliquée et d'Agriculture Coloniale (Paris). 2:331-37.
1947 Arvores e Plantas Uteis. São Paulo: Companhia Editora Nacional.

Lévi-Strauss, Claude
1964 Tristes Tropiques. New York: Atheneum.
Lima, Pedro E. de
1950 Os Indios Waurá. Boletim do Museu Nacional. Antropologia 9.
Lothrop, Samuel K.
1946 Indians of the Paraná Delta and La Plata Littoral. In: Handbook of South American Indians, edited by Julian H. Steward. Bureau of American Ethnology Bulletin 143, Vol. 1:177-99.
McGovern, William Montgomery
1927 Jungle Paths and Inca Ruins. New York: Grosset & Dunlap.
Magalhães [de Gandavo], Pero de
1922 The Histories of Brazil. Vol. 2. New York: The Cortes Society.
Métraux, Alfred
1928 La Civilisation Matérielle des Tribus Tupi-Guarani. Paris: Librairie Orientaliste Paul Geuthner.
1942 The Native Tribes of Eastern Bolivia and Western Mato Grosso. Bureau of American Ethnology Bulletin 134.
Métraux, Alfred and Herbert Baldus
1946 The Guayakí. In: Handbook of South American Indians, edited by Julian H. Steward. Bureau of American Ethnology Bulletin 143, Vol. 1:435-44.
Miller, Leo E.
1917 Up the Orinoco to the Land of the Maquiritares. The Geographical Review 3:258-77, 356-74.
Murphy, Robert F.
1960 Head Hunter's Heritage. Social and Economic Change among the Mundurucú Indians. Berkeley: University of California Press.
Murphy, Yolanda, and Robert F. Murphy
1974 Women of the Forest. New York: Columbia University Press.
Nimuendajú, Curt
1952 The Tukuna. University of California Publications in American Archaeology and Ethnology 45.
Nordenskiöld, Erland
1924 The Ethnography of South America Seen from Mojos in Bolivia. Comparative Ethnographic Studies 3. Göteborg.
Oberg, Kalervo
1953 Indian Tribes of Northern Mato Grosso, Brazil. Smithsonian Institution, Institute of Social Anthropology Publication 15.
O'Neal, Lila
1949 Basketry. In: Handbook of South American Indians, edited by Julian H. Steward. Bureau of American Ethnology Bulletin 143, Vol. 5:69-96.
Oviedo y Valdés, Gonzalo Fernández de
1851 Historia General y Natural de las Indias, Islas y Tierra Firma de la Mar Océano. Vol. 1. Madrid.
Reichel-Dolmatoff, Gerardo
1971 Amazonian Cosmos. The Sexual and Religious Symbolism of the Tukano Indians. Chicago: University of Chicago Press.
Reichel-Dolmatoff, Gerardo, and Alicia Reichel-Dolmatoff
1956 Momil. Excavaciones en el Sinu. Revista Colombiana de Antropologia 5: 109-333.
Rogers, David J.
1963 Studies of *Manihot esculenta* Crantz and Related Species. Bulletin of the Torrey Botanical Club 90(1):43-54.
1965 Some Botanical and Ethnological Considerations of *Manihot esculenta*. Economic Botany 19:369-77.

n.d. Botanical Considerations on the Origin of *Manihot esculenta*. Paper presented at the annual meeting of the American Anthropological Association, New York, 1971.

Rogers, David J., and Henry S. Flemming
1973 A Monograph of *Manihot esculenta*—with an Explanation of the Taximetric Methods Used. Economic Botany 27(1):1-113.

Roth, Walter Edmund
1924 An Introductory Study of the Arts, Crafts, and Customs of the Guiana Indians. Bureau of American Ethnology Annual Report 38(1916-1917): 25-745.

Roumain, Jacques
1942 Contribution à l'Étude de l'Ethnobotanique Précolombienne des Grandes Antilles. Bulletin du Bureau d'Ethnologie de la République d'Haiti 1.

Rydén, Stig
1942 Notes on the Moré Indians, Rio Guaporé, Bolivia. Ethnos 7:84-124.

Schery, Robert W.
1952 Plants for Man. New York: Prentice-Hall.

Schmidt, Carlos Borges
1951 A Mandioca. Boletim de Agricultura. Ano de 1951:N.° unico: 73-128. São Paulo.

Schmidt, Max
1914 Die Paressi-Kabiši. Baessler-Archiv 4:167-250.

Schultz, Harald
1959 Ligeiras Notas Sôbre os Makú do Paraná Boá-Boá. Revista do Museu Paulista 11:109-32.

Schwerin, Karl H.
1970 Apuntes sobre la Yuca y sus Orígenes. Asociación Venezolana de Sociología, Boletin Informativo 7:23-27.
n.d. The Bitter and the Sweet. Some Implications of Techniques for Preparing Manioc. Paper presented at the annual meeting of the American Anthropological Association, New York, 1971.

Simpson, George Gaylord
1940 Los Indios Kamarakotos. Revista de Fomento 3:22-660. Caracas.

Siskind, Janet Louise
1968 Reluctant Hunters. Ph.D. dissertation, Columbia University, New York.

Southey, Robert
1822 History of Brazil. Vol. 1. 2nd ed. London.

Spath, Carl D.
n.d. The Toxicity of Manioc as a Factor in the Settlement Patterns of Lowland South America. Paper presented at the annual meeting of the American Anthropological Association, New York, 1971.
1973 Plant Domestication: The Case of *Manihot esculenta*. Journal of the Steward Anthropological Society 5(1):45-67.

Staden, Hans
1928 The True History of His Captivity. London: George Routledge and Sons.

Sturtevant, William C.
1969 History and Ethnography of Some West Indian Starches. In: The Domestication and Exploitation of Plants and Animals, edited by Peter J. Ucko and G. W. Dimbleby, pp. 177-99. Chicago: Aldine Publishing Company.

Taylor, Douglas
1972 Interpretations of Some Documentary Evidence on Carib Culture. In: Aspects of Dominican History, pp. 8-17. Dominica, W.I.: Government Printing Division.

Villas-Boas, Claudio, and Orlando Villas-Boas
1955 Atração dos Índios Txukamãi, In: S. P. I.–1954. Relatorio das Atividades do Serviço de Proteção aos Índios durante o Ano de 1954, edited by Mario F. Simões, pp. 79-88. Rio de Janeiro.
Wagley, Charles
1964 Amazon Town. New York: Random House.
Whiffen, Thomas
1915 The North-West Amazons. London: Constable and Company, Ltd.

# PART IV
# ANTHROPOGENIC PLANTS AND COMMUNITIES

## INTRODUCTION

Adherence by a culture to general principles for manipulating vegetation does not preclude the creation of new plant forms or communities. In fact, following rules like "take only the part you will use" is precisely the process which leads to selection for larger leaves or seeds characteristic of domesticated plants.

The changes in plant forms and vegetational patterns affected by human activities follow a continuum of interactions. While some plant populations are collected, no further modification occurs. Others may be assisted through weeding, pruning, thinning, etc., procedures that enhance growth and reduce competition but do not necessarily result in genetic changes or dependence on humans. Transplanting and tending plants may lead to dependence, especially if a new habitat is maintained. Finally, the most extensive human intervention is domestication, which results in genetic changes that usually affect useful plant parts and produce total dependence upon humans for a plant's very existence. In the absence of humans, these plants simply would become extinct.

The North American continent is an ethnobotanical laboratory for the study of human impact on vegetation. Archaeologists have saved plant remains from their excavations for almost a century, and a long tradition of natural history has given even greater antiquity to observations of Indians using plants in numerous ways.

Gathering, gardening, and field horticulture form an evolutionary sequence of ecological subsistence techniques. Since the first Asian hunters crossed into Alaska and to this day, the collecting of natural plant products has remained an important economic pursuit in North America. During the activities related to gathering and processing these plants, some were accidentally introduced to new areas. As Black discusses in her paper, even in the most northern regions of the forests in the Northeast, the Indians have been agents for extending the range and for perpetuating plants accidentally introduced by the wind, birds, and other vectors into potentially disadvantageous environments. What the Algonquins do to maintain disjunct populations of sweetflag, butternut, chokecherry, and strawberry, other Indians undoubtedly did elsewhere for millennia, with many other plants.

The first domesticated plants were introduced into the continental United States from tropical Mesoamerica. Through contact between nomadic hunting and gathering bands thousands of years ago, these plants were added to the existing subsistence economies; gardening and gathering continued simultaneously. Until recently, the general consensus was that all of these cultigens entered the Southwest by several routes and some of them, notably corn, squash (*Cucurbita pepo*), and beans (*Phaseolus vulgaris*) moved eastward from there.

This model has been revised in light of recent discoveries and radiocarbon dating. Corn, squash, and bottle gourd (*Lagenaria siceraria*) were introduced into the Southwest between 3000 and 4000 years ago. Later common beans came and even later a series of plants—tepary beans, lima beans, runnerbeans, jackbeans, cotton, grain amaranth—mainly dependent upon irrigation in the Sonoran Desert were brought to southern Arizona. Meanwhile, it now appears that the bottle gourd and two varieites of squash, *C. pepo* and the smaller gourd squash *C. pepo* variety *ovifera*, were introduced from northeastern Mexico into the East independent of the Southwest and perhaps 1000 years earlier than in the Southwest. The research of Kay at Phillips Spring, Missouri, Watson on the Green River in Kentucky, and Chapman on the Little Tennessee support this contention. The cultivation and domestication of native plants followed these introductions. Other tropical cultivars did come from the Southwest, but much later in time. Corn appears only in the first few centuries B.C., and common beans from the same source appear shortly before A.D. 1000. In other words, the plant geography of introduced plants is quite complex and, as the papers in this volume reveal, the cultivation and domestication of native plants is equally intriguing.

Following the introduction of squashes and gourds, several indigenous ruderal plants were cultivated. Cowan discusses maygrass in this context. It is found in archaeological sites in areas beyond its known natural range. Sumpweed was also first recognized growing beyond its natural range, but unlike maygrass, human selection did create plants unknown in nature with larger seeds and in other habitats. Yarnell also discusses the sunflower as undergoing a similar process of domestication.

Despite its relative importance in the last 3000 years of Eastern U.S. prehistory as carefully documented by Asch and Asch, the sumpweed rapidly lost favor, and today the domesticated species is known only from archaeological contexts. Quite the opposite was the fate of the sunflower, except that many people who grow it do not realize that American Indians first domesticated it.

Archaeological evidence suggests that the Mexican domesticates and the native cultigens were grown in gardens that did not require extensive clearing of the natural vegetation and that were an ecologically complex, albeit human dominated, plant community.

It was not until 1000 years ago that large fields were cleared for planting corn and other crops. Although gathering and garden plots continued, an extensive anthropogenic landscape was created with the evolution of field horticulture based upon corn, beans, and squash.

# PLANT DISPERSAL BY NATIVE NORTH AMERICANS IN THE CANADIAN SUBARCTIC

*M. Jean Black*
University of North Carolina at Greensboro

The reciprocal nature of the relationships between humans and plants is recognized today as integral to the study of ecological anthropology, cultural ecology, and ethnobotany. It is so central to our thinking that its importance cannot be overstated. Students who came into contact with Volney Jones during their formative years were fortunate to have been exposed to this basic assumption of ecology before it was widely recognized. One area of plant and man relations in which Jones has had some interest is that of the role played by native American Indian populations as agents of plant dispersal. This question not only touches upon his interests in the ecological nature of the relationships but it also reflects his conception of ethnobotany as both relying upon and contributing to our knowledge of botany and of anthropology. In this paper some examples of American Indian influence on native flora and some suggestions concerning the nature of this influence are offered, with speculations about its influence on our own scientific, botanical traditions. The reciprocal relationships involve certain species of plants, some native Americans in the Canadian subarctic, and contemporary Euro-American and Euro-Canadian botanists.

## PLANT DISPERSAL BY MAN IN NORTH AMERICA

In 1931, Melvin Gilmore published a short paper in which he suggested that the American Indian had played an important but hitherto little recognized role in the distribution of some species of plants. Listing some examples derived from his own observations, he described the occurrence of six plant species at stations far beyond what is generally regarded by plant geographers as their normal distribution or range. Gilmore's examples came from the plains and prairie regions of the central part of North America. He suggested that isolated stations of the six species represented either intentional or unintentional introductions by groups of Indians such as the Pawnee, as they moved westward into the plains away from their former locations in the prairies or woodlands. The resulting discontinuity of the plant distributions often involves distances of hundreds of miles and was a key element in Gilmore's conclusions. It was this discontinuity which made the suggested human intervention recognizable in the first place. He also limited his discussion to plants we would not normally consider to be weedy and therefore appear to have become established under somewhat unusual circumstances, such as human intervention. He assumed rather logically that any group of people accustomed to using these plants might be inclined to bring them along when traveling. He did not assume that subsequent planting was necessarily intentional, but it may have been in some cases.

While some discontinuities of the sort described by Gilmore are found in the plants used by the Indians of Canada's eastern subarctic, this discussion is not limited to these plants. Included are other species which appear to be on the borders of their normal range or distribution as recognized by contemporary authorities. Others show some discontinuity but of less spectacular magnitude than those in Gilmore's study. Reasons for including the latter type will be discussed below.

The ethnobotanical knowledge of two linguistic and cultural groupings in the eastern subarctic, the Algonquin and the Cree, has been described and discussed in some detail (Black 1973). This provides us with a substantial list of plants for consideration.

## THE ALGONQUIN AND CREE ENVIRONMENTS

The Algonquin are located within the drainage of the Ottawa River, primarily in Quebec Province. These bands are particularly suitable for this discussion. Although the several bands share a common cultural tradition, their recent history has resulted in some of the bands being located within one defined forest type, the Boreal Forest, while others are located to the south in the Great Lakes-St. Lawrence Forest (Rowe 1972). Collectively, the Algonquin might be described as occupants of an interface between the two forest types. This is particularly interesting for our discussion of the northern distribution of some plant species beyond their accepted ranges.

Two bands in particular figure in the discussion. The first is River Desert, a relatively large band located quite near the northern extent of the Great

Lakes-St. Lawrence Forest. Many species of plants associated with this forest type are at or near their northern limits at the River Desert reserve. The other prominent Algonquin band is that of the Barrière Indians located approximately 100 miles north and west of River Desert. This band is well within the Boreal Forest. Also in this forest type and within the Ottawa drainage are two other Algonquin bands, Grand Lake Victoria and Simon Lake. Grand Lake is approximately 40 miles to the west and north of Barrière band's location by road. Simon Lake is north of Grand Lake about 24 miles on the same road.

The Cree bands included in the discussion are all located in the St. Maurice River drainage rather than the Ottawa. They are all within the Boreal Forest type. Manouan is nearest to River Desert being located approximately 100 miles to the northeast as the crow flies. There is no direct access between the two bands. Weymontaching is 50 to 60 miles northeast of Manouan with access between the two bands by road and water. Obedjiwan is 70 to 80 miles northwest of Weymontaching and about 200 miles north and east of River Desert. The locations of all these Algonquin and Cree bands can be visualized as occupying points along a large circle with River Desert on the most southern point of the circumference and Obedjiwan on the northern point.

## DISTRIBUTION OF SPECIES IN THE STUDY AREA

The northern limits of a number of important trees and other smaller plant species characteristic of the Great Lakes-St. Lawrence Forest are to be found in the vicinity of the River Desert reserve. Observations made during field work, however, indicate the difficulty in generalizing about limits that form the basis for any definition of a forest or vegetational type of zone. Trees generally regarded as southern are sometimes found growing in small numbers in isolated locations far to the north. Plants in such outlying stations frequently do not reproduce, presumably because of an inability to flower or set fruit in such places. When these isolated stations are the result of natural seed dispersal from the center of the species' primary distribution, this situation can persist for indefinite periods of time. Although each outlying population has a definite lifespan and does not reproduce itself, new outliers may be started from time to time by seed dispersed from the centers of main distribution. An example of such a plant in this area of eastern Canada is red pine (*Pinus resinosa* Ait.). Individuals of the species are found scattered north of the largest populations. Although some individual trees reach considerable size, they do not appear to be successful in reproducing themselves. Each new individual tree, then, is the result of seed dispersed from the south. This characteristic may sometimes aid in the identification of species which have been spread by human intervention. If we cannot identify other vectors responsible for the dispersal (the wind, in the case of red pine) then human intervention is suggested. However, even when a species in an outlying station does set fruit, other factors might prevent its dispersal or successful germination and growth unless some element of human activity is present. The nurture or cultivation of wild plant species may be indicative of the intervention of humans in the actual dispersal of that species. A list of plant species of some interest in this context has been compiled by examining the ethnobotanical knowledge of the Algonquin and Cree (Black 1973).

## Sweet Flag (*Acorus calamus* L.)

Sweet flag (*Acorus calamus* L.) is perhaps the most interesting species on this list. It is one of the plants listed by Gilmore, who mentions several places west of its main distribution where it grows in patches near old Indian village sites or camping places. He offers his opinion that it was intentionally planted by the Indians in those locations (Gilmore 1931:91). Sweet flag is a very highly regarded medicine plant which was widely used throughout the eastern part of North America (Black 1973:213-15). The Algonquin and Cree in western Quebec value it highly and it is known to the members of every band surveyed. However, it was found growing locally at only three of the reserves: River Desert, Simon Lake, and Weymontaching. Individuals in the other bands know of these locations and may sometimes travel considerable distances in order to obtain it from members of the bands or to dig it for themselves. According to Fernald (1950:385), it grows from Prince Edward Island to Montana and Oregon and southward to Florida and Texas. Gilmore (1931:91) observes that its greatest frequency is on the Atlantic seaboard. Marie-Victorin (1964:845) observes that sweet flag is not a boreal plant and is not found very far north in the Laurentides of Quebec, nor is it found around the Gulf of St. Lawrence. The two most extreme northerly stations he mentions in Quebec are at Ste. Anne-de-la-Pocatière and Rivière du Loup on the St. Lawrence River below Quebec City. These he believes to be the result of introductions. He also mentions that this plant has been used by Europeans as well as Indians. Fernald (1950:385) also points out that it has been introduced and naturalized in Europe, where it is usually sterile. All three authorities mention the infrequency of its reproduction except by means of its rhizomes.

The distances from Simon Lake to River Desert and Weymontaching to River Desert are both approximately 200 miles. Both Simon Lake and Weymontaching are well within the Boreal zone and beyond the normal distribution of this plant. The two extreme northern stations mentioned above are at about the same latitude as Simon Lake and Weymontaching but are within the Great Lakes-St. Lawrence forest type, which extends farther north on the south shore of the St. Lawrence River than it does in the interior. The presence of sweet flag in the Boreal zone so far removed from other known stations and in close proximity to Indian communities is notable, and the dispersal of its rhizomes through human agency is likely. Sweet flag is of such great value to the people of this area that the motives for transference scarcely need explanation. The implanting of the rhizomes may have been a conscious attempt to insure a supply of this plant in a convenient location.

### Butternut (*Juglans cinerea* L.)

The second plant is somewhat different. Butternut (*Juglans cinerea* L.) is a tree typical of the Great Lakes forest. It is found at River Desert in a few groves and is known to the people at that reserve as the source of an edible nut. It is not reported in the ethnobotanies of the other Algonquin or Cree bands, none of whom occupy areas within the known range of the species. In fact, River Desert appears to be at the northernmost extent of its range. It is to be expected that the numbers of trees would be fewer here than to the south and the distribution of groves discontinuous. However, in the distribution of the trees at River Desert there appears to be a relationship between the locations of groves and of former and present human habitation. The largest group of trees found there were located in an area which had been occupied by a number of family groups within recent memory. Conversely, no trees were noted growing which were not in the vicinity of homes or old house sites. Although this may be circumstantial evidence, it should be pointed out that Gilmore (1931:91) made a similar suggestion for the closely related black walnut (*Juglans nigra* L.) in southern Nebraska. Gilmore attributes the planting of black walnut to the Pawnee some 150 to 200 miles west of its normally recognized distribution. This involves a greater distance than would be the case in Quebec where butternut may be found growing 30 to 50 miles to the south of River Desert. It is unlikely that the activities of rodents could account for this discontinuity nor the action of rivers or streams. The suggestion that humans were the vector for dispersal seems likely. Rather than extending the plant hundreds of miles only a relatively short distance is involved. This tends to obscure the discontinuity which might appear superficially to be part of the normal distribution of butternut to the plant geographer unfamiliar with its association with human habitations.

### Canada Plum (*Prunus nigra* Ait.)

Direct ethnographic evidence exists for the next two plants on the list. Like butternut, Canada plum (*Prunus nigra* Ait.) is near its northern limit at River Desert, and some authorities place its northern limits even farther to the south and east (Hosie 1973:248). It was found growing on the reserve at River Desert, however, where it is valued for its fruit. Its distribution there is very sporadic, and in all cases noted it was found near present or former house sites. One individual informed me that she had intentionally planted the seeds near her house several years ago. Although that house has been unoccupied for some time, the trees have survived and continue to bear fruit which she claims as hers. In this case there is direct evidence of intentional introduction of a plant beyond its normal distribution.

### Chokecherries (*Prunus virginiana* L.)

Chokecherries (*Prunus virginiana* L.) reportedly grow in isolated spots as far north as James Bay (Hosie 1973:242). However, since they prefer open woods, fields, and roadsides, they appear to be more common in the Great Lakes-St. Lawrence forest zone than in the Boreal zone. They are common at River Desert and for approximately 30 miles or so to the north, but occurrences thereafter become somewhat rare and isolated in the vicinities of the other Algonquin and Cree bands. Chokecherries were actually observed growing at only one other band reserve, that of the Barrière band in the upper Ottawa drainage. Informants knew of their occurrence at a few other locations in the Boreal forest zone at isolated spots along streams or rivers. Since birds and other wildlife are the commonly recognized vectors for seed dispersal for the cherries, these isolated stations might all have been considered natural occurrences had it not been for the direct information provided by informants at Barrière. The chokecherry trees observed growing there were planted at that location many years ago by a woman who had moved into Barrière territory from River Desert after her marriage. It was her stated desire to grow the trees in order to make preserves, wine, and medicine. Although her house is no longer standing, the clearing is apparent, and her trees continue to bear cherries which are used by the women of the band today. This case in particular suggests that we might look for the intentional human dispersal of plant species even within their recognized normal distributions. This would be particularly true of plants which were highly desirable, were in short supply, or were very sporadic in their distribution. It is not suggested that the people in this area necessarily extended the distribution of chokecherries northward, as was the case in the previous three plants, but rather that they may have been responsible for increasing the supply of the cherries within its natural distribution. The relationships implied between plants and humans are the same in either case.

### Wild Strawberry (*Fragaria virginiana* Duchesne)

Wild strawberry (*Fragaria virginiana* Duchesne) grows at River Desert, where it is abundant in old pastures and fields. It also grows north of that reserve. All the Algonquin bands know the plant, use it, and have some access to it within their territories although the habitat is somewhat restricted in the Boreal zone where clearings are less common. The Cree also report use of it and have some access to it. The northern limits for wild strawberry according to Fernald (1950:802) are in southeast Labrador, central Quebec, and northern Ontario. The Cree bands are located along the northern limits of its range. In that area, the few plants found growing were along roadsides and some small clearings made by pulp and paper companies for construction of buildings. The presence of strawberries in these locations was presumably due to the activity of birds. However, in the Obedjiwan reserve, one individual was cultivating wild strawberry in a garden plot. No other plants were seen within approximately 50 miles of that garden. It cannot be assumed that this person had brought the plant to his garden; he was not available for interviews. However, he was responsible for providing a suitable spot for the plant to grow even if birds had been the agents of dispersal. It is unlikely the plant would have survived without the presence of humans at that location. The presence of the species at that station may or may not have been due to the more usual means of seed dispersal in combination with the intervention of humans.

### Other Plants

Several other plants might have a place on this list. These include, provisionally, *Amelanchier* spp. (serviceberry or sugar plum), which was observed at both River Desert and Obedjiwan but not in the intervening area. Hawthorn (*Crataegus* spp.) grows at River Desert but is at or near its northern limits there. Prickly gooseberry (*Ribes cynosbati* L.) was found growing at River Desert also north of its accepted range or at its extreme limit. It grew behind an abandoned house at only one observed location. Finally, a species of wild rice (*Zizania palustris* L.) grows in small amounts and sporadically in the vicinity of River Desert. However, the activity of ducks or other vectors in the dispersal of these species cannot be discounted, and without additional evidence for human intervention they remain tantalizing possibilities.

## CONCLUSIONS

Gilmore pointed out possible human intervention when discontinuity of considerable distance is found in the distribution of a plant species. This is illustrated here in the case of sweet flag. The presence of a discontinuity is not a necessary precondition for human intervention in the distribution of plant species in every case, however. The examples given here from ethnographic data in western Quebec would indicate that dispersal through human agency may take place over relatively short distances and may thereby give the appearance that the normal distribution is somewhat more extensive than actually is the case. This appearance may have in turn influenced the conclusions of plant geographers and the authors of floras, leading them to include the extended range as the normal range. This leads us to the conclusion that the American Indian has had a direct influence on our perceptions of the flora in North America and through this has directly affected contemporary scientific botany.

## REFERENCES

Black, M. Jean
1973 Algonquin Ethnobotany: An Interpretation of Aboriginal Adaptation in Southwestern Quebec. Ph.D. Dissertation, Anthropology Department, University of Michigan.

Fernald, Merritt L. (ed.)
1950 Gray's Manual of Botany. 8th ed. New York: American Book Company.

Gilmore, Melvin R.
1931 Dispersal by Indians a Factor in the Extension of Discontinuous Distribution of Certain Species of Native Plants. Michigan Academy of Science, Arts and Letters, Papers 12:89-94.

Hosie, Robert C.
1973 Native Trees of Canada. 7th ed. Ottawa: Department of the Environment, Canadian Forestry Service.

Marie-Victorin, Frere
1964 Flore Laurentienne. 2nd ed. Montreal: University of Montreal Press.

Rowe, J. S.
1972 Forest Regions of Canada. Ottawa: Department of the Environment, Canadian Forestry Service.

# THE PREHISTORIC USE AND DISTRIBUTION OF MAYGRASS IN EASTERN NORTH AMERICA: CULTURAL AND PHYTOGEOGRAPHICAL IMPLICATIONS

*C. Wesley Cowan*
Museum of Anthropology, University of Michigan

Archaeologists conducting research in temperate eastern North America have only recently begun to realize the potential contributions of the careful collection and interpretation of plant remains from archaeological contexts. Paleoethnobotany, the study of the direct interrelationships between past human and plant populations, has received tremendous impetus from this newfound interest, and already much new information has been added to our understanding of extinct subsistence and environmental systems. In addition, paleoethnobotany is one of the few natural sciences that can directly contribute to knowledge of the processes leading to plant domestication. Studies of contemporary domesticates cannot elucidate evolutionary trends since we can only observe the end products, not their precursors; in this regard, the study of archaeological plant remains provides the necessary evidence.

While much information concerning the evolution of domesticated plants has been provided through studies of Old World species or New World plants from south of the continental United States, in recent years a considerable amount of attention has been devoted to the processes of domestication of plants native to temperate eastern North America. From a historical perspective, interest in these processes stems directly from the early paleoethnobotanical work of Melvin R. Gilmore and Volney H. Jones.

In 1931 Gilmore published a rather extensive paper dealing with collections of desiccated plant materials excavated by M. R. Harrington from Ozark Bluff-Dweller sites in northwestern Arkansas and southwestern

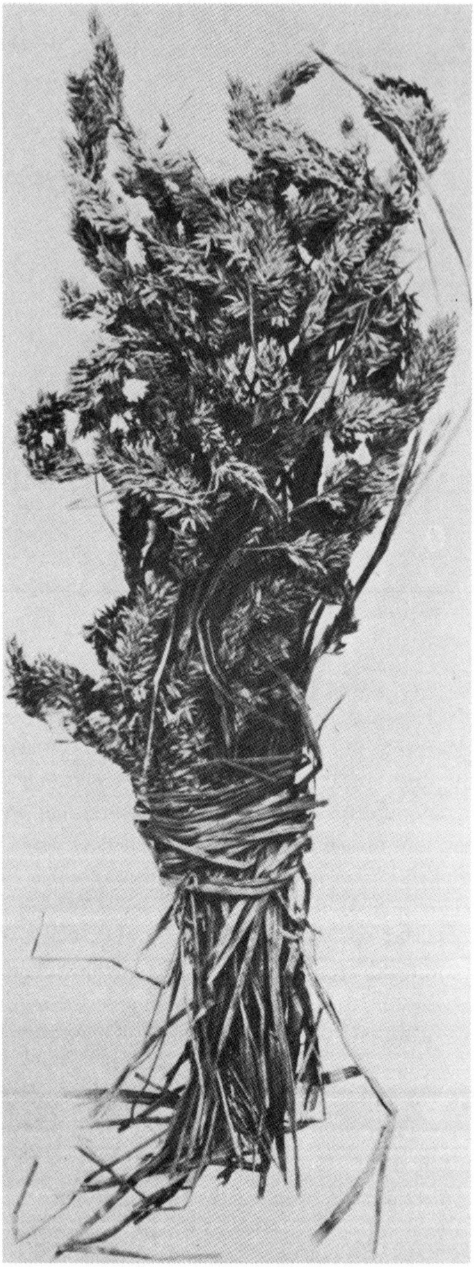

Fig. 1. Bundle of maygrass (*Phalaris caroliniana*) inflorescences from Indian Bluff Rockshelter. Benton County, Arkansas. Collected by M. R. Harrington. (Photograph Courtesy of the Museum of the American Indian, Heye Foundation)

Missouri (Gilmore 1931). Gilmore was of course impressed with the remarkable preservation of these ancient plants but was also struck by the presence of several species that appeared to be the remains of previously unknown eastern North American cultigens. Commenting on the presence of such genera as *Iva*, *Chenopodium*, *Ambrosia*, and *Phalaris* he noted

> But besides the well-known staple crops of corn, beans and squashes and pumpkins and sunflowers, which were cultivated by those ancient peoples and many other tribes from that time down to the present, there is also evidence that the ancient Ozark Bluff-Dwellers also had certain other species of plants not cultivated at the present time. [ibid.:85]

To my knowledge, this was the first suggestion that peoples indigenous to eastern North America actually domesticated plant species native to the temperate region.

Following the lead provided by Harrington, W. S. Webb and W. D. Funkhouser of the University of Kentucky began to search in the summer of 1929 for similar rock shelter sites in the mountainous region of eastern Kentucky. Their efforts were quickly rewarded with the discovery of such dry overhangs, and soon reports of their research began to appear (Funkhouser and Webb 1929, 1930; Webb and Funkhouser 1936).

Perhaps the most significant site excavated by the University of Kentucky during this era was the Newt Kash Hollow Shelter in Menifee County (Webb and Funkhouser 1936). Beneath the dry dust floor of this overhang, hundreds of items of perishable material culture were collected from deep storage pits and areas of compacted vegetal material that the excavators interpreted as "beds" or sleeping places. No systematic attempt was made to collect plant remains other than items of material culture, but a large sample of the bedding material was retained for future analysis. Undoubtedly Webb and Funkhouser were familar with Gilmore's work on the Ozark plant remains and were anxious to have the Newt Kash plants examined.

In January of 1936 Webb submitted the sample of bedding to the Ethnobotanical Laboratory at the University of Michigan for analysis. Gilmore, failing in health, turned the materials over to a newly hired assistant, Volney H. Jones, for analysis.

The report that resulted from Jones' examination stands as a landmark in the history of paleoethnobotany. In contrast to Gilmore's highly romanticized accounts of Ozark plant utilization, Jones' publication ". . . had all the characteristics of a well-developed scientific report. He was thorough in his description of the material, resourceful in his use of comparative materials . . . and discerning in drawing conclusions and constructing hypotheses" (Schwartz 1967:102).

Although most of the plants from Newt Kash were typical members of the deciduous forest, the complex of possible cultigens identical to that found by

Gilmore was also reported. By analyzing samples of paleofeces from the site, Jones was able to report with certainty that *Iva*, *Chenopodium* and *Phalaris* were common dietary components. He suggested that these plants may have been cultivated by the Newt Kash inhabitants, basing this assertion on the facts that *Iva* and *Chenopodium* seeds from the site were considerably larger than wild varieties and that *Phalaris* was not a present member of any plant community in Kentucky.

Although the early work of Gilmore and Jones on this complex of possible temperate cultigens raised more questions than it answered, it laid the groundwork and provided a stimulus for future research. However, with the rise of the Great Depression and the shift in interest towards excavation of large, more spectacular open-air sites, the rock shelter areas slipped for a time into obscurity.

As interest in the elucidation of prehistoric subsistence systems in eastern North America began to emerge in the mid to late 1960s, the rock shelter plant remains once again gained prominence, and the question of the domestication and cultivation of the native temperate region plants was revived. Payne and Jones (1962) studied archaeological and modern populations of giant ragweed achenes and concluded that the *Ambrosia* found in the Ozark shelters was not cultivated, since it fell well within the range of variability noted in modern forms.

After several years of research with sumpweed populations from a variety of archaeological contexts, Yarnell (1972) was able to show a gradual increase in the size of the achenes spanning a period of three millenia. Coupled with the occurrence of sumpweed achenes in archaeological sites far outside the natural distribution of the plant, Yarnell concluded that *Iva annua* var. *macrocarpa* represented an extinct, cultivated plant.

Recently, Asch and Asch (1977) reexamined populations of archaeological *Chenopodium* from the Ozark and Kentucky shelters in hopes of solving the question of cultivation of this genus. No evidence of morphological differences between present-day and archaeological forms could be discovered, and the authors were forced to conclude that there was no positive basis to conclude that chenopods were cultivated.

Maygrass (*Phalaris caroliniana*) is the last of the four plants originally noted by Gilmore and Jones as possible temperate cultigens. Much new information that directly contributes to Gilmore and Jones' early observations concerning the use of this plant has become available in the last few years. This paper reviews this evidence and takes up the question originally posed by these early ethnobotanists—"Was maygrass a cultivated plant?"

The following discussion is divided into six basic sections. The first presents information regarding the botanical characteristics of the plant, its geographic distribution, and its habitat preferences. The second section reviews archaeological evidence for the use of maygrass in eastern North America, while the third addresses the role of maygrass in prehistoric economic systems. A fourth section presents a brief discussion of possible ethnographic accounts of maygrass utilization in eastern North America. The question of maygrass cultivation is taken up in the fifth section, and the final section attempts to integrate the data presented in previous sections.

## BOTANICAL CHARACTERISTICS OF MAYGRASS

Maygrass (*Phalaris caroliniana* Walt.) is a grass in the subfamily Festucoideae in the tribe Phalaridae (Hitchcock 1950). There are approximately 15 species in the genus *Phalaris* with Old and New World distribution. In the Old World, most of the species are centered around the Mediterranean area, with one species, *Phalaris minor*, extending as far east as India. In the New World, species are located on both the North and South American continents. One species, *Phalaris arundinacea*, has a distribution that extends over the northern latitudes of both the Old and New Worlds.

Besides *P. arundinacea*, three other species are located in North America. *Phalaris californica* and *P. lemonii* are restricted to California and southern Oregon, while *P. caroliniana* is distributed over the southern half of the United States and northern Mexico (Anderson 1961:16-17). All four species seem to have followed different evolutionary lines (ibid.).

Maygrass is an annual (Fig. 2) with erect stems ranging from 30 cm to 1 m in height (Anderson 1961:78; Hitchcock 1950:552). As with all members of the genus, flowers are densely packed on a terminal panicle or inflorescence ranging in length from 1 to 7 cm. Maygrass has a shallow, fibrous root system, and unlike many grasses does not have the ability to reproduce vegetatively from underground stolons or above ground rhizomes. It is thus dependent upon reproduction from seed produced from one year to the next.

As suggested by its name, the plant begins to flower around the first or second week in April to the first week in May. Like most of the undomesticated grasses, maygrass has an indeterminate type of inflorescence–the flowers at the extreme tip of the inflorescence develop first, with flowers below successively reaching anthesis. Because of this fact, mature seeds are not present along the entire length of the inflorescence at the same time. An examination of 20 specimens of *P. caroliniana* on file at the University of Michigan Herbarium indicates that the terminal florets (the portion of the inflorescence containing the grain) mature first, and upon ripening, fall free from the plant. Maturation of these florets seems to begin around the first week in May with subtending florets ripening as late as the last week in June. By the first or second week in July, most of the florets have completely matured and have fallen from the inflorescence.

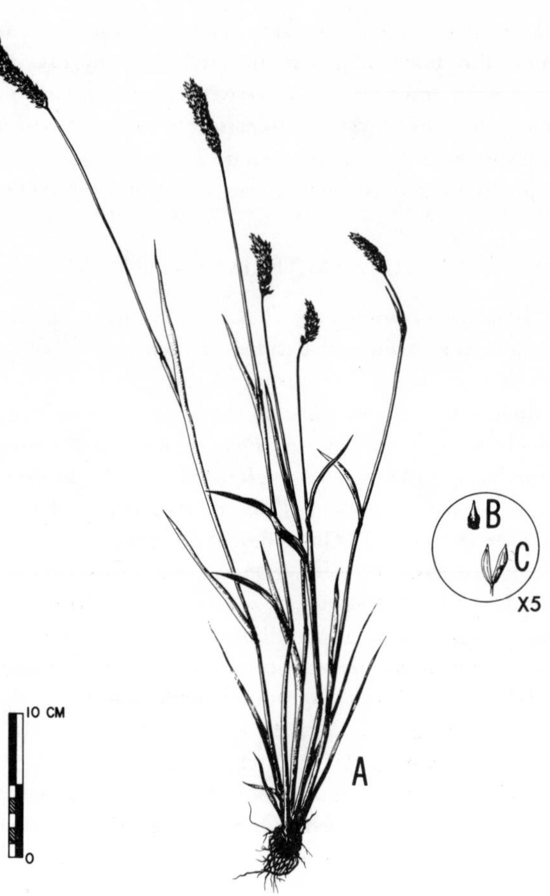


Fig. 2. Maygrass, *Phalaris caroliniana* Walt. A) plant, B) floret, C) caryopsis.

The small grains of maygrass are reddish in color, and range in length from 1.8 to 2.3 mm, and from 0.9 to 1.8 mm in width (Anderson 1961:78; Radford et al. 1964:122). From an economic standpoint, the grain is free from the surrounding lemma and palea, although it is not easily threshed. No information regarding the nutritional content of the grains could be located; however, such data are available for the grains of canary grass (*P. canariensis*), a cultivated member of the genus common to many parts of the Old World. While slight differences between the two species almost certainly exist, it is assumed that there is probably general compatibility between the two.

As indicated in Table 1, the nutritional content of canary grass seed is quite comparable with that of a variety of other cultivated cereals. While this information is reliable from a chemical standpoint, it should be remembered that all of the constituents may not be available for human digestion. Maygrass grains that have been found in paleofeces from Salts and Mammoth caves, Kentucky are still tightly enclosed in the florets in which they developed (Richard A. Yarnell, personal communication), and thus the nutritional benefit derived must have been minimal.

According to Hitchcock (1950) and Anderson (1961) maygrass is now distributed across the southern half of North America and northern Mexico. Although the maps provided by these two authors indicate the broad distribution of maygrass in eastern North America, they are fairly spotty in their coverage. Accordingly, during the winter of 1977 additional information regarding the distribution of maygrass was collected from 19 eastern North American museums and university herbaria.[1] The county by county distribution of maygrass garnered from this survey is presented in Figure 3. This information confirmed the observations of Anderson and Hitchcock regarding the phytogeography of maygrass and helped fill in gaps in their original descriptions. In addition, at least two apparent disjunct occurrences were recorded—one in Muskingum County, Ohio (Braun 1967:138) and another in southern Middlesex County, Massachusetts (specimen on file at the University of Illinois). Both specimens were collected from disturbed habitats (the Ohio specimen from a cultivated garden; the Massachusetts specimen from along a railroad track), but it is not known if maygrass has naturalized in either state.

In general, the natural distribution of maygrass in eastern North America includes the Coastal Plain and Lower Piedmont of the southeastern states, the Ozark Plateau, and the forest-prairie transition areas of Missouri and Kansas. Except for two isolated collections from Tennessee, in the southeast, maygrass does not seem to occur much north of the Fall Line. Significantly, maygrass does not seem to be a member of any natural plant community in the Blue Ridge, Ridge and Valley, Appalachian or Central Lowlands physiographic provinces.

[1] Questionnaires were received from the following herbaria: The University of Alabama; Auburn University; the University of Arkansas; the University of Florida; Florida State University; the University of Georgia; Emory University; the University of Illinois; Southern Illinois University; the University of Indiana; Louisiana State University; Southwestern Louisiana State University; the Gray Herbarium of Harvard University; the University of Missouri; the University of Mississippi; the University of North Carolina; Duke University; the University of Tennessee; Virginia Polytechnic University; West Virginia University; the United States National Museum, Smithsonian Institution. In addition the author personally examined the collections at the University of Michigan and at the University of Kentucky. Dr. Dennis Anderson's notes on *Phalaris caroliniana* collected as part of his dissertation research were also made available.

TABLE 1

COMPARISON OF NUTRITIONAL QUALITIES OF VARIOUS GRAINS

| Species | Food Energy Cal/100 g | Water | Crude Protein | Fat | Carbohydrates |
|---|---|---|---|---|---|
| *Phalaris canariensis** (canary grass) | 334 | 9.49 | 15.96 | 6.13 | 52.46 |
| *Triticum vulgare* (wheat) soft red winter | 326 | 14.0 | 10.20 | 2.00 | 69.80 |
| *Hordeum vulgare* (barley) dehusked kernel | 339 | – | 12.00 | 2.00 | 68.00 |
| *Sorghum vulgare* (sorghum) | 332 | 11.00 | 11.00 | 3.30 | 71.30 |
| *Setaria italica* (foxtail millet) dehusked | 355 | – | 10.00 | 2.50 | 73.00 |
| *Zea mays* (corn) whole ground, unbolted | 420 | 12.00 | 11.00 | 5.00 | 87.00 |

*From Febrel and Carballido 1965. Table 5, all other information from Sheffer 1972: Fig. XXXII.

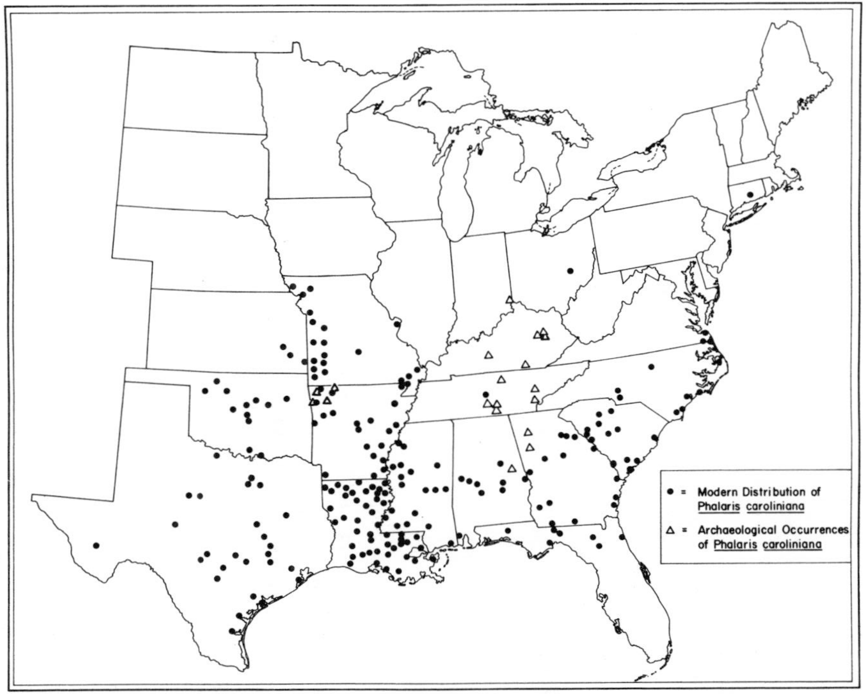


Fig. 3. Present-day and prehistoric distribution of maygrass in eastern North America. (*Phalaris caroliniana* is distributed all across the southern half of North America, but for the purpose of the present discussion only its distribution in Eastern North America has been examined. Dots and triangles are only indicative of counties where *Phalaris caroliniana* has been recorded either as a modern or prehistoric occurrence. For archaeological occurrences, the reader should consult Table 2.)

Information concerning habitat requirements for maygrass was also collected from the various herbaria. Judging from the data provided by them, maygrass is tolerant of a wide variety of habitats, a fact that is supported by its fairly broad distribution. Although the plant is capable of growth in numerous environmental settings, almost 60% of the specimens in the herbaria were collected from disturbed, open habitats. Approximately 10% were collected from the floodplains of streams and river; 9% from low, wet swales or depressions in open areas; 10% from open upland prairies (this is particularly true of the specimens from western Missouri); the remaining specimens were collected from wet marsh shores, open woods, and other, unspecified areas.

Based on this information, it can be concluded that maygrass is highly intolerant of shaded habitats and outside of the prairie areas of Missouri, is probably not a climax member of any plant community in eastern North America. Its decided preference for disturbed, open habitats and other waste places indicates that maygrass can safely be classified as an adventive weed.

It should be noted that at least two members of the genus *Phalaris* are grown for consumption by animals and man. An Old World species, *Phalaris canariensis* (canary grass), is a common constituent of commercial birdseed and is cultivated in Spain (Febrel and Carballido 1965), South Africa (Selschop and Wolfaardt 1952 as quoted in Anderson 1961:58) and England (Sturtevant 1919). Sturtevant (ibid.:428) also indicates that at one time the seeds of canary grass were also used as a food source by the inhabitants of the Canary Islands and Italy.

The reed-canary grass (*Phalaris arundinacea*) is commonly cultivated as a source of hay and pasturage in the mid-latitudes of eastern North America (Harrison and Davis 1966; Myers 1967).

## ARCHAEOLOGICAL EVIDENCE OF MAYGRASS UTILIZATION IN EASTERN NORTH AMERICA

Carbonized and desiccated reproductive structures of maygrass have been found in a variety of archaeological contexts in eastern North America (Table 2). Geographically, sites which have produced evidence of maygrass utilization range from Alabama to southwestern Indiana. Chronologically these sites embrace a period of over three millennia, from the terminal Archaic to the late Prehistoric period. A brief discussion of the evidence from each of these cultural/chronological periods is presented below.

### Late Archaic

Three sites from this cultural/temporal period have produced carbonized maygrass grains. Two sites (Bacon Bend and Iddins) on the Little Tennessee

TABLE 2
ARCHAEOLOGICAL OCCURRENCES OF MAYGRASS IN EASTERN NORTH AMERICA

| Cultural/Temporal Affiliations of Sites | Location | River Drainage | Physiographic Province |
|---|---|---|---|
| Late Archaic | | | |
| Bacon Bend | Monroe Co., Tenn. | Little Tennessee | Ridge and Valley |
| Iddins | Loudon Co., Tenn. | Little Tennessee | Ridge and Valley |
| Salts Cave | Edmonson Co., Ky. | Green River | Interior Low Plateau |
| Early Woodland | | | |
| Newt Kash Hollow | Menifee Co., Ky. | Licking River | Appalachian Plateau |
| Mammoth Cave | Edmonson Co., Ky. | Green River | Interior Low Plateau |
| Salts Cave | Edmonson Co., Ky. | Green River | Interior Low Plateau |
| Middle Woodland | | | |
| Icehouse Bottom | Monroe Co., Tenn. | Little Tennessee | Ridge and Valley |
| Owl Hollow | Franklin Co., Tenn. | Duck River | Interior Low Plateau |
| Shofner | Bedford Co., Tenn. | Duck River | Interior Low Plateau |
| Raus | Bedford Co., Tenn. | Duck River | Interior Low Plateau |
| Eoff I | Coffee Co., Tenn. | Elk River | Interior Low Plateau |
| Anawakee Creek Mound | Douglas Co., Georgia | Chatahoochee River | Piedmont |
| Garfield | Barton Co., Georgia | Etowah River | Piedmont |
| Late Woodland | | | |
| Rogers Shelters | Powell Co., Ky. | Red River | Appalachian Plateau |
| Haystack Rock Shelters | Powell Co., Ky. | Red River | Appalachian Plateau |
| Leonard Haag? | Dearborne Co., Ind. | Great Miami River | Central Lowlands |
| Fort Loudon | Monroe Co., Tenn. | Little Tennessee River | Ridge and Valley |

TABLE 2 (Continued)

| Cultural/Temporal Affiliations of Sites | Location | River Drainage | Physiographic Province |
|---|---|---|---|
| Mississippi | | | |
| Icehouse Bottom | Monroe Co., Tenn. | Little Tennessee River | Ridge and Valley |
| Nukaya | Talapoosa Co., Ala. | Talapoosa River | Coastal Plain |
| Eoff I | Coffee Co., Tenn. | Elk River | Interior Low Plateau |
| Fort Ancient | | | |
| W. S. Webb Memorial Shelter | Menifee Co., Ky. | Red River | Appalachian Plateau |
| Uncertain Affiliation | | | |
| Rock Creek Shelter | McCreary Co., Ky. | Cumberland River | Appalachian Plateau |
| Hooton Hollow Shelter | Menifee Co., Ky. | Red River | Appalachian Plateau |
| Buckner Hollow Shelter | Lee Co., Ky. | Kentucky River | Appalachian Plateau |
| Cloudsplitter Shelter | Menifee Co., Ky. | Red River | Appalachian Plateau |
| Green Bluff Shelter | Benton Co., Ark. | White River | Ozark Plateau |
| Indian Bluff Shelter | Benton Co., Ark. | White River | Ozark Plateau |
| Salt Bluff Shelter | Benton Co., Ark. | White River | Ozark Plateau |
| Bushwack Shelter | Benton Co., Ark. | White River | Ozark Plateau |
| Edens Bluff Shelter | Benton Co., Ark. | White River | Ozark Plateau |
| Gibson Shelter | Washington Co., Ark. | White River | Ozark Plateau |
| Putman Shelter | Washington Co., Ark. | White River | Ozark Plateau |
| Walden Shelter | Carrol Co., Ark. | White River | Ozark Plateau |
| Amos Shelter | Carrol Co., Ark. | White River | Ozark Plateau |
| Ash Cave | Madison Co., Ark. | White River | Ozark Plateau |
| Fort Loudon | Monroe Co., Tenn. | Little Tennessee River | Ridge and Valley |

river in Monroe and Loudon Counties, Tennessee have produced small quantities of maygrass grains. Although the analysis of the Bacon Bend site is incomplete, carbonized maygrass grains have been identified from seven of the nine features examined (Jefferson Chapman and Andrea Brewer: personal communications). Other plant food remains recovered include hickory, walnut, and acorn nuts, grape, *Chenopodium* sp., and sumac. Significantly, carbonized squash rinds (*Cucurbita pepo*) have also been identified from these features. Two radiocarbon assays from the site fix its occupation at somewhere around 2400 B.C.

One carbonized maygrass grain has been recovered from a pit feature at the Iddins site. A wide variety of carbonized plant food remains, including nuts of several species, squash, and gourd were also recovered. Chronometric analysis of carbonized plant materials from this site indicate an occupancy dating somewhere around 1705 and 1255 B.C. (ibid.).

One site in the Green River drainage of western Kentucky has also yielded evidence of maygrass utilization. Continuing investigation of the prehistoric use of Salts Cave in Edmonson County by Patty Jo Watson and her team of researchers has shown that by at least 1500 B.C. the grains of maygrass were an important seasonal food source. Analysis of flotation and paleofecal samples from the cave has led Yarnell and Watson to assert that maygrass was never an important, stored commodity but was heavily used during its period of maximum availability (Yarnell 1969, 1974; Watson 1974).

## Early Woodland

Three Kentucky sites have produced excellent evidence for maygrass utilization by Early Woodland populations. Maygrass was particularly prevalent in the desiccated materials from the Newt Kash Hollow shelter in the Licking River drainage of Menifee County (Jones 1936). Along with the culms of little bluestem grass (*Andropogon scoparius*), entire maygrass plants were an important constituent of the "beds." Jones also indicated that sheaves of the inflorescences were also present, and in the paleofeces he examined " . . . seeds of this grass were common and secondary only to those of marsh-elder (sumpweed)" (ibid.:152). Subsequent radiocarbon dating of some of the perishable material from Newt Kash yielded dates of 700 B.C. and 650 B.C. ± 300 (Crane 1956).

Both Salts and Mammoth caves in the Green River drainage have produced abundant evidence of maygrass utilization by the Early Woodland groups who entered them. Analysis of flotation and paleofecal samples has shown that the seeds were a common component of the Early Woodland diet (Yarnell 1969, 1974; Stewart 1974; Marquardt 1974; Watson 1974).

## Middle Woodland

Numerous Middle Woodland sites in the Little Tennessee, Duck and Elk river drainages of Tennessee have produced carbonized grains of maygrass. Several Owl Hollow Phase sites (Owl Hollow, Shofner, and Raus) in Bedford, Coffee, and Franklin counties have produced quantities of maygrass grains from various pit and midden contexts (Andrea Brewer and Charles Faulkner: personal communication). One feature from the Shofner site produced 245 carbonized grains in the 25% of the sample examined. A large sample of radiocarbon assays indicate a temporal span of these sites from between A.D. 200 to 500.

Two carbonized maygrass grains have been recovered from the McFarland phase Eoff I site in Coffee County. McFarland phase sites seem to have been occupied between 200 B.C. and A.D. 200 (McCollough and Faulkner 1976: 234-37).

Seventeen carbonized maygrass grains were recovered from a rock-filled fire basin at the Icehouse Bottom site in Monroe County, Tennessee, which Chapman (1973) assigned to the Middle Woodland occupation of the site.

## Late Woodland

Two sites in the Red River drainage of eastern Kentucky, and one in the Great Miami River drainage of southwestern Indiana have produced evidence of *Phalaris* utilization. Uncarbonized grains and inflorescences of maygrass have been recovered in small quantities from the Rogers and Haystack rock shelters in Powell County, Kentucky. Five radiocarbon dates from the Rogers shelters fix the time of occupancy at somewhere between A.D. 500 and A.D. 700 (Cowan 1973). No dates are available from the Haystack shelter, but cultural materials recovered from the site are identical to those found at the Rogers shelters (Cowan 1976). Thin limestone-tempered ceramics from both these sites are attributable to the Newtown phase of the Middle Ohio Valley (Ochler 1973). Other plant food remains include a variety of nuts, squash, gourd, sunflower, sumpweed, and *Chenopodium bushianum*.

The Leonard Haag site in Dearborne County, Indiana has also produced numerous carbonized maygrass grains. The exact chronological placement of these seeds is not clear since they were not found in association with any explicitly definable cultural entity. Crawford (Gary Crawford: personal communication) believes that they could be associated with either the Late Woodland Newtown component of the site or with the younger, Fort Ancient Madisonville component. Significantly, a carbonized cluster of 52 grains was found in one of the features at the site. The Newtown occupation probably began somewhere around A.D. 500 to 600, with the Madisonville occupation occurring somewhere between A.D. 1000 to 1300 (Reidhead 1976).

A single carbonized grain of maygrass was also recovered from a Late Woodland pit from the site of Fort Loudon in Monroe County, Tennessee on the banks of the Little Tennessee River. The majority of the occupation at this site is attributed to the historic Cherokee and British inhabitants of the fort, and comparatively little is known about earlier occupations (Carl Kuttruff: personal communication).

### Mississippi

Two sites in Tennessee, one in the Little Tennessee drainage and one in the Duck River drainage have produced carbonized maygrass grains. One carbonized grain was recovered from a large pit feature at the Icehouse Bottom site on the Little Tennessee River (Chapman 1973), and two were found in a feature at the Eoff I site in Coffee County on the Duck River (Andrea Brewer: personal communication). In addition to these Tennessee sites, seven carbonized grains were identified from a pit feature at the Nukaya site in Talapoosa County, Alabama (Gary Crawford: personal communication). All of these sites range in age from A.D. 1000 to 1200.

### Fort Ancient

A small quantity of desiccated maygrass inflorescenses was discovered in a small pit feature at the W. S. Webb Memorial rock shelter in Menifee County, Kentucky in the Red River drainage. This small, single component site seems to have been sporadically occupied by a few Fort Ancient people. Ceramics from the site are typical of the Madisonville Focus, and probably date to somewhere between A.D. 1000 and 1300 (Cowan 1975).

### Sites of Uncertain Cultural Affiliation

In addition to the sites discussed above, there is a considerable group of other sites that have produced maygrass remains whose occupations cannot be adequately fixed in time. Four of these sites are located in eastern Kentucky and ten in the Ozark region of Arkansas.

A tied bundle of maygrass inflorescences was discovered by an amateur archaeologist in a rock shelter on a tributary of the Cumberland River in McCreary County, Kentucky. Along with a wooden bowl and a length of rope, the maygrass was found accompanying the desiccated body of a mature human male. While Woodland materials were also located in the same overhang, the exact chronological placement of the maygrass bundle cannot be ascertained.

A single inflorescence of maygrass was recovered from the interior of a fragment of a large woven fiber bag from the Hooton Hollow rock shelter in the Red River drainage of Menifee County, Kentucky. While Haag (1974) has maintained that this site was occupied by Adena groups, this assertion is not tenable. Although Woodland and Fort Ancient artifacts were recovered, no typical Adena artifacts were found.

One inflorescence of maygrass was recently discovered in a looter's back-dirt pile at the Buckner Hollow shelter in the Kentucky River drainage of Lee County, Kentucky. This site was partially excavated by the University of Kentucky during the early part of this century (Funkhouser and Webb 1929), but at the present time it is impossible to ascertain the exact chronological placement of the site. Much of the material recovered by the University of Kentucky is attributable to the Woodland and Fort Ancient periods.

A single maygrass inflorescence has also been recovered from the backdirt of looters of the Cloudsplitter rock shelter in Menifee County, Kentucky (Cowan and Wilson 1977). Unfortunately this site has not been professionally excavated, and at present it is impossible to determine the cultural/temporal affiliation of these materials.

A number of rock shelter sites excavated by the Museum of the American Indian, Heye Foundation, and the University of Arkansas during the 1920s and 1930s yielded abundant amounts of *Phalaris caroliniana*. Information regarding this material is scarce, but notes made by M. R. Gilmore during the 1930s indicate that maygrass was found at the following Ozark sites: Benton County—Green Bluff shelter, Indian Bluff, Salt Bluff, Bushwack shelter, Edens Bluff; Washington County—Gibson shelter, Putnam shelter; Carrol County—Walden shelter, Amos shelter; and Madison County—Ash Cave. An examination of Gilmore's extensive notes on the vegetal remains from these sites indicates that actual bundles of tied maygrass inflorescences were discovered at the Gibson, Putnam, Indian Bluff, and Edens Bluff shelters. The occupation of the Edens Bluff site has been dated by two radiocarbon assays (A.D. 630 ± 120 and A.D. 1080 ± 110, according to Crane and Griffin 1968), but the cultural affiliation of these materials remains unclear.

Four carbonized maygrass grains were recovered from a circular bell-shaped pit at Fort Loudon in Monroe County, Tennessee. Kuttruff (personal communication) has tentatively identified this feature as Woodland in origin, but has not yet assigned it a position within this broad tradition.

## THE ROLE OF MAYGRASS IN PREHISTORIC EASTERN NORTH AMERICAN ECONOMIC SYSTEMS

Even though maygrass has been recovered from numerous archaeological sites in eastern North America, its role in prehistoric economic systems remains an enigma. In part, this is due to the nature of the paleoethnobotanical data base. The carbonized seeds of maygrass that have been recovered from open sites undoubtedly represent only a small fraction of what was originally present, and interpretations based on these meager samples are justifiably subject to sharp criticism.

Our best direct evidence of maygrass utilization comes from various analyses of prehistoric coprolites from Newt Kash Hollow, and Salts and Mammoth caves, Kentucky (Jones 1936; Yarnell 1969, 1974; Stewart 1974; Marquardt 1974). The importance of maygrass in the feces from Newt Kash has already been described. Of the 154 coprolites that have been examined from Salts and Mammoth caves, 27% contained varying amounts of maygrass grain. Even though from a statistical standpoint, the sample size from Salts and Mammoth caves is satisfactory, it is important to remember that the majority of the coprolites were deposited by prehistoric mirabilite miners. If a sexual division of labor existed in these Green River societies such that only men participated in mining activities, then the samples that have been analyzed possibly represent a biased sample of the population. In addition, as Watson (1974) has indicated, far too little is known about the movement of food waste in the human digestive tract; just how many meals does one coprolite represent?

Based on his initial analysis of 100 Salts Cave coprolites, Yarnell (1969) felt that maygrass was an important late spring-early summer food source. However, the sporadic occurrence of maygrass grains in these specimens indicated that the plant was not present in sufficient quantities to suggest its importance as a stored commodity. Based on his examination of these and subsequent samples, Yarnell (1974:122) has also suggested that the cave was not heavily utilized during the spring and early summer. Since this is the time of ripening of maygrass grains, once again we must reckon with the distinct possibility of sampling error; the feces may not be truly representative of the yearly subsistence base.

Data from flotation samples from a small Late Archaic to Early Woodland campsite in the vestibule of Salts Cave provided additional information regarding the importance of maygrass in the prehistoric economic systems of these people (Yarnell 1974). Maygrass, along with the seeds of a number of other small grains showed a steady increase through time in the stratigraphic column. In fact, in terms of seeds per gram of material examined, the grains of maygrass were more abundant than those of sunflower and sumpweed except in two levels. This undoubtedly reflects the differential dispersal rates of the larger achenes of sunflower and sumpweed versus those of maygrass. Whereas the achenes of the latter species are retained in their fruiting disks, those of maygrass fall free much more easily. If, however, these figures are indicative of the amount of plant foods being processed, then it appears that maygrass was heavily utilized, and it, like sunflower and sumpweed, formed a significant portion of the diet.

Unfortunately, data from the other archaeological sites where maygrass has been discovered may not provide an accurate reflection of the economic importance of maygrass. Except in a few cases, the sample of *Phalaris* grains from any one site is negligible. The discovery of a cluster of 52 charred maygrass grains at the Haag site in Indiana and the recovery of 245 grains from a small sample of pit fill from the Shofner site in Tennessee, seem to indicate that maygrass was harvested in appreciable quantities, but data from other sites are inconclusive.

In part, the deficiencies in our understanding of the relative importance of maygrass are due to our lack of knowledge concerning the method(s) by which maygrass was prepared. Grains that have been discovered in paleofeces from dry caves and rockshelters reveal that no attempt was made to separate the grain from the surrounding chaff; all grains from feces are still enclosed by the surrounding lemma and palea. A similar lack of preparation has been noted for the seeds of sunflower, sumpweed and *Chenopodium* (Jones 1936; Yarnell 1969). Perhaps all were simply boiled to make a thick gruel. On the other hand, the presence of carbonized maygrass grains in numerous sites (including Salts Cave) suggest that fire may also have been directly applied to the grains to prepare them for consumption. Obviously these differences in method of preparation seriously affect the amount of maygrass grain that might become part of the archaeological record.

In spite of these exigencies, certain observations concerning the role of maygrass in the yearly subsistence base can be made. Maygrass begins to ripen during early May and by the end of June is no longer available for harvesting. In temperate eastern North America the natural availability of plant foods in the late spring and early summer months is characteristically low. Theoretically, by this time stored food reserves from the previous fall would have been exhausted. In such a situation, the collection of maygrass would help even out a low point in the yearly cycle of resource availability and, if present in abundance, could provide food for future utilization.

## POSSIBLE ETHNOGRAPHIC ACCOUNTS OF MAYGRASS UTILIZATION IN EASTERN NORTH AMERICA

None of the early travelers and explorers in the southern portions of eastern North America who observed native plant usage described a plant that may be positively identified as *Phalaris caroliniana*. There are, however, at least two tantalizing accounts that may indicate use of the plant.

Captain John Smith, in describing the foods of the Indians of coastal Virginia in the year 1612 noted the following:

> *Mattoume* which groweth as our bents do in meddows. The seed is not much unlike to rie, though much smaller. This they use for a dainty bread buttered with deere suet. [Smith 1612 (Arber Edition 1884:58)].

Clearly what Smith observed was a species of grass. In current botanical usage, bent grass refers to any of a number of species of the genus *Agrostis.* None of the *Agrostis* species bear any resemblance to maygrass, but it is not clear whether Smith meant that the grass he saw simply grew in large stands, or whether it actually resembled bent grass.

Later, in the first two decades of the 18th century, Le Page Du Pratz noted of the Natchez Indians of Louisiana that

> They also make food of two grains, of which one is called *choupichoul*, which they cultivate without difficulty, and the other is the *widlogouill*, which grows naturally and without any cultivation. These are two kinds of millet which they hull in the same way as rice. [Du Pratz 1758, quoted in Swanton 1911:76].

In the same context Du Pratz referred to the *choupichoul* as the "belle dame sauvage"—a common European name for members of the genus *Atriplex.* Recently, Asch and Asch (1977) have suggested that *choupichoul* might also have been a member of the genus *Chenopodium.* Unfortunately, Du Pratz left no other clue as to the identity of *widlogouill.* Since he indicated that the plant was similar to millet, and that the grains were hulled in the same manner as rice, *widlogouill* may have been a member of the grass family.

It should be remembered that neither Smith nor Du Pratz were trained botanists. Maygrass had not yet been introduced to the early botanists of the day, and it was not until 1753 that Linnaeus first described the Mediterranean *Phalaris canariensis* (Anderson 1961:4). Walter's description of the holotype for Carolina canary grass (maygrass) was not published until 1788 (Walter 1788). Thus even if Smith and Du Pratz had been trained as botanists, they would have been describing a plant that was unknown to the rest of the botanical world.

These brief references to the use of a New World grass as a source of food certainly do not constitute impeccable evidence for the use of maygrass but nonetheless should be noted.

## MAYGRASS AND THE QUESTION OF CULTIVATION

The survey of archaeological occurrences of maygrass in eastern North America demonstrates conclusively that the small grains were utilized as a food source on at least a seasonal, if not a regular basis. Moreover, its discovery with some of the earliest tropical and native cultigens reported from eastern North America indicates an intimate association with an early horticultural complex. We may now return to the question posed at the beginning of this paper—"Was maygrass, like *Iva* and *Helianthus*, another distinctly eastern cultigen?" Unfortunately, there is no easy answer to this question.

If the plant that Smith and Du Pratz observed being utilized during the early 15th and mid-16th centuries was indeed maygrass, then we must conclude that *Phalaris* was not a cultigen. Both authors indicate that the plant they saw grew wild. As noted previously, there is a distinct possibility that this was not *Phalaris*. In light of the uncertainty attendant in their identifications, we must turn to other evidence.

As a number of authors have noted, evidence for domestication might be shown by (1) genetic change in a ruderal ancestor, or (2) prehistoric extension of the range of the wild form (Yarnell 1965; Flannery 1965; Asch and Asch 1977). To date there is no indication of any change in the gross morphology of maygrass over the three millennia of its recorded utilization. While a change in seed size has been cited by Yarnell (1972) as evidence of domestication of sumpweed, all archaeological grains of maygrass fall well within the published range of modern populations.

If maygrass became a true domesticate, dependent upon man for the completion of its lifecycle, we might also expect to see at least two biological changes in the plant. An increase in the length of the inflorescence, with a concomitant decrease in natural variability, is one of these. Unfortunately, inflorescences have not been preserved except in a few rock shelter sites, but the specimens that have been recovered exhibit a range of variability that can be duplicated by modern populations (Fig. 4).

Another expected change in maygrass would be the loss of its indeterminate form of inflorescence. Since not all of the grains in the inflorescence mature at the same time, it is virtually impossible to harvest the plant at a time when all grains are present. Because of this factor, maximum productivity could be realized only if the indeterminate inflorescence is lost and all grains ripen at once. This is true of other domesticated grasses such as corn, wheat, barley, etc. Although I have never examined the maygrass from Salts or Mammoth caves, florets recovered from the Late Woodland Rogers rock shelters in Powell County, Kentucky occur in both fully mature and immature forms. This may indicate that when the inflorescences were collected they contained both mature and immature florets. If such was the case, then it must be concluded that the indeterminate form of inflorescence was present in prehistoric times.

At the present time our best evidence for the cultivation, but not domestication, of maygrass comes from numerous archaeological sites where grains have been recovered far outside the natural distribution of the plant (Fig. 3). Moreover, many of these, particularly those sites in the Appalachian Plateau region of Kentucky, and the Ridge and Valley province of eastern Tennessee

Fig. 4. Maygrass inflorescences from the Newt Kash Hollow shelter, Menifee County, Kentucky, showing the range of variation that typifies the collection (UMMA-EL Cat. No. 16441).

are situated in areas that during prehistoric times would have been dominated by dense deciduous forests. As we have seen, maygrass is poorly adapted to such low-light habitats, requiring strong sunlight for population maintenance. In addition, maygrass seems to prefer disturbed habitats; in densely forested areas, such niches could have been provided through cultural activities such as opening up the canopy for garden plots. Since maygrass has been found associated with other members of a garden complex in these areas, we may conclude that its necessary habitat requirements were most likely met by man-induced disturbance.

In spite of such an assertion, the question may legitimately be raised—"Such habitats are plentiful today; why hasn't maygrass invaded these?" At the present time, an answer to this question cannot be given; however, the answer may be intimately associated with the temperature requirements of the plant. The seeds of most plants of temperate zones require some exposure to cold soil conditions in order to break physiological dormancy. At the same time, if the temperature is too cold, the seed dies. This may partially explain the present-day southern distribution of maygrass; winter soil temperatures further north may inhibit germination of the grain. If experimental research proves this to be the case, then the prehistoric distribution of the plant in these more northern areas may indicate that cultural mechanisms maintained viable seeds over periods of cold. The ameliorating climate on the interior of caves and rock shelters, or other habitation structures, might be ideal localities for this type of storage. Such a proposition is supported by the discovery of sheaves of maygrass in Ozark and eastern Kentucky rock shelters, found stored away with the seeds of other cultigens.

The northern archaeological distribution of maygrass might also be explained by an actual extension of the natural range of the plant during the past. As cultural disturbance of the landscape increased through time, many new areas would have become available for the invasion of the plant. Later, with the introduction of a plethora of European weeds, maygrass may not have been able to compete successfully and was forced back into its original range. Again, however, such a proposition needs to be verified through experimental research and is one that at present does not carry a great deal of explanatory power.

## CONCLUSIONS

Based on current archaeological and botanical evidence, there is no indication that maygrass, like sumpweed and sunflower, ever became a true domesticate. At the same time, the presence of maygrass in a variety of archaeological contexts far north of its modern distribution requires an explanation. At present, the best answer seems to be that maygrass was introduced into these northern areas by man. The early association of maygrass with a series of manipulated tropical and indigenous eastern North American cultigens indicates that at least initially it was a welcome, and probably protected, garden adventive.

If the growth of maygrass is indeed inhibited by colder northern winter temperatures, then its continued existence in areas outside its natural distribution would be dependent upon interference by man. By providing a suitable micro-climate (in this case actually storing the seed) for the retention of biologically viable grains, man could have maintained an otherwise disjunct distribution. As we have seen, maygrass was in use for at least three thousand years in northern latitudes, thereby strengthening such a hypothesis. This suggestion also carries with it the implicit assumption that maygrass was intentionally propagated from seed. In the absence of any observed morphological change in the plant, it must also be concluded that the selective pressures necessary to induce genetic change were minimal. As Asch and Asch (1977:24) have recently noted, when plant species adapted to disturbed habitats are introduced to garden plots, the "habitat under cultivation may be quite similar to the habitat of the neglected weed." Under such circumstances one might not expect to see a morphological change in the plant.

Before we can accept with certainty that maygrass was actually cultivated, more information needs to be amassed concerning the biology of the plant. Although I have tentatively suggested that the present range of maygrass may in part be explained by winter temperature regimes, there could easily be other explanations as yet undiscovered, since so little is actually known about the natural history of the plant. Such aspects as the frequency of occurrence of the plant, size of local populations, range of variability within these populations, regional variation in flowering and fruiting cycles, and mutation rates and loci of mutations (if any) remain to be investigated.

**Acknowledgments**

A great number of individuals have contributed directly to this undertaking. Thanks are extended to Dr. Richard A. Yarnell and Mr. Gary Crawford of the University of North Carolina for information regarding *Phalaris* utilization in their files, as are Drs. Jefferson Chapman and Charles Faulkner of the University of Tennessee. Ms. Andrea Brewer Shea, also of the University of Tennessee provided much of the information regarding *Phalaris* utilization in the Owl Hollow and the McFarland Phase sites in Tennessee. Thanks are also extended to Mr. Charles Long of Georgetown, Kentucky, for allowing me to examine a bundle of *Phalaris* in his private collection.

My sincere thanks are also extended to the curators of the various herbaria who kindly responded to a questionnaire that was sent out from the University of Michigan Ethnobotanical Laboratory. Without their cooperation the distributional map included in this paper could never have been constructed, and the data concerning habitat preferences of *Phalaris* would have been incomplete.

Dr. Richard I. Ford is to be thanked for his constant encouragement and support throughout all phases of the preparation of the manuscript, and for allowing the questionnaire to be sent under postage provided by the University of Michigan Museum of Anthropology.

Ms. Shelley K. G. Cowan is to be acknowledged for her excellent photograph of the maygrass pictured in Figure 4, as are the cartographic and artistic skills of Ms. Jane Mariouw, illustrator for the University of Michigan Museum of Anthropology, who provided Figures 2 and 3.

Finally my warmest regards and thanks are extended to Mr. Volney H. Jones, Curator Emeritus of the University of Michigan Museum of Anthropology. Although unaware of its final destination, Mr. Jones willingly opened his and M. R. Gilmore's files concerning *Phalaris* utilization during the preparation of the manuscript. His advice and encouragement in approaching a problem of considerable personal interest are deeply appreciated.

## REFERENCES

Anderson, Dennis E.
1961 Taxonomy and Distribution of the Genus *Phalaris*. Iowa State Journal of Science 36(1):1-96.

Asch, David L. and Nancy B. Asch
1977 Chenopod as Cultigen: A Re-evaluation of Some Prehistoric Collections From Eastern North America. Midcontinental Journal of Archaeology 2(1): 3-45.

Braun, E. Lucy
1967 The Vascular Flora of Ohio: Vol. 1, The Monocotyledoneae. Columbus: The Ohio State University Press.

Chapman, Jefferson
1973 The Icehouse Bottom Site, 40MR23. Reports of Investigations Number 13. Department of Anthropology, University of Tennessee.

Cowan, C. Wesley
1973 Prehistoric Plant Utilization at the Roger's Rock Shelters, Powell County, Kentucky. Unpublished Paper Presented at the 30th Annual Meeting of the Southeastern Archaeological Conference, Memphis, Tennessee.
1975 An Archeological Survey and Assessment of the Proposed Red River Reservoir in Wolfe, Powell, and Menifee Counties, Kentucky. Report submitted to the National Park Service Interagency Archeological Services Office, Atlanta, Georgia by the University of Kentucky Museum of Anthropology.
1976 Test Excavations in the Proposed Red River Lake: 1974 Season. Report submitted to the National Park Service Interagency Archeological Services Office, Atlanta, Georgia by the University of Kentucky Museum of Anthropology.

Cowan, C. Wesley, and Frederick T. Wilson
1977 An Archaeological Survey of the Red River Gorge Area. Frankfort, Kentucky: The Kentucky Heritage Commission.

Crane, H. R.
1956 University of Michigan Radiocarbon Dates I. Science 124:664-72.

Crane, H. R. and James B. Griffin
1968 University of Michigan Radiocarbon Dates XII. Radiocarbon 10:61-114.

Febrel, J. and A. Carballido
1965 Estudio Bromatologico del Alpisto. Annales de Bromatologia 17(4):345-60.

Flannery, Kent V.
1965 The Ecology of Early Food Production in Mesopotamia. Science 147:1247-56.

Funkhouser, W. D. and W. S. Webb

1929 The So-Called "Ash Caves" in Lee County, Kentucky. University of Kentucky Reports in Archaeology and Anthropology 1(2):37-112.

1930 Rock Shelters of Wolfe and Powell Counties, Kentucky. University of Kentucky Reports in Archaeology and Anthropology 1(4):239-306.

Gilmore, Melvin R.

1931 Vegetal Remains of the Ozark Bluff-Dweller Culture. Papers of the Michigan Academy of Science, Arts and Letters 14:83-102.

Haag, William G.

1974 The Adena Culture. In: Archaeological Researches in Retrospect, edited by Gordon C. Willey, pp. 119-145. Winthrop Publishers, Inc., Cambridge Massachusetts.

Harrison, Carter M. and John F. Davis

1966 Reed Canarygrass for Wet Lowland Areas of Michigan. Cooperative Extension Service, Michigan State University, Bulletin E-517.

Hitchcock, A. S.

1950 Manual of the Grasses of the United States. United States Department of Agriculture Miscellaneous Publication No. 200. Washington, D.C.: United States Government Printing Office.

Jones, Volney H.

1936 The Vegetal Remains of Newt Kash Hollow Shelter. In: Rockshelters in Menifee County, Kentucky, edited by W. S. Webb and W. D. Funkhouser. University of Kentucky Reports in Archaeology and Anthropology 3(4): 147-65.

Marquardt, William H.

1974 A Statistical Analysis of Constituents in Human Paleofecal Specimens from Mammoth Cave. In: Archeology of the Mammoth Cave Area, edited by P. J. Watson, pp. 193-202. New York: Academic Press.

McCollough, Major C. R. and Charles H. Faulkner

1976 Normandy Archaeological Project Volume 3. Department of Anthropology, University of Tennessee, Reports of Investigations Number 16.

Myers, R. Maurice

1967 Forage Plants. Western Illinois Bulletin 46(3):1-56. Series in the Biological Sciences Number 5.

Oehler, Charles

1973 Turpin Indians. The Journal of the Cincinnati Museum of Natural History 23(2).

Payne, Willard W., and Volney H. Jones

1962 The Taxonomic Status and Archaeological Significance of a Giant Ragweed from Prehistoric Bluff Shelters in the Ozark Plateau Region. Papers of the Michigan Academy of Science, Arts, and Letters 47:147-63.

Radford, Albert E., H. E. Ahles, and C. R. Bell

1964 Manual of the Vascular Flora of the Carolinas. Chapel Hill: The University of North Carolina Press.

Reidhead, Van A.

1976 Optimization and Food Procurement at the Prehistoric Leonard Haag Site, Southeastern Indiana: A Linear Programming Approach. Ph.D. Dissertation, Indiana University. Ann Arbor, Michigan: University Microfilms.

Sheffer, Charles

1972 Notes on Agronomy with Special Reference to Wheat, Barley, Sorghum, and the Millet. Unpublished Ph.D. Preliminary Paper, on file at the Department of Anthropology, University of Michigan.

Schwartz, Douglas W.

1967 Conceptions of Kentucky Prehistory: A Case Study in the History of Archaeology. Studies in Anthropology Number 6. Lexington: University of Kentucky Press.

Smith, Captain John
1612 A Map of Virginia. In: Captain John Smith, Works 1608-1631, edited by Edward Arber. Birmingham, England: The Editor, 1884.
Swanton, John R.
1911 Indian Tribes of the Lower Mississippi Valley and Adjacent Coast of the Gulf of Mexico. Bureau of American Ethnology Bulletin 43.
Stewart, Robert B.
1974 Identification and Quantification of Components in Salts Cave Paleofeces 1970-1972. In: Archeology of the Mammoth Cave Area, edited by P. J. Watson, pp. 41-47. New York: Academic Press.
Sturtevant, E. Lewis
1919 Sturtevant's Notes on Edible Plants, edited by V. P. Hendrick. New York State Department of Agriculture Annual Report (1918-19) 27(2).
Walter, Thomas
1788 Flora Caroliniana. Londini: J. Fraser.
Watson, Patty Jo
1974 Theoretical and Methodological Difficulties Encountered in Dealing with Paleofecal Material. In: Archeology of the Mammoth Cave Area, edited by P. J. Watson, pp. 239-42. New York: Academic Press.
Webb, W. S., and W. D. Funkhouser
1936 Rockshelters in Menifee County, Kentucky. University of Kentucky Reports in Archaeology and Anthropology III(4):101-67.
Yarnell, Richard A.
1965 Early Woodland Plant Remains and the Question of Cultivation. Florida Anthropologist 18:77-82.
1969 Contents of Paleofeces. In: Prehistory of Salts Cave, Kentucky, edited by P. J. Watson, pp. 41-54. Illinois State Museum Reports of Investigations 16.
1972 *Iva annua* var. *macrocarpa*: Extinct American Cultigen? American Anthropologist 74:335-41.
1974 Plant Food and Cultivation of the Salts Caves. In: Archeology of the Mammoth Cave Area, edited by P. J. Watson, pp. 113-22. New York: Academic Press.

# DOMESTICATION OF SUNFLOWER AND SUMPWEED IN EASTERN NORTH AMERICA

*Richard A. Yarnell*
Department of Anthropology and
Research Laboratories of Anthropology
University of North Carolina, Chapel Hill

Only two plants are generally considered to have been domesticated aboriginally in eastern North America, but neither case is universally accepted. There are statements suggesting that the giant cultigen sunflower (*Helianthus annuus* var. *macrocarpus* Ckll.) originated in the American Southwest, though there is little to support this contention. The cultigen status of sumpweed (*Iva annua* var. *macrocarpa* Jackson) has been questioned because there has never been a recorded observation of the plant which produced the giant archaeological seeds. Apparently the variety became extinct in late prehistoric or early historic times, perhaps to our own economic detriment.

I am indebted to Charles B. Heiser, Jr., to Raymond C. Jackson, and especially to Volney H. Jones for providing information about these plants and their significance and for the stimulation to undertake analyses of old and new archaeological collections that are pertinent to these plants' evolution under domestication. Jones' analyses of archaeological plant remains from the East and also from the Southwest and his ethnographic research have provided much of the original framework which led to and oriented many other studies. His seminal work on materials from Newt Kash Hollow rock shelter (1936) provided fundamental data and laid the groundwork for further investigations of this nature. Among other results, it demonstrated considerable antiquity for sunflower and sumpweed remains and indicated that these plants preceded maize as cultigens in the East.

Sunflower, especially in full bloom, has a rather distinctive appearance, whereas sumpweed is quite an ordinary-looking plant; but these two species are similar in some respects. They are members of the same tribe of the family Asteraceae (Compositae). The seeds are similar in color, texture, flavor, and high oil content. Their nutritional values probably do not differ significantly. Sunflower seeds are ordinarily about 50% longer than sumpweed seeds from the same general time period, but widths are similar except for the cultigen forms, after about A.D. 1000, when sunflower seeds surpassed sumpweed seeds in all dimensions (see Tables 1 and 2).

As in all members of the family, the fruit is an achene which consists of a single seed enclosed in a dry, indehiscent pericarp. Sunflower and sumpweed achene pericarps are thin, with a fine longitudinally oriented structure and surface texture. Sumpweed pericarps are considerably thinner than those of sunflower and have a finer texture. Substantial variation is exhibited in the achenes of both species, but variation in shape and external surface of sumpweed achenes from a single population, or even from a single plant, are especially marked.

Habitat tolerances for both species are rather broad so long as there is little competition with other plants. Habitat disruptions are generally to their advantage, whereas developing ecological succession eliminates them. Soil moisture preference is for mesic to moderately hydric conditions for sumpweed and mesic to moderately xeric conditions for sunflower. This may be the reason for the unexpectedly small size of sunflower achenes from the Boyd site in northwestern Mississippi where soil moisture is likely to have been relatively high. Sumpweed achenes from the same site are slightly larger than expected for the estimated date of A.D. 500 (John M. Connaway 1977: personal communication) and are approximately equal in size to the sunflower achenes. (For a more extended description of habitat preference and tolerance and for further relevant information, see Heiser 1951, 1954, 1955, 1969, and 1976 for sunflower; see Jackson 1960, Black 1963, and Yarnell 1972 for sumpweed.)

The wild form of the common sunflower appears to be native to the general region around the Colorado Plateau. Unlike sumpweed, it is not native to eastern North America. Nevertheless, all of the archaeological evidence of the early stages of sunflower domestication has been recovered from Kentucky and Tennessee and a few adjacent locations. Modern sunflowers are abundantly represented throughout the greater Midwest and in the Great Plains by large ruderal weed forms growing in highly disturbed habitats. It is possible that they represent an ancestral type intermediate between wild and cultigen varieties, but it is just as likely that the ruderal forms are feral descendents of early domesticated forms. It might be more realistic to view all of the eastern forms of *Helianthus annuus* as members of a polymorphic evolving and interbreeding complex, but a certain amount of genetic isolation from ruderal forms would be necessary for the ultimate development of the giant monocephalic cultigen sunflower.

By early historic times, the cultigen sunflower in the East was distributed from eastern Texas to North Carolina, Quebec, and North Dakota. There is no evidence for aboriginal sunflower husbandry from Alabama, Georgia, Florida, or South Carolina. Although there is no archaeological evidence of sunflower husbandry in Mexico or the Southwest, there are historical and ethnographic references to cultigen sunflower in those areas (Heiser 1951, 1955).

Sumpweed is apparently native from southern Illinois southward to Mississippi and northeastern Mexico and westward along river valleys into the Great Plains (Jackson 1960; Black 1963). Archaeological evidence indicates that the early stages of its domestication took place in Kentucky, southern Illinois, and eastern Missouri, probably in the region of sunflower domestication and at the same time. Like sunflower it exhibits a continuing increase in seed size from approximately 1500 B.C. to late prehistoric times, when the cultigen sumpweed with giant achenes disappeared without a trace. The greatest extent of its distribution was during the Mississippian period when it was grown from Arkansas and Iowa to western North Carolina. This is a much more restricted range than that of prehistoric cultigen sunflower.

Tables 1 and 2 summarize the available metric data from the more significant and better documented collections of sunflower and sumpweed seeds and achenes. For purposes of comparison these tables also present the available metric data for modern sunflower and sumpweed.

The original wild ancestral sunflowers probably had mean achene lengths of 4.5 to 5.0 mm. This estimate is based on the modern achene length range of 4.0 to 5.5 mm as given by Heiser (1954) for *H. annuus lenticularis*. Modern ruderal sunflower achene length ranges from 4.0 to 7.0 mm, and modern cultigen achenes range from about 6 mm to more than 20 mm in length. Mean achene length of measured samples from modern varieties ranges from 7 to 15 mm.

While the modern ruderal sunflower achenes are intermediate in size between wild and modern cultigen achenes, the early archaeological achenes are intermediate in size between the ruderals and the modern cultigens but with considerable overlap in size. This indicates that the early archaeological sunflower remains are from populations which were in early, but not initial, stages of domestication, even if achenes of the ancestral stock were as large as those of modern ruderals. However, we have no basis for assuming that achenes from sunflowers in the initial stages of domestication were as large as those from modern ruderals on the average. In fact, they may have had

TABLE 1

SUNFLOWER ACHENES (*Helianthus annuus*)[a]

| Modern | | | Range of L and W | | |
|---|---|---|---|---|---|
| wild | | | 3.0-5.5 x? | | |
| ruderal | | | 4.0-7.0 x? | | |
| cultigen | | | 6.3-20+ x 3.2-12+ | | |
| **Archaeological** | **N** | **Mean LxW=LW** | **Range of L and W** | **Dating** | **Data Source[b]** |
| *Late Archaic to Early Woodland* | | | | | |
| [c] Higgs, TN | 24 | 7.8x3.1=24 | 5.7-10.1x1.8-4.2 | 900 B.C. | Shea |
| [c] Salts Cave J4:10-11 | 3 | 6.1x3.0=18 | 5.6-6.8x2.5-3.1 | 1500 B.C.? | |
| [c] Salts Cave J4:5-8 | 7 | 7.2x3.3=24 | 6.1-8.1x2.3-3.9 | 1000 B.C.? | |
| [c] Salts Cave J4:4 | 47 | 7.5x3.3=25 | 6.0-9.7x1.8-4.2 | 500 B.C.? | |
| Salts Cave feces | 1000 | 7.4x3.2=24 | 5.3-9.8x2.2-4.8 | 700-300 B.C. | |
| Mammoth Cave cadaver | 80 | 7.0x3.1=22 | 5.1-8.6x2.3-4.0 | | |
| [c] Mammoth Cave-Nelson | 39 | 10.4x4.7=49 | 7.4-12.4x3.8-5.7 | ? | |
| Newt Kash Hollow, KY | 14 | 8.6x3.4=29 | 7.0-11.1x3.0-4.2 | ? | Cowan |
| *Middle to early Late Woodland* | | | | | |
| [c] Owl Hollow, TN | 9 | 7.4x4.6=34 | 6.5-7.8x3.5-5.6 | A.D. 200-600 | Shea |
| [c] Boyd, MS | 10 | 7.3x3.4=25 | 5.9-8.4x2.3-3.1 | ca. A.D. 500 | |
| Hooten Hollow, KY | 4 | 9.0x4.0=36 | 8.6-9.2x4.0 | ? | Cowan |
| Haystack Shelter, KY | 2 | 9.0x4.0=36 | | A.D. 500-700 | Cowan |
| Rogers Shelters, KY | 11 | 8.6x4.1=35 | 6.6-10.0x3.0-5.0 | A.D. 500-700 | Cowan |

TABLE 1 (Continued)

| Archaeological | N | Mean LxW=LW | Range of L and W | Dating | Data Source[b] |
|---|---|---|---|---|---|
| *Mississippian* | | | | | |
| [c] Great Oasis, IA | 4 | 11.9x4.7=56 | 11.7-12.4x3.6-5.6 | A.D. 880-1070 | L. Blake |
| [c] Mitchell, SD | 76 | 10.3x? | 6.6-17.0x? | A.D. 985-1125 | D. Benn |
| [c] Wilford, MS | 16 | 9.8x4.2=41 | 8.5-11.3x3.3-5.3 | | |
| [c] Steed Kisker, MO | 6 | 10.8x5.3=58 | 9.7-12.7x4.3-7.0 | | |
| [c] Paul McCulloch, MO | 1000 | 12.6x6.5=82 | 7.7-16.9x3.6-9.6 | A.D. 1100-1200 | Waselkov |
| [c] Turner-Snodgrass, MO | 3 | 11.5x6.6=76 | 10.5-13.1x4.5-5.7 | A.D. 1300 | L. Blake |
| [c] Campbell, I., OH | 299 | 11.5x6.8=78 | 9.1-14.1x4.4-9.8 | | |
| [c] Cramer, OH | 20 | 11.6x6.7=78 | 10.0-13.4x5.0-9.1 | | |
| *Ozark Bluff Dwellers* | | | | | |
| Harrington | 9 | 9.3x4.8=45 | 8.2-10.1x4.4-5.2 | | |
| Dellinger | 3 | 9.3x5.6=52 | 8.9-10.0x4.7-6.2 | | |
| Indeterminate | 10 | 11.4x7.0=80 | 10.4-12.7x5.9-7.7 | | |

[a]Modified from Heiser (1953) and Yarnell (1977); all dimensions are mm.
[b]Measured by Heiser or Yarnell unless specified.
[c]Dimensions derived by conversion from dimensions of carbonized seeds or achenes.

TABLE 2
SUMPWEED ACHENES (*Iva annua* L.)[a]

| Modern | N | Mean LxW=LW | Range of L and W |
|---|---|---|---|
| S. F. Blake (1939) | ? | | 2.3-3.8x1.8-2.6 (mid range: 3.0x2.2=7) |
| R. C. Jackson (1960) | ? | | 2.0-4.5x? (2 varieties) |
| St. Louis RR Yard | 60 | 2.9x2.5=7 | 2.3-3.8x1.7-3.2 (width distortion=0.3) |
| Heiser field | 277 | 2.8x2.2=6 | 2.0-4.0x1.5-3.4 (12 plants; LW=5-10) |
| Herberger garden | 330 | 3.5x2.2=8 | 2.8-4.6x1.7-3.0 (1 plant: 1882 achenes) |
| Heiser greenhouse | 818 | 3.7x2.6=10 | 2.4-5.4x1.6-4.0 (9 plants; LW=7-14) |
| Heiser polyploid | 201 | 4.4x3.0=13 | 3.0-6.0x2.0-4.8 (6 plants; LW=11-18) |

| Archaeological | N | Mean LxW=LW | Range of L and W | Dating | Data Source[b] |
|---|---|---|---|---|---|
| *Middle Archaic* | | | | | |
| [c] Koster, IL-Level 6 | 20 | 3.0x2.1=6 | 2.4-3.6x1.7-2.5 | 3500 B.C. | Asch |
| *Late Archaic to Early Woodland* | | | | | |
| [c] Salts Cave J4:7-11 | 23 | 3.5x2.5=9 | 2.5-5.7x1.8-3.2 | 1000-1500 B.C.? | |
| [c] Salts Cave J4:4-6 | 286 | 3.7x2.7=10 | 1.7-5.7x1.4-4.0 | 500-1000 B.C.? | |
| [c] Salts Cave misc. | 87 | 3.9x2.8=11 | 2.6-5.9x1.7-3.7 | 500-1000 B.C.? | |
| Salts Cave feces | 879 | 4.2x3.2=14 | 2.8-7.0x2.2-5.0 | 300-700 B.C.? | |
| Salts Cave cadaver | 53 | 4.5x3.7=17 | 3.3-6.1x2.8-4.5 | A.D. 10 | |
| Mammoth Cave cadaver | 40 | 4.0x3.1=12 | 2.5-5.7x2.5-3.8 | | |
| [c] 30 Acre I., TN | 4 | 4.5x3.2=14 | 4.0-5.2x2.9-3.7 | | Shea |
| [c] Collins, MO | 19 | 4.9x3.4=17 | 3.3-6.5x2.3-5.0 | ca. 600 B.C. | |
| Cloudsplitter, KY | 11 | 4.5x3.2=14 | 4.0-6.2x2.8-3.8 | ? | Cowan |
| Newt Kash Hollow, KY | 74 | 5.5x3.9=21 | 4.2-7.5x3.0-5.0 | ? | Cowan |

TABLE 2 (Continued)

| Archaeological | N | Mean LxW=LW | Range of L and W | Dating | Data Source[b] |
|---|---|---|---|---|---|
| *Middle to early Late Woodland* | | | | | |
| [c] Macoupin, IL | 193 | 4.8x3.3=16 | 2.8-8.5x1.7-5.4 | | Asch |
| [c] Apple Creek, IL | 19 | 5.4x3.6=19 | 4.4-6.2x2.9-4.8 | | |
| [c] Boyd, MS | 20 | 6.1x4.2=26 | 4.7-7.4x3.2-5.4 | ca. A.D. 500 | |
| [c] Stillwell, IL | 65 | 5.9x3.9=23 | 4.6-7.0x2.8-5.3 | | |
| Hooton Hollow, KY | 13 | 5.7x3.9=22 | 4.0-7.4x3.0-5.0 | ? | Cowan |
| Haystack, KY | 74 | 6.0x4.2=25 | 4.0-7.5x3.0-6.0 | A.D. 500-700 | Cowan |
| Rogers Shelters, KY | 19 | 6.2x4.2=26 | 5.4-7.7x3.0-5.0 | A.D. 500-700 | Cowan |
| *Mississippian* | | | | | |
| [c] Great Oasis, IA | 7 | 6.5x4.3=28 | 5.8-7.4x3.1-5.0 | A.D. 880-1070 | |
| [c] Friend & Foe, MO | 6 | 5.5x3.7=20 | 4.1-6.2x2.6-3.9 | A.D. 1100 | |
| [c] Paul McCulloch, MO | 19 | 7.0x4.5=32 | 5.5-8.8x3.9-5.3 | A.D. 1100-1200 | |
| [c] Turner-Snodgrass, MO | 33 | 7.3x4.5=33 | 6.0-8.7x3.6-5.3 | A.D. 1300 | |
| [c] Warren-Wilson, NC | 6 | 5.9x4.3=25 | 4.8-6.6x3.9-4.8 | A.D. 1250-1450 | |
| *Ozark Bluff Dwellers* | | | | | |
| Montgomery, MO | ? | | 4.8-9.3x3.2-5.7 | | S. Blake |
| Proether, MO | 10 | 7.0x5.2=36 | 6.0-8.0x4.5-6.0 | | Ford |
| Alred, 103, AR | 50 | 7.6x4.9=37 | 6.0-9.8x4.0-6.0 | | |
| Craddock 380, AR | 50 | 5.5x3.9=21 | 4.2-7.6x2.9-5.4 | early? | |
| Craddock 552, AR | 56 | 7.1x5.0=36 | 5.6-8.8x3.6-6.7 | | |
| Edens 1706, AR | 40 | 5.5x3.9=21 | 4.2-7.6x2.8-5.1 | early? | |
| Edens 980a, AR | 45 | 7.5x5.1=38 | 6.1-9.5x4.1-6.1 | | |

[a] Modified from Yarnell (1972,1977); all dimensions are mm.
[b] Measured by L. Blake, H. C. Cutler, and/or Yarnell (cf. 1972) unless specified.
[c] Dimensions derived by conversion from dimensions of carbonized seeds or achenes.

mean dimensions no larger than 5.0 x 2.5 mm at a date of perhaps 2000 B.C., which is more recent than the earliest evidence of plant husbandry in Kentucky and Missouri (Chomko and Crawford 1978).

The oldest collections of sunflower seeds and achenes are from central Kentucky and eastern Tennessee. Those from the Higgs site in Tennessee are rather long and narrow and are securely dated to about 900 B.C. (Brewer 1973; McCollough 1973). Several samples, uncarbonized as well as carbonized, dating from approximately the last 1500 years B.C., have been recovered from Mammoth Cave and Salts Cave in Mammoth Cave National Park, Kentucky (Watson 1969, 1974; Yarnell 1976).[1] For the most part mean length and width of achenes from these Kentucky caverns are 7.0 to 7.5 mm and 3.0 to 3.3 mm, with the product of length x width ranging from 22 to 25. The carbonized achenes recovered by Nelson (1917) from the Mammoth Cave vestibule and measured by Heiser are much larger with a converted mean length x width product of 49. This is larger than the mean size of any collection that is dated earlier than the Mississippian period and much larger than any Early Woodland sample. It would appear that Nelson's achenes are Late Woodland in age.

Sunflower collections that are Middle Woodland to early Late Woodland in age are mostly larger in mean size than those from older sites and smaller than those of Mississippian age. The size range of Mississippian achenes is roughly comparable to the size range for modern sunflowers. They are larger than most of the Indian varieties but smaller than some of the commercial varieties.

If one examines the statistics presented in Table 1, it should be clear that there was a rather consistent increase in size of sunflower achenes during the period of approximately the last 3000 years of prehistory in eastern North America. Data from pre-Mississippian collections are available mainly from Kentucky, but some of the Ozark Bluff Dweller collections probably are pre-Mississippian. (Additional measurements of uncarbonized sunflower achenes from these sites will soon be made.) Metric data for the collections from Kettle Hill Cave in Ohio would be helpful, and pre-Mississippian collections from Illinois and Missouri are needed.

The record of prehistoric development of cultigen sumpweed is similar to that of sunflower in many respects, but there are some differences. There are good pre-Mississippian collections from Illinois and from Missouri

---

[1] In order to convert measurements of carbonized sunflower seeds and achenes to estimates of original achene size, it is necessary to increase achene length and width by 11% and 27% respectively, in accordance with results obtained by Heiser (1953), and to increase seed length and width by 30% and 45% (or more) respectively, in accordance with averaged results obtained by Waselkov and Yarnell.

(Klippel 1972), but there is very little prehistoric sumpweed from Ohio, perhaps reflecting the paucity of published results of flotation sample analysis from there.

For sumpweed we have good data indicating that the initial pre-domestication achene dimensions were approximately 3 x 2 mm. My own reconstruction of the mean achene dimensions for carbonized seeds from a late Middle Archaic level of the Koster site in the lower Illinois Valley is 3.0 x 2.1 mm (2.18+10%+0.6x1.65+10%+0.3) (Asch, Ford, and Asch 1972). We have data from modern sumpweed plants that were subjected to a variety of treatments by Heiser in Indiana and by Lee Herberger in Michigan (see Table 2). Plants grown in a cultivated field produced achenes with mean dimensions of 2.8 x 2.2 mm. Garden grown plants that were watered and fertilized produced achenes with means of 3.5 x 2.2 mm. Achenes from watered and fertilized greenhouse plants average 3.7 x 2.6 mm. Heiser produced tetraploid sumpweed by treatment with colchicine. These plants, which were watered and fertilized in the greenhouse, produced achenes with mean dimensions of 4.4 x 3.0 mm. Watering and fertilizing probably were minimal to nonexistent aboriginally in the region of sumpweed husbandry, so it seems likely that the continuing increase in sumpweed seed size exhibited by the sequence of archaeological collections was primarily due to polyploidy and selection in favor of genes responsible for larger seed size. Jackson points out that achene size is a quantitative trait (personal communication, 1978).

It appears that sumpweed domestication resulted in achene size expansion from an original length x width product of 6 up to a value of more than 30 in the Mississippian period. Given the observed, but seldom recorded, increase in achene thickness, the total increase in seed size probably was on the order of 1000%, an amount that is comparable to the prehistoric increase in seed size for sunflower. Some of the dates are uncertain, and the continuing size increases of sunflower and sumpweed achenes as indicated by the metric data are somewhat irregular; however, it seems clear from an overall examination of the archaeological record that the prehistories of sunflower and sumpweed involved parallel sequences of continuing change from essentially wild or weed forms to developed cultigen forms which were a part of the Mississippian plant husbandry complex.

## Acknowledgments

The contributions of Volney Jones to whatever value this study may have cannot be overestimated. In addition, the study could not have been made without the generous contributions of David Asch, Nancy Asch, David Benn, M. Jean Black, Leonard Blake, F. A. Calabrese, John M. Connaway, Wesley Cowan, Hugh Cutler, Hester Davis, Roy Dickens, Charles Faulkner, Richard I. Ford, Charles B. Heiser, Dale Henning,

Lee Herberger, Michael Hoffman, Peggy Hoffman, Raymond C. Jackson, Walter Klippel, Richard Marshall, Major McCollough, Louise Robbins, Andrea Brewer Shea, Stuart Struever, Gregory Waselkov, and Patty Jo Watson.

## REFERENCES

Asch, Nancy B., Richard I. Ford, and David L. Asch
1972 Paleoethnobotany of the Koster site: The Archaic Horizons. Illinois State Museum Reports of Investigations 24.

Benn, David W.
1974 Seed Analysis and Its Implications for an Initial Middle Missouri Site in South Dakota. Plains Anthropologist 19:55-72.

Black, Meredith
1963 The Distribution and Archaeological Significance of the Marsh Elder, *Iva annua* L. Papers of the Michigan Academy of Science, Arts and Letters 48:541-47.

Blake, S. F.
1939 A New Variety of *Iva ciliata* from Indian Rock Shelters in the South-Central United States. Rhodora 41:81-86.

Brewer, Andrea
1973 Analysis of Floral Remains from the Higgs Site (40Lo45). In: Excavation of the Higgs and Doughty Sites, I-75 Salvage Archaeology, by M. C. R. McCollough and C. H. Faulkner, pp. 141-44. Tennessee Archaeological Society Miscellaneous Papers 12.

Chomko, Stephen, and Gary W. Crawford
1978 Plant Husbandry in Prehistoric Eastern North America: New Evidence for Its Development. American Antiquity 43:405-08.

Heiser, Charles B., Jr.
1951 The Sunflower among the North American Indians. Proceedings of the American Philosophical Society 95:432-48.
1953 The Archaeological Record of the Cultivated Sunflower with Remarks Concerning the Origin of Indian Agriculture in Eastern North America. Manuscript. Files of the Author.
1954 Variation and Subspeciation in the Common Sunflower, *Helianthus annuus*. American Midland Naturalist 51:287-305.
1955 The Origin and Development of the Cultivated Sunflower. The American Biology Teacher 17 (May):161-67.
1969 The North American Sunflowers (Helianthus). Memoirs of the Torrey Botanical Club 22:1-217.
1976 The Sunflower. Norman: University of Oklahoma Press.

Jackson, Raymond C.
1960 A Revision of the Genus *Iva* L. University of Kansas Science Bulletin 41: 793-876.

Jones, Volney H.
1936 The Vegetal Remains of Newt Kash Hollow Shelter. In: Rock Shelters in Menifee County, Kentucky, edited by W. S. Webb and W. D. Funkhouser. University of Kentucky Reports in Archaeology and Anthropology 3(4): 147-67.

Klippel, Walter E.
1972 An Early Woodland Period Manifestation in the Prairie Peninsula. Journal of the Iowa Archaeological Society 19:1-91.

McCollough, Major
1973 Supplemental Chronology for the Higgs Site (40Lo45), with an Assessment of Terminal Archaic Living and Structure Floors. Tennessee Archaeologist 29:63-68.

Nelson, Nels C.
1917 Contributions to the Archaeology of Mammoth Cave and Vicinity, Kentucky. American Museum of Natural History Anthropological Papers 22: 1-73.

Watson, Patty Jo
1969 The Prehistory of Salts Cave, Kentucky. Illinois State Museum Reports of Investigations 16.

Watson, Patty Jo (ed.)
1974 Archaeology of the Mammoth Cave Area. New York: Academic Press.

Yarnell, Richard A.
1972 *Iva annua* var. *macrocarpa*: Extinct American Cultigen? American Anthropologist 74:335-41.
1976 Early Plant Husbandry in Eastern North America. In: Cultural Change and Continuity, edited by C. E. Cleland, pp. 265-73. New York: Academic Press.
1977 Native Plant Husbandry North of Mexico. In: Origins of Agriculture, edited by C. A. Reed, pp. 862-75. The Hague: Mouton.

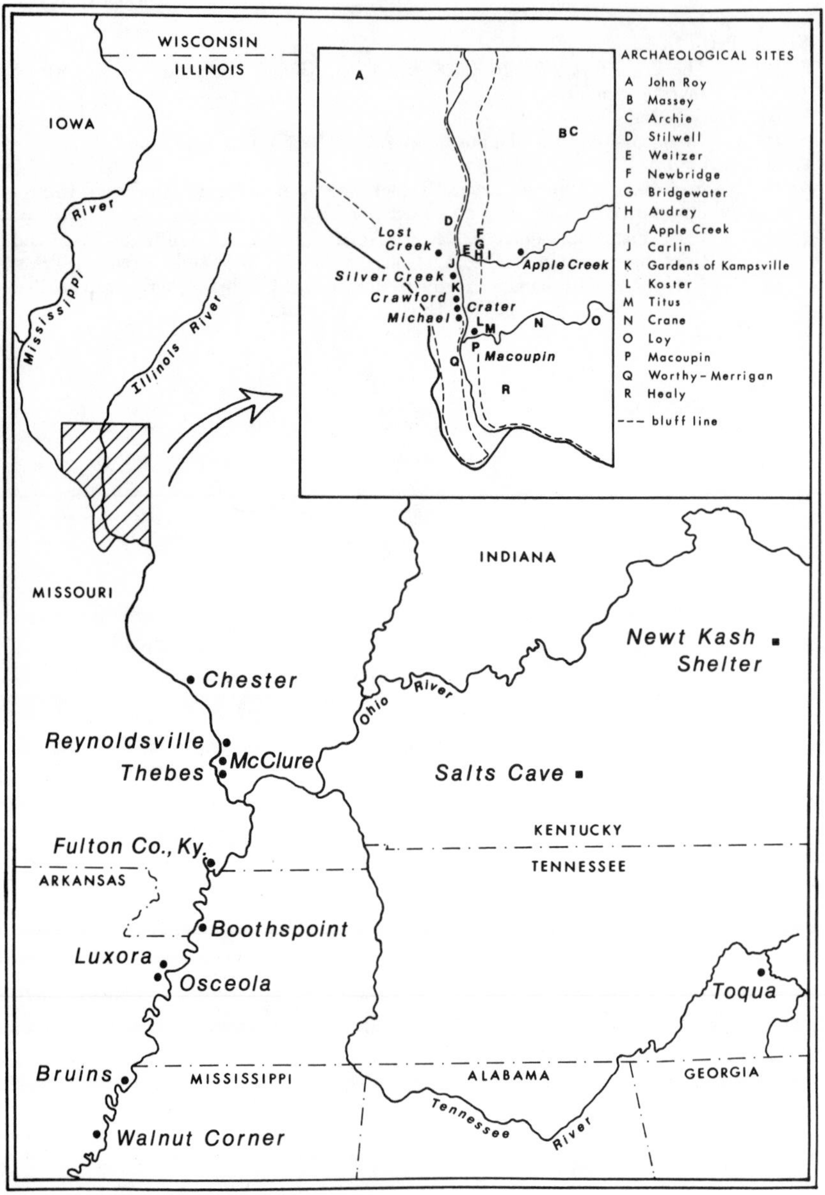


Fig. 1. Location of field collections and archaeological sites.

# THE ECONOMIC POTENTIAL OF *IVA ANNUA* AND ITS PREHISTORIC IMPORTANCE IN THE LOWER ILLINOIS VALLEY

*David L. Asch and Nancy B. Asch*
Northwestern University Archaeological Program

In eastern North America, the transformation from man's sole dependence on hunting and gathering of wild resources to major reliance on maize agriculture was a process spanning more than 3000 years. Volney Jones has made important contributions to the concept of a pre-maize "Eastern Agricultural Complex" involving largely indigenous annual seed plants. Native species with the best evidence for cultivation are sunflower and an obscure plant *Iva annua* L. (sumpweed, marsh elder). Either *Iva* was not a component of any historic Indian crop complex or else its use was not recorded by Europeans. However, its achenes have now been found in many prehistoric archaeological contexts. Two observations support *Iva*'s early status as a cultivated plant and a domesticate: (1) The prehistoric achenes are usually larger than modern wild achenes, and (2) they have been found in abundance at sites like Newt Kash Shelter, Kentucky, which may lie beyond the modern continuous geographic range of the species (Jones 1936). There is also evidence from the Koster site in the lower Illinois Valley that *Iva* was the first of the native annual seed plants to be singled out for extensive exploitation. Except for the sunflower, the native components of the Eastern Agricultural Complex had been almost entirely supplanted by the Mesoamerican triad of maize, squash, and beans at the time of European entry into North America.

In this paper, two approaches to the study of *Iva annua* are employed. (1) An investigation is made of its economically significant attributes (including the wild distribution, abundance, season of availability, productive potential, palatability, and nutritional value), its potential complementation and conflict with other premodern wild resources, and characteristics that may help to understand how it was brought under cultivation and eventually domesticated. The potential economic value of the species is not obvious; in fact it has been totally overlooked as a resource by modern wild food collectors. An understanding of the rational basis of prehistoric subsistence strategies requires extensive knowledge of plant and animal resources under realistic conditions of human exploitation. (2) The prehistoric record of *Iva*'s utilization is presented for one region–the lower Illinois Valley–and compared with other trends in plant exploitation from the region. The emphasis on archaeology of a single region offers some control over the environmental and cultural contexts of prehistoric change.

## PALATABILITY AND NUTRITIONAL VALUE

When *Iva* was first discovered in archaeological contexts–at Ozark Bluff Dweller sites in Arkansas and Missouri–its status as a food item seemed dubious. M. R. Gilmore and W. E. Safford considered the achenes of *Iva* to be unpalatable as well as lacking in starches (Gilmore 1931), and Gilmore suggested that *Iva* may have been used as medicine or perfume rather than food. With the discovery of quantities of achenes in human paleofeces at Newt Kash Rockshelter (Jones 1936) and Salts Cave (Watson et al. 1969), its use as food was reasonably established. Nevertheless, its reputation as an inferior food item has persisted. Black (1963), for instance, suggested that its inferiority gives a probable reason for its disappearance once maize was introduced to eastern North America.

An *Iva* "seed" is similar in appearance and structure to a tiny sunflower seed. Technically, it is termed an achene, and it consists of a kernel (the true seed) attached at one place to a thin dry shell (the pericarp). Unprocessed achenes are unpalatable because of the objectionable odor and taste of the shell, and because the tough, fibrous, indigestible shell is about 45% of the total achene weight. The soft, oily kernel is nonfibrous; it has a pleasant, somewhat nutty flavor. We found that boiling or roasting the achenes eliminates the objectionable odor and flavor. The shells do not split readily and are difficult to remove. Pounding or grinding the achenes will crush the soft kernels, making it impossible to winnow out the shells or to float them off in water. A separation method that proved successful involved the following steps: (1) boiling the achenes for several minutes, which causes many shells to split open partially and weakens the rest; (2) drying the boiled achenes to reharden the kernels; (3) rubbing the material between the hands to separate kernels from shells; (4) winnowing the shells. Not all the shells will split, but the rubbing and winnowing can be repeated until the continued abrasion causes unacceptable losses of kernels. With this method we were able to remove two-thirds of the shells, and the resulting product was quite palatable.

Yarnell (1977) states that *Iva* in the Salts Cave feces was unprocessed or only slightly roasted. However, with the method of shell removal described above, the free kernels would be digested while the remaining achenes with shells would pass through the digestive tract whole. Thus, fecal remains could give the mistaken impression that *Iva* was unprocessed. Yarnell also notes that the shell is often missing from carbonized *Iva*, and we too have observed this. Is the shell destroyed by burning or could its absence be due to processing? Evidence for the latter alternative comes from the early Late Woodland Newbridge site in the lower Illinois Valley. One of the pit features had served for a time as a refuse dump in which *Iva* was thrown along with other debris. Subsequently, the garbage was covered with sterile soil, and the pit was converted into a hearth. Heat from the hearth carbonized the garbage beneath without destroying delicate materials as direct burning does. Many rarely preserved materials were carbonized but were otherwise in perfect condition, including folded sections of a woven skirt or bag, *Apios* tubers (groundnuts), cattail roots, twisted fibers, and bundles of grass. Of the 300 *Iva* achenes recovered, only one-third retained their shells. In this instance at least, processing is the only plausible explanation of why the shells were missing.

Despite earlier questioning of its food value, *Iva annua* has proved to be very nutritious. A sample of achenes was sent to Analytical Bio Chemistry Laboratories, Inc., Columbia, Missouri, for determination of nutritional composition. Shells and kernels were each subjected to proximate analysis (Table 1). For the kernels there was also an amino acid analysis (Table 2) and determination of selected minerals and vitamins (Table 3).

As expected, the crude fiber in the shells approaches half of the shell weight. Spiller and Shipley's (1977) review of dietary fiber studies leads one to suspect that much of the remaining carbohydrate in the shell as well as some of the shell's lipids, minerals, and proteins are also indigestible. Without milling, any nutrients of the shell could scarcely be utilized, and as noted above, Salts Cave paleofeces apparently did not contain ground achenes. Consequently, we do not believe that *Iva* shells were a source of nutrients.

*Iva* kernels are a concentrated source of food energy because of their high fat and low moisture content. They provide somewhat fewer Calories per 100 g than nuts, about the same number of Calories as sunflower kernels, and considerably more than starchy roots or starchy seeds like *Chenopodium* and acorns (Table 1). An adult who requires 2500 Cal/day of food energy can satisfy this requirement by eating 467 g of *Iva* kernels.

The protein content of *Iva* also is high. However, protein quality is poor (Table 2), as measured by the essential amino acid in shortest supply compared with the FAO (1973) reference pattern. Lysine is the most limiting

TABLE 1

PROXIMATE COMPOSITION (g/100 g) OF *IVA ANNUA* AND SELECTED FOODS

| | Cal./ 100 g* | Water | Protein † | Fat | Carbo-hydrate‡ | Fiber | Ash | Source § |
|---|---|---|---|---|---|---|---|---|
| Oily Seeds (kernels) | | | | | | | | |
| Sumpweed, *Iva annua* kernels | 535 | 5.06 | 32.25 | 44.47 | 10.96 | 1.46 | 5.80 | A |
| shells | — | 6.47 | 5.07 | 2.81 | 36.84 | 45.18 | 3.63 | A |
| Sunflower, *Helianthus annuus* | 560 | 4.80 | 24.00 | 47.30 | 16.10 | 3.80 | 4.00 | B |
| Squash, *Curcurbita pepo* | 553 | 4.40 | 29.00 | 46.70 | 13.10 | 1.90 | 4.90 | B |
| Starchy Seeds | | | | | | | | |
| Lambsquarters, *Chenopodium album* | 320 | 13.40 | 13.30 | 5.60 | 45.90 | 14.60 | 7.20 | C |
| Field corn, *Zea mays* | 348 | 13.80 | 8.90 | 3.90 | 70.20 | 2.00 | 1.20 | B |
| Tubers and Roots | | | | | | | | |
| Duck potato, *Sagittaria latifolia* | 123 | 66.90 | 4.40 | 0.80 | 24.90 | 1.00 | 2.00 | D |
| Groundnut, *Apios americana* | 109 | 70.70 | 4.10 | 1.00 | 18.60 | 3.50 | 2.10 | D |
| Cattail rootstock flour, dried, *Typha latifolia* | 367 | 7.60 | 6.90 | 3.10 | 79.80 | ¶ | 2.60 | E |
| Nuts | | | | | | | | |
| Shagbark hickory, *Carya ovata* | 696 | 2.20 | 11.00 | 72.70 | 10.60 | 1.50 | 2.00 | F |
| Black walnut, *Juglans nigra* | 621 | 2.90 | 24.10 | 58.50 | 10.80 | 1.00 | 2.70 | F |
| White oak, *Quercus alba* | 221 | 47.30 | 2.80 | 3.30 | 43.90 | 1.30 | 1.40 | F |
| Red oak, *Quercus rubra* | 299 | 38.20 | 3.40 | 12.90 | 42.10 | 1.90 | 1.50 | F |

TABLE 1 (Continued)

| | Cal./ 100 g* | Water | Protein † | Fat | Carbo-hydrate‡ | Fiber | Ash | Source § |
|---|---|---|---|---|---|---|---|---|
| Other Plant Food | | | | | | | | |
| Pumpkin flesh, *Cucurbita pepo* | 26 | 91.60 | 1.00 | 0.10 | 5.40 | 1.10 | 0.80 | B |
| Lambsquarters greens, *Chenopodium album* | 43 | 84.30 | 4.20 | 0.80 | 5.20 | 2.10 | 3.40 | B |
| Meat | | | | | | | | |
| Deer (raw, lean), *Odocoileus* sp. | 126 | 74.00 | 21.00 | 4.00 | .00 | .00 | 1.00 | B |
| Raccoon (roasted), *Procyon lotor* | 255 | 54.80 | 29.20 | 14.50 | .00 | .00 | 1.50 | B |
| Wild duck (raw), *Anas platyrhynchos* | 233 | 61.10 | 21.10 | 15.80 | .00 | .00 | 1.10 | B |
| Catfish (raw), *Ictalurus* sp. | 103 | 78.00 | 17.60 | 3.10 | .00 | .00 | 1.30 | B |
| Pink heel-splitter, *Proptera alata* | 77 | 76.50 | 9.50 | 0.80 | 7.80 | 1.10 | 4.30 | G |

*Using Atwater conversion factors (Watt & Merrill 1963).

†Computed as N x 5.30 for oily seeds and nuts, N x 6.25 for other foods.

‡Nitrogen-free extract.

§A = Analytical Bio Chemistry Laboratories, Columbia, Missouri; B = Watt & Merrill 1963; C = Spinner & Bishop 1950; D = Winton 1932, Vol. 2; E = Claassen 1919; F = Wainio & Forbes 1941; G = Parmalee & Klippel 1974.

¶Fiber not reported; included in carbohydrate.

TABLE 2
AMINO ACID COMPOSITION (mg/g of nitrogen) AND
AMINO ACID SCORE OF *IVA ANNUA* KERNELS

| | *Iva* Kernels | | *Iva* Kernels | Suggested Level* | Score† |
|---|---|---|---|---|---|
| Nonessential amino acids | | Essential amino acids | | | |
| Alanine | 210 | Histidine | 115 | ‡ | – |
| Arginine | 638 | Isoleucine | 209 | 250 | 84 |
| Aspartic acid | 557 | Leucine | 348 | 440 | 79 |
| Glutamic acid | 1265 | Lysine | 150 | 340 | 44 |
| Glycine | 271 | Total S-containing | 196 | 220 | 89 |
| Hydroxyproline | 0 | Methionine | 94 | – | – |
| Proline | 235 | Cystine/2 | 102 | – | – |
| Serine | 245 | Total aromatic | 424 | 380 | 112 |
| | | Phenylalanine | 281 | – | – |
| | | Tyrosine | 143 | – | – |
| | | Threonine | 153 | 250 | 61 |
| | | Tryptophan | 110 | 60 | 183 |
| | | Valine | 270 | 310 | 87 |

*Source: FAO (1973:63).
†Amino acid level of *Iva* (per g N) expressed as a percentage of reference pattern level.
‡Required by infants; no evidence that adults require histidine.

amino acid; there is even less lysine per gram of nitrogen than in maize (Orr and Watt 1957). Fortunately, the low quality of *Iva* protein is more than compensated for by the high ratio of protein to food energy. Correcting for protein quality and basing computations on the protein and energy requirements published by FAO (1973), one finds that *Iva* protein is available at a safe level for an individual of any age, sex, or body weight who meets his caloric requirements. That is to say, protein requirements would be met if one ate only *Iva*; or eaten in combination with other plant foods, *Iva* would compensate in part for protein deficiencies in the other foods.

*Iva* is an unusually good source of several vitamins and minerals. For comparison, data on mineral and vitamin composition of several foods are given in Table 3. The table includes some foods that would not have been available prehistorically but which should be comparable to similar prehistoric foods. *Iva* would have been one of the best prehistoric sources of calcium, being rivaled only by greens, by mussels and by bones. Of the foods listed in Table 3, only squash seeds have a comparable iron content, and even beef liver has a lower concentration. However, the nonheme iron of a plant may be in a chemical form that is less available than the heme iron of meats (FNB 1974: 93). Phosphorus and potassium are abundantly available in *Iva*, and human diets under normal health conditions should not be deficient in these elements.

TABLE 3

MINERAL AND VITAMIN CONTENT (mg/100 g edible portion) OF *IVA ANNUA* AND SELECTED FOODS*

| | Calcium | Phosphorus | Iron | Potassium | Thiamine | Riboflavin | Niacin |
|---|---|---|---|---|---|---|---|
| Oily Seeds (kernels) | | | | | | | |
| Sumpweed, *Iva annua* | 290 | 1300 | 11.4 | 780 | 2.13 | 0.75 | 13.1 |
| Sunflower, *Helianthus annuus* | 120 | 837 | 7.1 | 920 | 1.96 | 0.23 | 5.4 |
| Squash, *Cucurbita pepo* | 51 | 1144 | 11.2 | — | 0.24 | 0.19 | 2.4 |
| Starchy Seeds | | | | | | | |
| Quinoa, *Chenopodium quinoa* | 131 | 424 | 6.8 | — | 0.52 | 0.31 | 1.6 |
| Buckwheat, whole grain, *Fagopyrum esculentum* | 114 | 282 | 3.1 | 448 | 0.60 | 0.15 | 4.4 |
| Nuts | | | | | | | |
| Pecan, *Carya illinoensis* | 73 | 289 | 2.4 | 603 | 0.86 | 0.13 | 0.9 |
| Black walnut, *Juglans nigra* | Trace | 570 | 6.0 | 460 | 0.22 | 0.11 | 0.7 |
| Other Plant Food | | | | | | | |
| Pumpkin flesh, *Cucurbita pepo* | 21 | 44 | 0.8 | 340 | 0.05 | 0.11 | 0.6 |
| Lambsquarters greens, *Chenopodium album* | 309 | 72 | 1.2 | — | 0.16 | 0.44 | 1.2 |
| Meat | | | | | | | |
| Venison, lean, *Odocoileus* sp. | 10 | 249 | — | — | 0.23 | 0.48 | 6.3 |
| Beef, carcass, total edible utility grade, *Bos taurus* | 11 | 172 | 2.8 | 100 | 0.08 | 0.17 | 4.5 |
| Beef, liver | 8 | 352 | 6.5 | 281 | 0.25 | 3.26 | 13.6 |
| Beef, kidneys | 11 | 219 | 7.4 | 225 | 0.36 | 2.55 | 6.4 |
| Duck, domesticated, total edible, *Anas platyrhynchos* | 10 | 176 | 1.6 | — | 0.08 | 0.19 | 6.7 |
| Catfish, *Ictalurus* sp. | — | — | 0.4 | 330 | 0.04 | 0.03 | 1.7 |
| Pink heel-splitter, *Proptera alata* | 370 | 812 | 12.5 | 41 | — | 0.3 | 2.0 |

*Data are from Watt & Merrill (1963) except for *Iva* (Analytical Bio Chemistry Laboratories, Columbia, Missouri), quinoa (White et al. 1955) and pink heel-splitter (Parmalee & Klippel 1974).

Except for yeast, partially defatted sunflower kernels, and rice bran, the thiamine levels of *Iva* are higher than any unprocessed food listed in *Composition of Foods* (Watt and Merrill 1963). *Iva* is one of the best plant food sources of riboflavin, a vitamin which also is concentrated in liver and kidney and well represented in venison. Niacin levels in *Iva* are very high. In addition, the amino acid tryptophan is converted to niacin in the body in a ratio of 1 mg niacin equivalent for 60 mg of tryptophan consumed (FNB 1974:69-70). Tryptophan conversion nearly doubles the effective availability of niacin in *Iva*.

To summarize, few plant foods compare with *Iva* in any one of the analyzed minerals and vitamins, and scarcely any approach *Iva* across the entire suite. Oil seeds commonly have higher than average quantities of these vitamins and minerals, and sunflower kernels are the prehistoric food most similar to *Iva* in these nutrients. Sunflower is most similar as well in its proximate composition.

Vitamin A and Vitamin C levels of *Iva* kernels were not tested because they are generally negligible in mature seeds. In prehistoric diets, Vitamin A and Vitamin C could have been provided by organ meats, greens, squash flesh, fruits, and berries.

## DISTRIBUTION, ABUNDANCE, AND SEASONALITY

Floodplain localities where *Iva annua* grows have been massively disturbed by Euroamerican settlement, levee construction and land drainage, cultivation and cattle grazing, and plantings for wildlife conservation. Little detailed information exists about premodern vegetation associations in which *Iva* occurred. We have therefore attempted first to determine the habitat factors controlling the modern distribution and abundance of *Iva* and then to evaluate the limited descriptions of early historic or relatively unmodified floodplain vegetation in conjunction with our knowledge of these factors. Since *Iva* occurs in early stages of plant succession, its ecological position in man-modified habitats should have parallels to its occurrence in naturally disturbed habitats.

According to Steyermark (1963:1536) the habitats of *Iva annua* in Missouri are "alluvial soils along streams, borders of ponds and sloughs, river bottom prairies and meadows, low fields in valleys, along roadsides and railroads." Jackson (1960) recorded habitat information from 115 herbarium sheets of *Iva annua*, and all but three descriptions are consistent with Steyermark's generalization (Yarnell 1972).

Our own observations in the Illinois and Mississippi River drainages between central Illinois and central Arkansas indicate that presently the species occurs in habitats that are (1) flooded in the spring and often wet throughout the growing season; (2) on recently disturbed soil; (3) open, growing where there is a cover of short grasses but less successful where it competes with other weeds and tall grasses. It is tolerant of open conditions maintained by mowing or cattle grazing.

*Iva annua* does not grow everywhere these general conditions are met. For example, it does not grow on soil wet enough to support the semiaquatic, perennial species of *Polygonum*, nor is it likely to occur in stands with the following common floodplain weeds: water hemp (*Amaranthus*, the dioecious species), cocklebur (*Xanthium*) or giant ragweed (*Ambrosia trifida*). *Iva* does not occur in the annual herb zone of river shorelines or in the beds of dried-up backwater lakes, except at their margins.

*Iva* is mainly an edge species occurring between permanently wet and somewhat better drained soils, and its extension in either direction is limited by better adapted plant competitors. Thus, its distribution is linear, following microtopographic drainage contours. Stands are frequently only a few plants wide, and a 5-meter-wide stand would be exceptionally broad.

Within the narrow zone where it competes successfully with other species, *Iva* commonly occurs in dense stands in which it is the tallest plant. As shown in Table 4, the density of *Iva* is variable in stands where it is dominant. When multiply-branched and tall (sometimes over 2 m), plants are more widely spaced due to competition for light; where plants are short and unbranched due to mowing, grazing, or a late start in the growing season, there may be several hundred *Iva* plants per square meter.

Judging from modern conditions, the following characteristics of *Iva* distribution and abundance have implications for wild gathering: (1) It is concentrated in certain open habitats of large river valleys, with only minor occurrence elsewhere. (2) Its distribution changes with each year's pattern of flooding and with plant succession. (3) It commonly grows in dense stands, which makes harvesting efficient, but (4) the total area of a stand is small. We have never seen a continuous stand that occupied more than a fraction of a hectare.

There are a few records of *Iva annua* in more natural settings of bottomland prairies (now virtually extinct), especially at the fluctuating margins of floodplain prairie lakes (Turner 1934, 1936; Hus 1908; Mackenzie 1902; Gates 1940). These botanical records do not provide adequate information about the extent of mowing and grazing and other disturbance of the bottomland prairies, which is unfortunate since *Iva* is favored rather than reduced by such activities.

Sampson (1921) and Schaffner (1926) recognize three major associations of lowland grasses in the Eastern prairie region. An *Andropogon gerardi* (big bluestem) association in which *Iva annua* is unlikely to occur occupies high ground that is seldom or never flooded. A *Spartina pectinata* (sloughgrass)

TABLE 4
SIZE, DENSITY, AND YIELD OF *IVA ANNUA* IN 1 $m^2$ PLOTS

| Plot # | Location | Number of Plants | Mean Height (cm) | Achene Weight (g) | Yield* (kg kernels/ha) | Comments† |
|---|---|---|---|---|---|---|
| 1 | Chester, Randolph Co., Ill. | 72 | 54 | 41 | 226 | Seed loss less than 10%. |
| 2 | Same as Plot I | 81 | 65 | 40 | 220 | Seed loss less than 10%. |
| 3 | Luxora, Mississippi Co., Ark. | 28 | 149 | 84 | 462 | Most plants very green. |
| 4 | Fulton Co., Ky. | 97 | 114 | 14 | 77 | Ca. 20% seed loss. |
| 5 | Thebes, Alexander Co., Ill. | 27 | 133 | 30 | 165 | Ca. 75% seed loss. Weedy. |
| 6 | Reynoldsville, Union Co., Ill. | 625 | 42 | 61 | 336 | Ca. 35% seed loss. Mowed. |
| 7 | McClure, Union Co., Ill. | 113 | 60 | 74 | 407 | Ca. 5% seed loss. Mowed. |
| 8 | Bruins, Lee Co., Ark. | 115 | 98 | 85 | 468 | Ca. 25% seed loss. |

*Assuming that kernels are 55% of the achene weight.
†Observations of seed loss not considered very reliable.

association grows on wet, poorly aerated alluvium, and a *Panicum virgatum* (switchgrass) association sometimes occurs where the other two major associations intergrade. *Spartina* grows rapidly in the spring and early summer, usually forming dense, almost pure stands with deep shade beneath (Weaver 1954:31-33) Any *Iva* germinating in dense stands would be shaded out because its growth spurt occurs later than that of *Spartina*. However, Weaver states that many tall forbs may be found where *Spartina* stands are thinner.

Turner (1934) studied two *Spartina* prairies in the lower Illinois and adjacent Mississippi valleys and found that *Iva annua* was very common in both, occurring in 82% of 50 one-meter-square plots from the Illinois Valley prairie and in 36% of 110 Mississippi Valley plots. Turner specifically stated that the Illinois Valley prairie was grazed. Apparently, the Mississippi Valley prairie was not grazed, but it probably was not completely undisturbed since *Ambrosia bidentata* (ragweed) was as common as *Iva*.

We have observed *Iva* in two essentially undisturbed lowland areas dominated by low-growing rice cutgrass (*Leersia oryzoides*). This association is found in conditions even wetter than a *Spartina* association. Rice cutgrass does not grow tall enough to compete effectively with *Iva* for light.

However, the distribution of *Iva* was controlled by other factors since it exhibited microtopographic zonation and was not dispersed over a wide area within the association.

Turning now to the seasonal growth pattern of *Iva annua*, Ungar and Hogan (1970) in Nebraska found that germination occurs during April when temperatures average 11°C. *Iva* plants have a burst of growth late in the summer which culminates in flowering at the end of August. Since flowering is photoperiodically timed, achenes in a region ripen at about the same time regardless of the size or location of the plants. In an average year, *Iva* achenes are ripe and ready to drop around October 15 in the lower Illinois Valley (later in regions to the south). In 1977 cold weather in late September and rain in early October caused most seeds to drop by the end of the first week in October. Wind and rain always shorten the potential harvest season, so even in good years the effective harvest season for *Iva* is probably not longer than two weeks.

## HARVESTING

A study of the harvest potential of *Iva* was conducted in the fall of 1977. Objectives of the study were to determine under optimal conditions the productivity per unit area and the harvest rate. Investigations were concentrated on the best stands for two reasons. (1) We suspected that *Iva* could not be collected efficiently enough to justify harvesting it for its supply of food energy. To test this, it would be necessary to harvest *Iva* under the most favorable conditions. (2) Stands of *Iva* vary widely in their harvest potential and it is reasonable to assume that prehistoric collectors concentrated, as we did, on the more productive stands.

Initially the harvest was planned for the lower Illinois Valley region where the archaeological data for this report were obtained. However, an unanticipated early seed drop forced us to make the collections farther south where the plants matured later. Several stands were harvested in the Mississippi Valley between southern Illinois and central Arkansas (Fig. 1).

Eight one-square-meter plots from seven localities were selected and carefully harvested to provide data on maximum productivity per unit area (Table 4). Plant size in the plots varied greatly. However, where *Iva* is dominant, the number of plants is inversely correlated with plant size, and we did not observe a consistent effect of stand type on harvest potential.

Dried-up plants are the easiest to strip. A harvester would probably avoid green plants whose fruiting spikes break off when the achenes are stripped because it is difficult to thresh a bulky collection containing many spikes with achenes tightly attached. Seed loss is inevitable in a plot of dry plants, and some had fallen in all of the study plots. Thus, our data should be regarded as the measure of an economic variable—harvest yield—rather than a biological measure of seed production.

Achene yields in the study plots ranged from 14 g to 85 g/m$^2$. Estimating that 55% of the achene weight is kernel, this represents a mean of 295 kg/ha of kernels or 470 kg/ha in the best harvest. In terms of food energy, the mean return per hectare is equivalent to 1.6 x $10^6$ Cal, which is enough to supply the total annual food energy requirements (at 2500 Cal/day) of 1.7 adults.

Stands selected to measure harvest rates were dense enough that the time spent moving from one plant to the next was minimal. By fastening sacks at waist level, harvesters were able to strip seeds into them using both hands. A total of 20 timed collections were obtained from 7 stands (Table 5). Collection times ranged from 15 minutes to almost 3 hours. The maximum collection rate was 1.41 kg/hr for achenes, which is equivalent to 0.78 kg kernels/hr. At the Osceola and Walnut Corner localities a large proportion of the seeds

TABLE 5
HARVEST RATES FOR *IVA ANNUA*

| Collection Area | Collector | Time (min) | Achene Weight (g) | Kg Kernels/ hr | Date |
|---|---|---|---|---|---|
| Thebes, Alexander Co., Ill., near Plot #5 | A | 15 | 172 | 0.38 | 10-14-77 |
| | A | 15 | 152 | 0.33 | 10-14-77 |
| | A | 15 | 180 | 0.40 | 10-14-77 |
| | B | 15 | 169 | 0.37 | 10-14-77 |
| | B | 15 | 221 | 0.49 | 10-14-77 |
| | B | 15 | 345 | 0.76 | 10-14-77 |
| Reynoldsville, Union Co., Ill., near Plot #6 | A | 15 | 307 | 0.68 | 10-15-77 |
| | A | 15 | 273 | 0.60 | 10-15-77 |
| | B | 15 | 216 | 0.47 | 10-15-77 |
| | B | 15 | 205 | 0.45 | 10-15-77 |
| McClure, Union Co., Ill. near Plot #7 | A | 15 | 353 | 0.78 | 10-15-77 |
| | B | 15 | 332 | 0.73 | 10-15-77 |
| Boothspoint, Dyer Co., Tenn. | A | 15 | 137 | 0.30 | 10-31-77 |
| | B | 15 | 227 | 0.50 | 10-31-77 |
| Osceola, Mississippi Co., Ark. | A | 170 | 544 | 0.11 | 11-01-77 |
| | B | 170 | 1603 | 0.31 | 11-01-77 |
| Bruins, Lee Co., Ark. near Plot #9 | A | 150 | 1099 | 0.24 | 11-02-77 |
| | B | 150 | 1650 | 0.36 | 11-02-77 |
| Walnut Corner, Phillips Co., Ark. | A | 45 | 127 | 0.09 | 11-02-77 |
| | B | 55 | 216 | 0.13 | 11-02-77 |

had fallen, and achene collection rates dropped below 0.2 kg/hr. The 15-minute collections were made working at top speed, and they may approach the average rate for a skilled harvester. The mean for these collections is 0.94 kg achenes/hr, equivalent to 0.52 kg kernels/hr or 2800 Cal/hr. In other words, the return from one hour of harvesting meets an adult's daily energy requirement.

To provide a perspective for these harvest statistics, information on a few other seed plants is given in Table 6. The yield of *Iva* is more or less comparable to domesticated sunflower (though not to high-yielding sunflower varieties under intensive cultivation), but it is much less productive than non-hybrid corn grown by nineteenth-century Indians of the Upper Missouri. *Iva* yields are close to teosinte, the postulated wild ancestor of maize, particularly if one takes into account *Iva*'s higher caloric content. The high yield per unit area for *Chenopodium bushianum* (an indigenous eastern North American species that was also collected or cultivated prehistorically) is based on small open plots in which the plants had maximum potential for vegetative growth. Thus it is probably unreasonable to extrapolate its production to a large stand where the chenopods would be competing with one another. Yields for the domesticated South American quinoa may provide a more reasonable approximation for large scale chenopod production.

Considering the difficulty of comparing harvest rates, which depend to an important extent on the skill of the harvester, it is surprising that the rates assembled in Table 6 are so similar. In this respect *Iva* ranks with some of the wild cereals which eventually became major world crops.

In sum, from the viewpoint of energetics it is feasible to harvest dense stands of wild *Iva*. The yield per unit area is similar to other wild plants that were taken into cultivation as dietary staples.

It is also important to assess the total harvest potential of *Iva*, since the major value of the plant was presumably as a storable source of food energy. If the collecting season lasted for two weeks (it could be shortened by inclement weather), if workdays were 8 hr long, and if kernels were collected at an average rate of 0.5 kg/hr, then one collector could harvest 56 kg of kernels in a season. This should be a generous estimate since an allowance for time traveling to and from *Iva* stands is not included. For a person consuming 2500 Cal/day, the seasonal harvest translates to a 120-day total energy supply.

For many reasons, these computations are far too high if they are taken to represent per capita harvest potential for a prehistoric community. (1) Collecting was probably done by women, while men engaged in other activities preparatory for the winter season. (2) No allowance was made for pre-storage processing. Drying would usually be necessary to prevent molding and to halt insect activity, and drying makes threshing and winnowing easier. Winnowing probably was performed prior to storage to reduce the bulk of

TABLE 6
YIELD AND HARVEST RATE OF SELECTED SEED PLANTS

| | Yield (kg/ha) | Harvest Rate (kg/hr) | Comments | Source |
|---|---|---|---|---|
| *Iva annua*, sumpweed (Mississippi Valley) | 295 | 0.52 | Kernels. Best yielding plot: 470 kg/ha. Mean harvest rate based on 14 15-minute collections. Best rate: 0.78 kg/hr. Harvested by hand-stripping. | Asch and Asch 1977 |
| *Helianthus annuus*, sunflower (U.S.) | 350 | –– | Kernels. Average yield of field crop in U.S., 1959. Assumes that shell is 46% of achene weight. | Martin and Leonard 1967 |
| *Zea mays*, maize (Upper Missouri Indians) | 1250 | –– | 19th century records. | Will and Hyde 1917 |
| *Zea mexicana*, teosinte (Guerrero and Valle de Bravo, Mexico) | 150-630 | 0.5 | Seed minus 50% allowance for roughage. Wild stands. Harvest rate based on estimate of 1 liter seed/hr. Harvested by hand-stripping. | Flannery and Ford 1972; Robson et al. 1976 |
| *Chenopodium bushianum*, lambsquarters seed (Lower Illinois Valley) | 1330-1740 | 0.83-1.12 | Clean seed. Yield based on two plots of 1 $m^2$, tall multiple-branched plants, open to sunlight. Harvest rate based on two 15-minute collections. | Asch and Asch 1977 |
| *Chenopodium quinoa*, quinoa seed (Andes, Altiplano region) | 350-800 | –– | Cultivated species. | Dendy et al. 1975 |
| *Triticum dicoccoides*, wild emmer wheat (Palestine) | 500-800 | 0.25-0.52 | Clean grain. Yield is estimated upper limit for rainy year. Harvest rates determined from two localities. Harvested by hand-stripping. | Zohary 1969; Ladizinsky 1975 |
| *Triticum boeoticum*, wild einkorn wheat (Diyarbakir, Turkey) | –– | 0.94-1.13 | Clean grain. Lesser rate by hand-stripping, greater rate using flint sickle. | Harlan 1967 |

material because uncleaned collections—containing stems, leaves, floral parts, and achenes—have more than five times the bulk of cleaned achenes. It might require almost as long to process *Iva* for storage as it does to make the collections. (3) There simply are not enough dense stands to support intensive collecting by a large task force. Today in most localities, one individual in a season could clean out the harvestable stands from several square kilometers of a river valley. In our opinion these restrictions on the total harvest would have been stringent enough to limit wild *Iva* to a small (though useful) contribution to the annual food supply.

## SUBSISTENCE STRATEGIES

The economic potential of *Iva* is not a property of the plant per se. Its potential depends as well on its attractiveness in relation to competing species, on the overall demand for food, on the organization of the settlement system, and on other system properties. A full-scale examination of relevant natural and cultural variables cannot be pursued in this paper. Of special importance, however, is the comparison of *Iva* with nuts and acorns, which were the major preagricultural source of plant-derived food energy in eastern North America. In the lower Illinois Valley region, hickories were preeminent among the nuts and acorns in the extent of utilization (Asch, Ford, and Asch 1972).

Zawacki and Hausfater (1969) estimated the premodern annual mast of oaks and hickories for a transect of the lower Illinois Valley and adjacent uplands. They obtained results suggesting an almost unlimited wild bounty. Their *lower bound* for annual yield in the upland forest, which we have tried to translate into quantities of edible food, is roughly equivalent to 700 kg/ha of hickory nutmeats and 300 kg/ha of oven-dry acorn meats—enough food to supply a human population exceeding 650 individuals/km$^2$.

There are two basic objections to the use of these data as an indication of nut and acorn potential: (1) They greatly overestimate the annual mast; (2) the aggregation and dispersion of the mast in time and space is a more critical limiting factor for human exploitation than the gross annual mast in a forest.

Certain unjustified procedural assumptions led Zawacki and Hausfater to overestimate the overall density of trees in premodern forests by a factor of three or more. Telford (1926) is a good alternative source on the composition and density of a variety of Illinois forest types.

To estimate the production of individual trees, Zawacki and Hausfater relied on statements from the *Woody-Plant Seed Manual* (U.S. Forest Service 1948:110) about what "thrifty" trees "often produce." Many factors, however, reduce production below the ideal. Five such factors are:

(1) Presence of less productive or nonproductive immature or overmature trees in a forest stand (Downs 1949).

(2) Degree of canopy development (Morris 1912; Sharp 1958; Verme 1953). Holding other factors constant, mast production is more or less directly related to canopy development. In a forest, with severe light competition, the mast is produced largely near the treetop. Open grown trees or trees at the forest margin have greater exposure to high intensity light and have more limb development. Consequently, they may produce mast at many times the rate of interior forest trees.

(3) Genetic or site variation (Sharp and Sprague 1967). Trees of the same size and age exhibit great and persistent individual variation in mast production.

(4) Insect damage (Korstian 1927; Christesen and Korschgen 1955). It is not uncommon for half of the mast to be infested by insects, rising to nearly 100% in years of light production.

(5) Regional synchronization (Downs and McQuilken 1944; Sharp and Sprague 1967; Thompson 1958). Trees bear heavily at irregular intervals, which are synchronized over broad regions by common weather patterns. Production of different species is somewhat complementary. Nevertheless in some years virtually all species fail.

Because acorns are an important deer food, several studies have sampled their masts on an areal basis. Two of the better studies whose conclusions should be fairly applicable for lower Illinois Valley forests were conducted in predominantly oak-hickory forests of southeast Ohio (Nixon et al. 1975) and Arkansas (Segelquist and Green 1968). Both obtained mean annual acorn masts on the order of 60 kg kernels/ha (oven-dry basis). Of course, an average year seldom occurs, and both studies recorded masts of less than 10 kg/ha within a span of 8 or 9 years.

The study of Nixon et al. (1975) provides virtually the only reliable quantitative data on areal yields of hickories. They recorded a mean of only 7.5 kg/ha of hickory nutmeats, production ranging over 9 years' time from less than 1 kg/ha to over 17 kg/ha. In this Ohio forest, hickories constituted 7.2% of total basal area among trees $6.0^{+}$ inches DBH (Nixon et al. 1968:300).

However large a mast may be, its availability ultimately is limited by the length of time it can be effectively harvested. Humans face serious competition with animals for the mast. In the fall, squirrels have a decided preference for hickory nuts (Smith and Follmer 1972), which they cut down while the nuts are too green for man to shake or knock down from the trees. In poor years few hickory nuts will escape the squirrels. Acorns are less favored by squirrels and remain on the ground longer. However, they will also disappear rapidly in poor years. Downs and McQuilken (1944) plot seasonal curves for the persistence of acorns on the ground.

Fallen leaves are another factor limiting the season of availability because they cause a drastic increase in the "search time" for nuts and acorns. In the lower Illinois Valley, leaves begin to accumulate on the ground by October 15. Ripened hickory nuts begin to drop in late September or early October, so the effective harvest season can be very short.

Our experiences with observing and collecting hickory nuts illustrate some of the problems of nut collecting. In 1975 we planned to conduct a survey of nut production in a near-virgin oak-hickory forest. The survey was soon abandoned, however, because nuts were so rare that the recovery rate was only four sound hickory nuts per man-hour. This was not a good year for the hickory mast, but a few trees on the forest edge were producing. Several collections were made under the open canopy portions of forest edge trees. Calculating harvest rates as if all nuts were sound, the maximum rate of nut-meat collection was attained under a pignut hickory (*Carya glabra*)–0.5 kg/hr. (Actually, half of the nuts under this tree were wormy.) We were trying to make complete pickups, so the rates were a little below what could have been achieved with a more hurried search. Nevertheless, the rates are far below what had been anticipated, certainly no better than collection rates for *Iva*. Retrospectively, we judge 3 kg/hr of hickory nutmeat to be an upper limit for collecting under a very heavily producing tree without insect damage and with ideal ground cover.

To summarize, a small percentage of all hickory trees have large yields while most produce very few nuts, and these are further decimated by squirrels. Human gatherers trying for a large harvest almost certainly would collect first under the high-yielding trees at the forest margin rather than collecting indiscriminately in a forest. In poor years the need to concentrate collecting effort on the best trees would be even greater. Consequently, estimates of nut production per unit area of forest appear to have little relevance for human gatherers. Due to destruction or removal of nuts by insects and squirrels, as well as the reduction of visibility on the ground after leaves fall, the supply of nuts for man is probably more limited by the length of the season of availability than by the area of forest.

*Iva* and hickory masts both have "patchy" distributions. For *Iva* the patches are narrow stands; for hickory nuts the patches are single high-yielding trees. The real labor cost of a harvest can be divided into two components: (1) time spent traveling to and from patches and (2) time spent harvesting within a patch. It may sometimes be more efficient to harvest *Iva* at a lower rate from a nearby stand than to travel to a higher yielding but distant nut tree. Thus, it would be misleading to expect that relative harvest rates (in patches) are the only important determinant of which species are collected.

Differences in the harvest season should also be considered in a comparison of *Iva* with hickory. Because hickory nuts are available before *Iva*, the need to collect *Iva* may depend on the success of the nut harvests. But it is unlikely that collectors would forego harvesting nuts in the expectation of collecting *Iva* since the *Iva* harvest can be decimated by unpredictably rainy, windy weather. Ordinarily some hickory nuts will still be available but in declining abundance when *Iva* is ready to harvest.

## CULTIVATION

In this section we discuss some models for the ecological setting of prehistoric *Iva* cultivation. First to be considered are manipulations which in effect aim at expanding the size or increasing the density of natural floodplain stands. Struever (1964) hypothesized that *Iva* and other indigenous pioneer seed plants were cultivated by simple horticultural techniques on mudflats of large midwestern river valleys during the Woodland period, and there is a little evidence for a similar practice during historic times. Le Page du Pratz (1758:1,316-17;III,9) described a simple form of floodplain cultivation by the Natchez and other Indians on seasonally flooded sandbanks in the Lower Mississippi lowlands. After the spring floodwaters fell and without any soil preparation, women and children planted a seed called *choupichoul* (probably a chenopod [Asch and Asch 1977]) on the sand banks using their feet to cover the seed.

However, the feasibility of mudflat cultivation of *Iva* is doubtful. It is continually dispersed by flooding into wet microhabitats where it does not succeed because of the competititon with other pioneer annuals whose seed is dispersed in great quantity. To grow *Iva* on parts of a mudflat where it is not found naturally, it would be necessary to somehow suppress the growth of quick germinating, fast growing competitors until it was tall enough to become dominant and shade them out. Weeding a field of *Iva* with a hoe or digging stick is impractical since it must be grown in continuous stands like wheat in order to be harvested economically.

Lowland *Spartina* (sloughgrass) communities, with a perennial grass cover, are a habitat in which the probabilities of managing *Iva* seem better than on a mudflat. *Spartina* grows tall and in pure stands it produces a dense shade that often is less than 1% of full sun, according to Weaver (1954). Still, Turner's (1934) observations suggest that potentially *Iva* competes well in this more stable situation. Mowing and cattle grazing in wet grassy areas favor *Iva* today. Prehistorically, a similar effect could be gained by cutting grass in the early summer to provide enough light for *Iva* to survive until its growth spurt occurs. Planting and germination of *Iva* in a *Spartina* meadow should not be difficult because *Spartina* dies down each fall and its basal shoots and leaves cover only 1-3% of the ground surface (Weaver 1954).

Under a system of floodplain management, little genetic change in *Iva* is expected. Since *Iva* is wind pollinated, "unconscious" or even deliberate selection in the harvested and sown seed would be counteracted by crossing with wild plants in and around the stand. Selection would also be opposed to the extent that stands were regenerated without sowing.

Since *Iva* ultimately did become domesticated–i.e., cultivated populations were established with genetic characteristics differentiating them from wild plants–some barrier to gene flow between wild and cultivated plants must have been established. Domestication could still occur in a floodplain management system if polyploid plants that did not cross freely with wild diploids were discovered and selected.

Another model for early cultivation which would erect a partial genetic barrier involves moving cultivation out of the floodplain. The discovery that *Iva* does not require wet soil to grow well was made accidentally. In the fall of 1972 we made large collections of *Iva* and stored them in our Kampsville, Ill., field laboratory. Record high floods the next spring entered the building and washed some of the achenes out the door. For the next two years a stand of *Iva* grew on the lawn before it was mowed down. The plants were not affected by the dryness of the lawn once the waters receded, and a second generation grew even though flooding the next spring did not reach the lawn. Wild *Iva* probably is restricted to floodplains because of its mode of seed dispersal by floodwaters rather than because of an inability to grow in open upland localities.

Once prehistoric gathering of wild *Iva* began, some achenes undoubtedly were lost at campsites, resulting in volunteer plants. However, it seems unlikely that the beginnings of *Iva* cultivation can be accounted for as a gradual shift from a gathered floodplain plant to a "weedy campfollower" to a "door garden" or "dump heap" cultigen. First, there is no evidence that *Iva* would have been an aggressive campsite pioneer even when its achenes were regularly introduced. After all, it is continually introduced today into dry, disturbed habitats (such as the tops of floodplain levees) without being an effective colonizer of these niches. Second, a few straggling plants at a campsite would scarcely provide a meal, and unless *Iva* grew as a dense stand it could not be effectively harvested. With its rather low productivity per unit area–on the order of 50 g kernels/m$^2$–*Iva* would have to occupy a large area in a campsite to have a significant economic impact.

Like wheat, in which single plants also yield a negligible amount of food, *Iva* is more appropriate as a field crop than as a small crop in a garden. Other early eastern North American cultigens are different in this respect. One or two squash or gourd plants growing in a camp would have perceptible value. A few sunflower plants would have made a contribution equal to many *Iva* plants (and would beautify the location with their flowers as well). Chenopods can be cut for greens through the growing season and still produce a seed harvest.

In sum, significant upland cultivation of *Iva* almost certainly would require planting in fields. Planting in upland prairies is unlikely because of the difficulty of removing sod with a hoe or digging stick. However, a simple swidden system on forest land should prove satisfactory for its cultivation. Immediately after preparation of a plot, *Iva* would face little weed competition. Also complete tree removal would not be necessary. We have seen reasonably large numbers of *Iva* in heavily grazed, semi-open floodplain woodlots, so it should be tolerant of light shade.

Conceivably, the early cultivation of *Iva* involved not merely its introduction into new habitats but a removal to localities beyond its geographic range. This would effectively isolate some cultivated populations from wild ones. Whatever its relation to the processes of cultivation and domestication, range extension has also been considered an important *proof* of cultivation. For these reasons, the modern and prehistoric natural range of *Iva annua* requires careful delimitation.

In floral manuals the natural distribution of the plant is generally described as Indiana to Nebraska, south to Mississippi and New Mexico (Steyermark 1963; Fernald 1950; Gleason 1952). Black's (1963) map of archaeological and modern herbarium records shows few modern records east of the Mississippi River, but archaeological *Iva* is well represented as far east as the vicinity of Asheville, North Carolina (Yarnell 1977). Yarnell follows some botanists in regarding the modern eastward occurrences of *Iva annua* as adventive, particularly since these records are from "artificially disturbed habitats." By implication, this reasoning assumes that *Iva* in its indigenous range occupies primary habitats. But, as discussed in the section on distribution, *Iva* appears to be a ruderal species wherever it occurs.

Southeastern United States lags behind other regions of the country in documentation of plant distributions. However, references to *Iva*'s occurrence there are accumulating, and we would like to suggest that the natural range of the species included much of the Southeast. In their manual for the flora of the Carolinas, Radford, Ahles, and Bell (1968) record *Iva annua* from 11 North Carolina counties, and they remark that it is locally abundant in fields of the lower piedmont and coastal plain. During a visit to the Toqua mound site in eastern Tennessee, we saw a thriving stand of *Iva annua* in the parking area (a grown-over floodplain field) used by the archaeologists. Since then Gary Crites (University of Tennessee) has searched for and found another large stand nearby. *Iva annua* is recorded as quail food from several localities of the Southeast (Landers and Johnson 1976), and it is particularly notable in the Black Belt of Alabama and Mississippi (Harper 1944:14,216).

To the north, evidence for *Iva* is sparse. It is not recorded in the *Flora of West Virginia* (Strausbaugh and Core 1964). For Indiana, Jackson (1960) gives only seven records, and Deam's (1940) intensive work confirms its infrequency. In Kentucky, we found that *Iva annua* is abundant in the Mississippi Valley but have not looked for it elsewhere in the state. McFarland (1942) cites it in his catalog of Kentucky's vascular flora, but Braun (1943) does not in hers. The modern distribution of *Iva* in Kentucky needs to be thoroughly documented because of the excellent archaeological records for it in the Salts Cave area and at Newt Kash. Wild size achenes appear to be illustrated in a photograph of the intestinal contents of the Salts Cave mummy, Little Al (Robbins, 1971), but according to Yarnell (1978, personal communication) this is due to incorrect scaling of the photograph.

Recently reported archaeological records of *Iva* from Middle Missouri tradition sites of South Dakota are well outside the modern continuous range of *Iva annua*, whose northern boundary occurs in Nebraska. Benn (1974) suggested that very small achenes (mean 1.75 mm x 1 mm) from the Mitchell site in southeastern South Dakota were *Iva annua* var. *caudata* (although the size is also consistent with *Iva xanthifolia*, a common weed of the northcentral Plains). Nickel (1977) referred small *Iva* achenes from the Bagnell site to *Iva xanthifolia*. From the Helb site in northcentral South Dakota, Nickel has also reported large achenes (5-7 mm long) which are more certainly *Iva annua*.

North Dakota has one modern record of a dense stand of *Iva annua* from a saline marsh (Stevens 1950). More significant is the identification of small quantities of *Iva annua* pollen (which is distinguishable from *Iva xanthifolia*) throughout the Holocene sedimentary record at Pickerel Lake, northeastern South Dakota (Watts and Bright 1968). For similar records from 12 Minnesota pollen sites, see McAndrews (1966), Watts and Winter (1966), and Birks (1976). Thus, in the past *Iva annua* may well have had a more extensive natural range.

## DOMESTICATION

Not every anomaly in the morphology of a prehistoric plant is the consequence of selection under cultivation. Alternative explanations may involve changes in the geographic clines of plant variability, phenotypic responses to different habitat conditions, inadequate documentation of variability in wild populations, or even misidentification of specimens. For eastern North America alone, *Ambrosia trifida* (giant ragweed), *Chenopodium bushianum*, and *Phytolacca americana* (pokeweed) can be cited as cases in which the occurrence of large-seeded prehistoric plants was accounted for by an explanation not involving domestication (Payne and Jones 1962; Asch and Asch 1977).

In this section, information about variability in modern wild populations of *Iva* is presented. No evidence was found to support the feasibility of making *wild* harvests of large achenes. Furthermore, it is shown that the archaeological collections of achenes from the lower Illinois Valley—except those dating to the Archaic—are clearly differentiated from modern populations by their larger size. Finally, a few comments are offered concerning the process of prehistoric selection in *Iva*.

Yarnell (1965) discussed the possibility that large-seeded collections could result from harvesting wild stands of large plants or through some other form of selection in the harvesting process. He concluded that this was unlikely but lacked observations on natural populations of *Iva* to fully substantiate his position. We measured achene lengths from 56 *Iva* plants whose height ranged between 15 cm and 2 m and found no correlation whatsoever between the two variables. Achenes from the upper and lower parts of 10 plants were also compared, but the differences were negligible. In general, variability among the 3-5 achenes within one of the many tiny flower heads on a plant is likely to be almost as great as variability in the entire plant.

For some species, delayed flowering results in larger than average seeds. One modern collection—from the Apple Creek population of Table 7—permitted a test of this possibility. Part of the stand had been mowed so late in the summer that plants flowered immediately without further production of leaves and stems. Achenes from the mowed plants were larger than achenes from unmowed plants by only 0.2 mm.

Table 7 gives achene measurements for several modern populations. There are small differences between population means. However, over the last eight years we have collected *Iva* from a wide variety of lower Illinois Valley habitats without ever finding a population whose mean achene length greatly exceeded 3.0 mm. There is some evidence of a modest north-south cline in achene size. The central Illinois collections average about 3.0 mm long, while those from Arkansas are close to 2.5 mm. According to Jackson (1960) var. *annua*, which is concentrated in the northern part of the geographical range, has achenes 2.5-4.5 mm in length, while var. *caudata* which is more common to the south has a slightly smaller range of 2-4 mm.

In the botanical literature precise definitions of range measurements and their sampling basis are usually not provided. The wild size range stated by Jackson is slightly too restrictive to encompass total achene variability but is much too large for the variation between population means. Among achenes harvested from 30 one-meter-square plots from six lower Illinois Valley localities, the largest achene from each plot averaged 4.2 mm long and the largest achene collected from any of the plots was 5.4 mm long.

Blake (1939) established the var. *macrocarpa* on the basis of achenes shown to him by Volney Jones from Ozark Bluff Dweller collections and

TABLE 7
*IVA ANNUA* ACHENE SIZE FROM MODERN COLLECTIONS

| Location | Number of Achenes | Mean (mm) | Standard Deviation | Range | Collection Date |
|---|---|---|---|---|---|
| Illinois | | | | | |
| Apple Creek (mowed) | 108 | 3.2 x 2.6 | .61 x .42 | 1.6-4.6 x 1.5-3.7 | 12-17-76 |
| (unmowed) | 138 | 3.0 x 2.4 | .48 x .41 | 2.0-4.6 x 1.6-3.6 | 12-17-76 |
| Crater | 204 | 2.8 x 2.1 | .33 x .30 | 1.9-3.8 x 1.2-3.3 | 10-31-71 |
| Lost Creek | 97 | 3.0 x 2.3 | .48 x .45 | 2.1-4.1 x 1.6-3.6 | 10-29-72 |
| Macoupin | 137 | 2.9 x 2.2 | .46 x .34 | 2.0-4.5 x 1.6-3.1 | 10-31-71 |
| McClure (mowed) | 104 | 2.8 x 2.3 | .41 x .34 | 1.8-4.0 x 1.6-3.2 | 10-15-77 |
| Michael | 102 | 2.8 x 2.0 | .37 x .29 | 2.1-3.6 x 1.5-2.9 | 10-31-71 |
| Silver Creek | 98 | 2.9 x 2.2 | .40 x .32 | 2.3-4.0 x 1.6-3.7 | 10-20-72 |
| Thebes | 112 | 3.0 x 2.3 | .39 x .31 | 2.2-4.1 x 1.6-3.3 | 10-14-77 |
| Arkansas | | | | | |
| Bruins | 110 | 2.5 x 2.0 | .37 x .31 | 1.9-3.7 x 1.5-2.8 | 11-02-77 |
| Osceola (green plants) | 124 | 2.5 x 2.0 | .23 x .21 | 1.9-3.0 x 1.5-2.5 | 10-12-77 |
| Tennessee | | | | | |
| Toqua | 113 | 2.8 x 2.3 | .37 x .31 | 2.1-3.9 x 1.6-3.1 | 10-27-76 |

from Newt Kash Shelter. Before deciding that the achenes were a domesticated variety of *Iva ciliata* (the species is now called *Iva annua*), Blake weighed the alternative possibilities that they were instead a distinct wild species now extinct or the wild ancestral form of the modern plant. *Iva annua* has seemed such an unlikely species to receive man's serious attention that one could scarcely avoid the occasional subversive thought that a simple explanation not involving cultivation and domestication might eventually emerge.

Nevertheless, the accumulating prehistoric data reinforce the hypothesis of a domestication from *Iva annua*. Achene measurements for archaeological collections from the lower Illinois Valley are given in Table 8, and Yarnell (1977) reports measurements of *Iva* from other regions. Although some specimens are within the range of modern variability, the mean dimensions of post-Archaic achenes are strikingly larger than modern population means. The fact that the earliest documented *Iva* from the Koster site are all small, whereas very small achenes are almost never found in later contexts, is favorable to the domestication hypothesis. Also supporting it are the records of prehistoric achene collections intermediate between the average size of modern *Iva annua* and the lower limit of 4.8 mm x 3.2 mm given by Blake for achenes of var. *macrocarpa*. Furthermore, as Yarnell (1972, 1977) has pointed out, there is tentative evidence of gradually increasing size through time, on the order of 1 mm per millennium for achene length.

Enough specimens were recovered from Newbridge to permit an examination of variability at that early Late Woodland site. Figure 2 plots the

TABLE 8
*IVA ANNUA* ACHENE SIZE FROM ARCHAEOLOGICAL SITES IN THE LOWER ILLINOIS RIVER VALLEY*

| Sites | Number of Measurements | Mean (mm) | Standard Deviation | Range |
|---|---|---|---|---|
| Middle Archaic | | | | |
| Koster Hor. 8 | 1L, 2W | 1.8 x 1.8 | —— x .57 | 1.8 x 1.3-2.2 |
| Late Archaic | | | | |
| Koster Hor. 6—all obs. | 284L, 340W | 2.3 x 1.9 | .32 x .29 | 1.4-3.3 x 1.2-2.9 |
| definite achenes | 19L, 22W | 2.2 x 1.8 | .33 x .36 | 1.6-3.0 x 1.3-2.6 |
| definite kernels | 17L, 18W | 2.2 x 1.6 | .28 x .19 | 1.6-2.5 x 1.4-2.1 |
| Titus Hor. 2 | 3 | 2.6 x 1.7 | .10 x .06 | 2.4-2.7 x 1.7-1.8 |
| Middle Woodland | | | | |
| Crane | 2L, 5W | 4.9 x 3.5 | .49 x .76 | 4.6-5.3 x 2.9-4.9 |
| Massey | 1 | 4.9 x 2.6 | —— | —— |
| Archie | 1 | 3.7 x 2.1 | —— | —— |
| Middle & Early Late Woodland | | | | |
| Macoupin | 205L, 248W | 5.0 x 3.3 | .76 x .57 | 3.0-8.6 x 1.8-5.4 |
| Apple Creek | 19 | 5.4 x 3.6 | —— | 4.4-6.2 x 2.9-4.8 |
| Early Late Woodland | | | | |
| Loy† | 3L, 2W | 5.1 x 3.7 | .60 x .57 | 4.4-5.8 x 3.2-4.1 |
| Carlin | 8 | 5.3 x 3.6 | .64 x .53 | 4.2-6.1 x 2.9-4.6 |
| Bridgewater | 9 | 4.7 x 2.9 | .52 x .25 | 3.7-5.7 x 2.4-3.2 |
| Newbridge | 224L, 306W | 5.5 x 3.7 | .68 x .51 | 3.2-8.1 x 2.1-5.6 |
| Stilwell | 65 | 5.8 x 3.9 | —— | 4.6-7.0 x 2.8-5.3 |
| Late Late Woodland | | | | |
| Koster-East | 10L, 19W | 6.1 x 4.0 | .59 x .52 | 5.4-7.2 x 3.1-5.0 |
| Carlin† | 1L, 2W | 5.7 x 3.7 | —— x .14 | 5.7 x 3.6-3.8 |
| Healy | 1 | 5.7 x 3.2 | —— | —— |
| Late Late Woodland/Mississippi | | | | |
| Loy† | 3 | 4.9 x 3.2 | .31 x .61 | 4.6-5.2 x 2.4-3.7 |
| Mississippi | | | | |
| Audrey | 2 | 5.2 x 3.4 | .20 x .20 | 5.0-5.4 x 3.2-3.7 |

*Correction for loss of pericarp is 0.7 mm L, 0.4 mm W. Correction for shrinkage after carbonization is 10% in each dimension. Koster Archaic and Titus not corrected for pericarp loss because (1) absence usually not definitely established and (2) standard correction too large for small achenes. Standard pericarp correction added to all Macoupin measurements since presence/absence of pericarp not noted for individual specimens. Apple Creek and Stilwell measurements reported by Yarnell (1977).

†Features constructed by later components also contained debris of earlier components. Loy mixing included Middle Woodland debris; Carlin late Late Woodland features also contained early Late Woodland sherds.

Newbridge frequency distribution for achene lengths, corrected for specimens lacking their shells and for shrinkage due to carbonization. Allowing for small-sample fluctuations and for some systematic error in recording measurements to the nearest 0.1 mm, it appears that the Newbridge frequency curve is

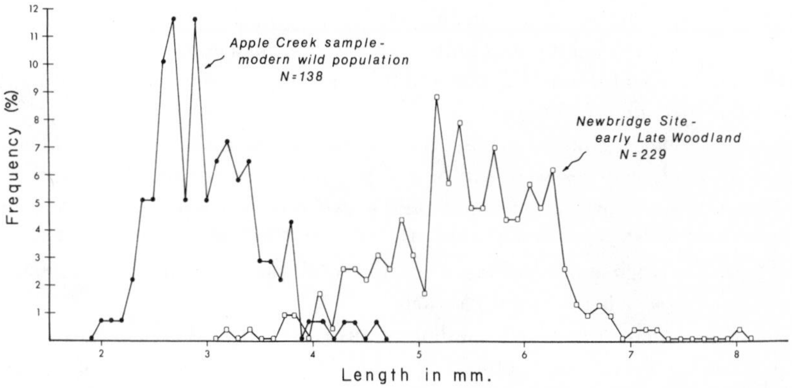


Fig. 2. Frequency distribution of *Iva* achene lengths from Newbridge site, compared with a modern wild population.

essentially unimodal. A few of the achenes are small enough to have been gathered from wild plants, but all are at least 3.2 mm long, which is larger than the mean size from any wild stand we have seen. Thus, it is almost certain that the smallest Newbridge specimens are immature or underdeveloped achenes from domesticated plants.

Historically, a permanent backwater lake was adjacent to Newbridge, and wild *Iva* undoubtedly grew along the shoreline. Nevertheless, Newbridge Indians relied exclusively on the domesticated, large-seeded crop. Similarly, the post-Archaic records of *Iva* in the lower Illinois Valley all suggest that a domesticated crop replaced rather than simply supplemented a wild *Iva* harvest.

Pollination of domesticated *Iva* by wild plants would threaten the maintenance of improved strains in regions like the lower Illinois Valley where wild strains are abundant. An expected consequence would be achenes of intermediate size and greater achene variability due to the heterogeneous genetic background. Achenes in prehistoric collections from the Illinois Valley are not as large as some that have been found (Yarnell 1977). However, they do not show evidence of increased variability. For the wild populations of Table 7, the coefficient of variation (cv) in achene length ranges from 0.09 to 0.19 (cv = standard deviation divided by mean). For the Newbridge prehistoric collection, the cv is only 0.12, and in fact none of the Illinois Valley prehistoric collections have a cv exceeding the wild values. This suggests that prehistoric cultivators maintained effective barriers against fertilization of *Iva* crops by wild plants or by wild-type cultivated plants.

In studies of prehistoric population variability and changes in achene size, there is a danger that corrections for shrinkage or missing shells will introduce systematic errors. On the basis of empirical investigations, Yarnell (1972) recommended a correction of 10% for shrinkage in carbonized specimens and the addition of 0.7 mm and 0.4 mm, respectively, to the length and width of specimens lacking a shell.

We conducted similar investigations. From the McClure collection (Table 7), 100 whole achenes were carbonized and measured for shrinkage, as were 140 Apple Creek whole achenes and 20 Crawford Creek kernels. Shrinkage in length and width was, respectively, 10.9% x 11.7%, 7.3% x 8.4%, and 9.4% x 8.5% of the original dimensions. Thus, the results are comparable to Yarnell's both for whole achenes and for kernels.

Except where noted, achene dimensions reported in this paper are based on Yarnell's correction factors to avoid confusing proliferation of correction systems. However, there remain serious questions about the validity of the corrections, particularly because they were made assuming that the shrinkage factor and shell correction terms are independent of specimen size. This assumption is demonstrably false for shell corrections. The relationship between the length of whole achenes and the kernels inside them was examined among wild collections. As Figure 3 shows, the smaller the specimen, the smaller the correction needed for a missing shell. However, we do not offer an amended correction procedure because these results imply that a correction computed from Newbridge Site achenes and kernels (0.6 mm length and 0.4 mm width) should have been much larger than it was. The need for further investigation of correction techniques is clearly indicated.

Motives for domestication are perhaps easier to understand than the process by which it was achieved. Economic advantages of greater achene size could be considerable if size increase was achieved without a reduction in the number of seeds produced per unit area. Table 9, based on wild collections, shows how a relatively small increase in the linear dimensions of achenes is accompanied by large increase in weight (see also Yarnell [1972:338]). Productivity per unit area would be a matter of greater concern to a cultivator than to a gatherer of wild *Iva* because of the cultivator's additional investments which are proportional to the area planted, i.e., preparation of the seed bed and care of young plants.

A second favorable effect of greater achene size is a reduction of the inedible shell as a percentage of total achene weight. Figure 4 illustrates this effect among wild achenes of varying size.

While advantages of domestication are apparent, *Iva* may have been a difficult plant to improve. Wind pollination would make it difficult to establish strains with desired characteristics. Deliberate selection would have been impeded by mixing of achenes from harvested plants. By comparison,

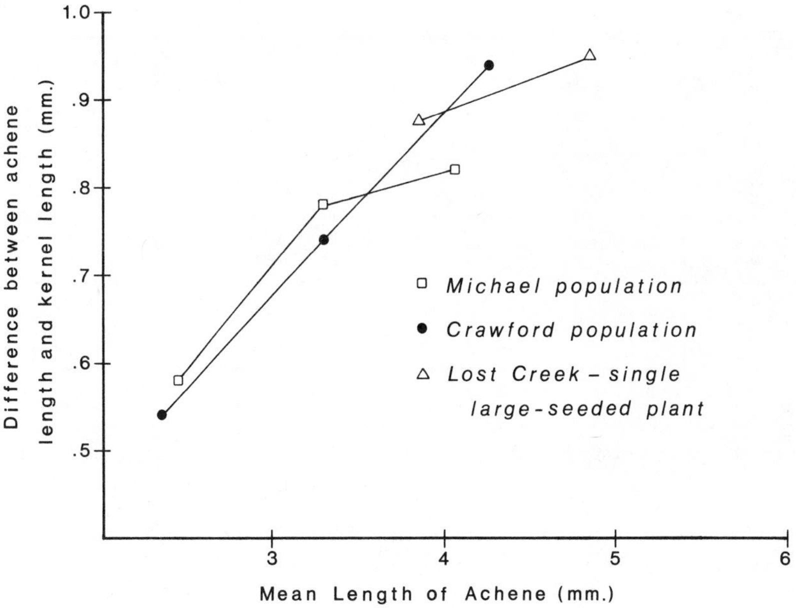


Fig. 3. Relationship of achene length and kernel length in wild *Iva*.

TABLE 9
*IVA* KERNEL WEIGHT VS. ACHENE LENGTH

| Mean Length 20 Achenes (mm) | Range (mm) | Mean Kernel Weight (mg) | Increase Factor | Source |
|---|---|---|---|---|
| 2.35 | 2.2-2.5 | 0.62 | 1.0 | Crawford population |
| 2.44 | 2.2-2.5 | 0.69 | 1.1 | Michael population |
| 3.30 | 3.1-3.5 | 1.25 | 2.0 | Michael population |
| 3.31 | 3.1-3.5 | 2.23 | 3.6 | Crawford population |
| 3.90 | 3.7-4.0 | 3.80 | 6.1 | Lost Creek single plant |
| 4.07 | 3.9-4.3 | 3.59 | 5.8 | Michael population |
| 4.27 | 4.1-4.6 | 3.66 | 5.9 | Crawford population |
| 4.86 | 4.5-5.3 | 6.67 | 10.8 | Lost Creek single plant |

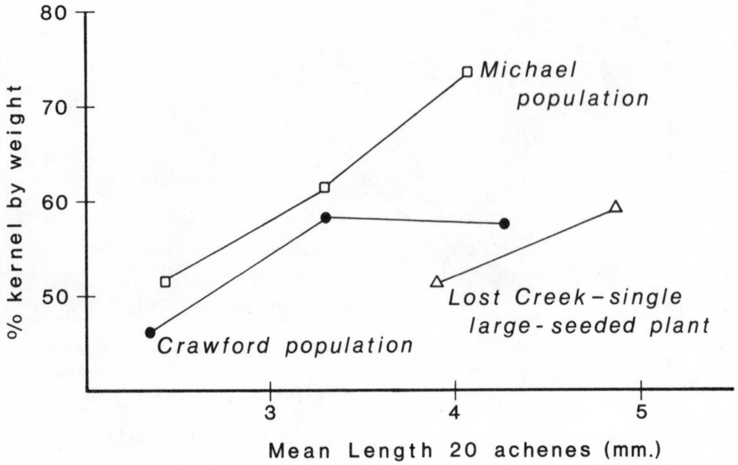


Fig. 4. Relationship of achene length and percentage kernel (by weight) in wild *Iva*.

selection could be more effective in plants like maize or squash for which seeds from individual plants remain segregated after harvesting.

Domestication could have occurred in several ways, ranging from "automatic selection" (in which character changes including greater seed size occur without purposeful human selection as the plant becomes adapted to its new cultivated habitat and new mode of dispersal [Harlan, deWet, and Price 1973]) to allopolyploidy involving the hybridization of *Iva annua* with an undetermined species. Only by combining a program of experimental domestication and related biological investigations with the developing archaeological record of variation in space and time is it likely that the domestication process for *Iva* will be significantly clarified.

## PREHISTORIC RECORD

Changes in subsistence systems more often involve shifts in emphasis than species replacement. For this reason it is necessary to develop quantitative archaeobotanical data and methods of sampling and interpretation which can detect subsistence changes and distinguish them from the many other factors that affect the archaeobotanical record.

In eastern North America, the record usually consists of carbonized plant fragments. Not only are carbonized food remains a small fraction of the quantities utilized, but the chances of preservation vary for structurally different types of plant tissue (e.g., dry and dense vs. moist and fleshy) and for plant parts with different economic significance (e.g., nutshell refuse produced each time a nut is processed vs. edible seeds which, as refuse, represent only the portion of seeds *not* consumed). Provisionally, we propose that variation in the percentages of different plants between archaeological assemblages can indicate changes in utilization. For example, in a comparison of a Late Archaic occupation from the Koster site with Woodland records at the nearby Macoupin site, we found that seed:nut ratios were 40 times greater at the Woodland site (Asch, Ford, and Asch 1972). This observation does not establish either the absolute or the relative economic importance of seeds and nuts at Koster or Macoupin, but it strongly suggests a change in emphasis. Admittedly, the reasoning involves an assumption that contexts of preservation and methods of recovery are similar enough between assemblages for major differences between archaeobotanical assemblages to be ascribed to differences in utilization.

Application of systematic and large-scale flotation recovery techniques at lower Illinois Valley sites has helped to control extraneous causes of archaeobotanical variation. With flotation it is possible (1) to recover all size classes of characoal fragments, (2) to sample a wide variety of contexts at sites, (3) to obtain statistically large samples of charcoal and the contexts in which it occurs. Flotation recovery eliminates the large idiosyncratic bias inherent in hand-collected samples. The smaller size classes of charcoal usually can be identified, and they constitute the bulk of all archaeological charcoal. Produced for the most part by everyday processing, cooking, and heating activities, these ubiquitous fragments provide a record of countless episodes of burning, not just the rare event and perhaps atypical kind of activity that leaves a large quantity of charcoal.

Table 10 summarizes results of archaeobotanical analyses for a series of components (see Fig. 1) that were excavated as part of an on-going research program in the lower Illinois Valley. Insofar as possible, the data are presented in the form of economically significant indices. For this paper, it is impossible to give an adequate description of cultural and environmental contexts or of various technical qualifications. The following synopsis of subsistence trends takes account of some of these complexities. However, at present these simple interpretations should be regarded as implicit hypotheses about the direction of subsistence change, and they are subject to further testing and refinement.

Before 5500 B.C. the archaeobotanical record at the Koster site (including occupations not summarized in Table 10) shows a low level of nut and seed utilization. At about 5500 B.C. nuts become very well represented at Koster, and their shell fragments are as abundant as wood charcoal. Seeds are still rare, but two achenes from Horizon 8C are the earliest documented occurrence of *Iva* in an archaeological context. *Iva* increases greatly in abundance and ubiquity in the major occupational unit dating 3800-2900 B.C. Seeds in general remain uncommon, but over 30% of them (including eroded and fragmented specimens) are wild-size *Iva annua*. In fact, the number of *Iva* specimens per kilogram of charcoal is greater than in any later archaeobotanical sample of Table 10, except at Newbridge where the abundance of *Iva* is inflated by a large number of achenes contained in one pit feature with exceptional preservation. It is not known whether the early small-sized *Iva* was cultivated or whether it was harvested from wild plants.

We do not have an Early Woodland record for the lower Illinois Valley, but data are available from several Middle Woodland sites. Squash (whose presence in eastern North America is now documented by 2000 B.C.) is ubiquitous in Middle Woodland contexts, although it is not an impressive component of flotation records due to the tiny size of most rind fragments and the small number of fragments in any single sample. Seeds are of variable but generally greater abundance, and there is a significant increase in representation of three starchy species that we believe were cultivated: *Chenopodium bushianum*, *Polygonum erectum* (knotweed), and *Phalaris caroliniana (?)* (maygrass). The shift to starchy seeds may have begun as early as 2000 B.C. at Koster (Hor. 4), but because of the small sample analyzed it is inconclusive at that early date. An abundance of starchy seeds is well documented at Salts Cave during the early part of the first millennium B.C. (Watson 1974).

The argument for cultivation of these three species in the lower Illinois Valley is indirect–namely, that the wild plants were not abundant enough to permit intensive harvest collection (or in the case of maygrass not even indigenous to the region). From our investigations of modern habitat relations of these species, we do not believe that extensive stands occurred either in naturally disturbed premodern habitats or on disturbed campsites. The impression of natural abundance given by Struever (1964, 1968) and Zawacki and Hausfater (1969) is based on the distribution of entire genera or on species other than the ones actually occurring in archaeological contexts.

Increasing achene size in *Iva* shows that domestication was underway at least by the Early Woodland period. However, in the lower Illinois Valley the percentage of *Iva* among seeds at Woodland sites is low. This implies not that utilization of *Iva* was declining but rather that its utilization remained at a plateau of importance while dependence on starchy seeds underwent a dramatic increase.

From the lower Illinois Valley, large seed masses have been recovered from Late Woodland contexts–*Chenopodium bushianum* and *Polygonum erectum* at Newbridge (Struever 1968), and these two species plus *Phalaris caroliniana (?)* at the upland John Roy site. These are early Late Woodland sites. From the Dinsmore mound group in southern Jersey County, Wadlow (1953:88) excavated a late Late Woodland pottery vessel that was half-filled with "seed hulls resembling smart-weed seed." This is probably *Polygonum erectum.* By the early Late Woodland period nuts undergo a sharp decline, possibly as dependence on cultivated plant foods became more securely established. Maize is absent or negligibly represented both in Middle and early Late Woodland times and first appears ubiquitously (though in low overall frequency) in the late Late Woodland period.

As agriculture based on maize as the staple cultigen became established, records of *Iva* continue in the lower Illinois Valley and elsewhere, and large *Iva* has even been reported from a proto-historic site in North Carolina (Yarnell 1972). The decline of *Iva* is not well documented. Too few late sites have been excavated with appropriate techniques for recovering small-scale plant remains, and archaeobotanical samples must be very large to detect a decline in a species never represented in large quantities.

Although the carbonized plant record cannot inform us about the actual extent of *Iva*'s utilization, it appears that less use was made of *Iva* and other native seed plants during Woodland times in the lower Illinois Valley than in the Salts Cave region of Kentucky where the paleofeces provide a direct record of human consumption (Watson 1974). Using semiquantitative methods, Yarnell estimated that *Iva* achenes were 19% of the seed bulk in a 100-feces sample from Salts Cave, and almost all of the abundant plant tissues in the feces were seeds. Stewart arrived at a lower estimate of the amount of *Iva* in the feces. On a completely quantitative basis, he determined that *Iva* achenes were only 3.8% of total seed weight in small subsamples from 38 Salts Cave fecal specimens. From nearby Mammoth Cave, he found only 2.5% by weight among the seeds in 27 fecal subsamples.

Flotation of sediments from Salts Cave vestibule and the interior cave floor yielded carbonized plant remains for comparison with the fecal record (Watson 1974). In the initial flotation series, seeds were recovered at a rate of 55 seeds/g of charcoal and *Iva* itself occurred at a rate of 3850 achenes/kg charcoal. In a one-meter-square block flotated from the vestibule, seeds in the upper part of the midden (levels 4 and 5) were recovered at a rate of 39 seeds/g charcoal and *Iva* at a rate of 484 achenes/kg charcoal. The rate of carbonized seed recovery from the Salts Cave midden is far greater than at Woodland sites in the lower Illinois Valley, except for the very small and undoubtedly unrepresentative John Roy sample. *Iva* is abundant at Newbridge where a single feature yielded most of the specimens. Otherwise, the relative abundance of *Iva* at Salts Cave is far higher than in the Illinois Valley.

This attempt at calibrating the lower Illinois Valley record in terms of Salts Cave data is very speculative, but it warns that the exceptionally well preserved archaeobotanical record from midwestern caves and rockshelters may not give a completely representative view of prehistoric plant utilization. If Salts Cave was visited only briefly by small groups who came to mine

TABLE 10
CARBONIZED PLANT REMAINS FROM LOWER ILLINOIS VALLEY ARCHAEOLOGICAL SITES (FLOTATION RECOVERED)

| | Site Location* | Feature or Midden | Number Flotation Samples | Wt Charcoal (g) | Number *Iva*/kg Charcoal | OILY SEEDS % *Iva*† | OILY SEEDS % Sunflower† |
|---|---|---|---|---|---|---|---|
| Early Archaic | | | | | | | |
| Koster Hor. 11-12 | VM | F+M | 129 | 313 | –– | –– | –– |
| Middle Archaic | | | | | | | |
| Koster Hor. 8C/8D¶ | VM | F | 118 | 499 | 4 | 6.5 | –– |
| Late Archaic | | | | | | | |
| Koster Hor. 6¶ | VM | F | 212 | 3001 | 94 | 41.1 | –– |
| Koster Hor. 4 | VM | F+M | 94 | 59 | –– | –– | –– |
| Titus Hor. 2 | VM | F+M | 44 | 177 | 17 | 10.7 | –– |
| Middle Woodland | | | | | | | |
| Archie | UP | F | 12 | 78 | 13 | 0.1 | –– |
| Crane | SV | F+M | 443 | 2772 | 2 | 0.1 | 0.2 |
| Loy | SV | F | 197 | 863 | 1 | 0.4 | 0.7 |
| Massey | UP | F | 55 | 494 | 2 | 0.1 | 1.2 |
| Early Late Woodland | | | | | | | |
| Bridgewater | VM | F | 2 | 132 | 68 | 0.4 | –– |
| Carlin | VM | F | 37 | 321 | 38 | 0.4 | –– |
| John Roy | UP | F | 8 | 88 | –– | –– | 0.01 |
| Loy‖ | SV | F | 31 | 233 | 20 | 10.8 | –– |
| Newbridge | VM | F+M | 230 | 1560 | 264 | 3.0 | 0.01 |
| Weitzer | FP | F | 103 | 179 | –– | –– | –– |
| Late Late Woodland | | | | | | | |
| Carlin‖ | VM | F | 21 | 85 | 23 | 1.6 | 4.0 |
| Healy | UF | F | 87 | 502 | 2 | 0.1 | 0.1 |
| Kos-E‖ (Early Bluff) | VM | F | 104 | 621 | 31 | 1.3 | –– |
| (Late Bluff) | | F | 70 | 211 | 24 | 0.8 | –– |
| Weitzer‖ | FP | F | 27 | 30 | –– | –– | –– |
| Late LW/Miss. | | | | | | | |
| Loy‖ | SV | F | 64 | 827 | 4 | 0.2 | 0.1 |
| Worthy-Merrigan | VM | F | 90 | 591 | –– | –– | –– |
| Mississippi | | | | | | | |
| Audrey | VM | F+M | 82 | 143 | 7 | 0.7 | –– |

*VM = Illinois Valley margin, FP = Illinois floodplain, SV = large secondary valley (Macoupin Creek), UF = dissected upland forest, UP = contact of small secondary valley and upland prairie.

†% of identifiable seeds, not including corn.

‡ *Chenopodium bushianum, Polygonum erectum*, or cf. *Phalaris caroliniana.*

§% of all charcoal fragments larger than 2 mm. Middle and early Late Woodland records not considered reliable due to low-level contamination by later components. Worthy-Merrigan summary excludes one feature with large charcoal sample (813 g) that was 78% corn.

TABLE 10 continued

| % Starchy Seeds‡† | Number Identifiable/ Number Total Seeds | Number Seeds/ g Nuts | Number Seeds/ g Charcoal | Nuts/ Other Charcoal# | % Units with Gourd, Squash | % Corn§ |
|---|---|---|---|---|---|---|
| 8 | 172/239 | 21.2 | 0.8 | 0.04 | -- | -- |
| 13 | 31/75 | 0.3 | 0.2 | 1.0 | -- | -- |
| 4 | 683/891 | 0.5 | 0.3 | 1.2 | -- | -- |
| 36 | 22/30 | 1.6 | 0.5 | 0.5 | -- | -- |
| 4 | 28/72 | 0.5 | 0.4 | 3.2 | -- | -- |
| 79 | 861/903 | 45.1 | 11.6 | 0.3 | 67 | -- |
| 82 | 5925/7787 | 4.5 | 2.8 | 1.7 | 30 | 0.03 |
| 39 | 280/503 | 0.9 | 0.6 | 1.7 | 21 | 0.002 |
| 77 | 998/1087 | 20.5 | 2.2 | 0.1 | 42 | -- |
| 85 | 2480/2710 | ? | 20.5 | ? | 50 | -- |
| 93 | 2744/2868 | 85.4 | 8.9 | 0.1 | 46 | -- |
| 99 | 44000/49200 | 5065.0 | 559.0 | 0.1 | -- | -- |
| 51 | 42/81 | 0.6 | 0.3 | 1.3 | 29 | 0.009 |
| 94 | 13900/15600 | 104.4 | 10.0 | 0.1 | 24 | 0.006 |
| 48 | 29/58 | 0.9 | 0.3 | 0.6 | 2 | -- |
| 36 | 124/168 | 12.6 | 2.0 | 0.2 | 14 | 0.5 |
| 81 | 712/914 | 22.7 | 1.8 | 0.1 | 23 | 8.2 |
| 69 | 1473/1908 | 29.5 | 3.1 | 0.1 | 45 | 0.003 |
| 89 | 627/814 | 52.0 | 3.9 | 0.1 | 11 | 0.8 |
| 100 | 33/41 | 75.9 | 1.4 | 0.02 | -- | -- |
| 96 | 1432/1846 | 3.1 | 2.2 | 2.4 | 23 | 2.2 |
| 57 | 165/207 | 1.2 | 0.4 | 0.4 | 2 | 0.7 |
| 86 | 153/259 | 9.9 | 1.8 | 0.2 | 1 | 5.7 |

¶A select sample of features whose contents have recognizable spatial organization. Seed preservation better in this subsample than in more disturbed features and midden.
#Based on counts of all fragments greater than 2 mm.
||Features constructed by later components at site also contain debris of earlier occupations, Features assigned to early components should be uncontaminated. Exception: Kos-E features assigned on basis of predominant pottery type.

gypsum or mirabilite, then it is conceivable that the visitants brought with them a high-energy low-bulk food supply. Certainly they were eating from a supply of stored food, since single fecal specimens contained foods harvested in both the spring and autumn (Watson 1974). *Iva* would be an ideal food for journeys because of its concentrated form, comparable to the sunflower seed balls carried by Hidatsa warriors and hunters of historic times (Wilson 1917:21).

Why did *Iva* ultimately fall into disuse? Perhaps it was never regarded as a primary food source since it had reached a plateau of utilization prior to the establishment of maize as the staple cultigen and even before the native complex of starchy seeds became well established. The Late Woodland decline in nut utilization may be indicative of a general reduction in dependence on oil seeds relative to starchy seeds. The uses of sunflower (*Helianthus annuus*) are most similar to *Iva* and it would have competed most directly with *Iva* for the attention of man. Yarnell (1972) suggests that as sunflower developed domesticated characteristics such as large achenes and a single large disk it replaced *Iva* because it became a superior cultigen.

The development of a polycultural crop system in eastern North America may also have played a part in *Iva*'s decline. *Iva* is most efficiently grown and harvested in a monoculture. But the Mesoamerican cultigens squash, maize, and beans, which were not all present in eastern North America until about A.D. 1000, are admirably adapted to growing in the same field, along with occasional chenopods and a few sunflowers planted at the edge of the field. A stand of *Iva* in such a field would only interfere with the individual attention given to the other kinds of plants during the growing season, and it would be partially shaded, hence adding little or nothing to the productivity of the plot.

Why was *Iva* once cultivated? Most of the information and ideas in this study are more relevant to how it could have been accomplished than why it occurred at all. A plant like squash with its multiple uses requires almost no attention to yield a return, and its early spread seems to require little justification. *Iva*, however, requires persistence to plant, to harvest, and to process. It fills the stomach but seems to have no unusual or uniquely valuable attributes. Thus, the early cultivation of *Iva* probably involved—broadly speaking—an element of necessity. It would have been cultivated to relieve a scarcity of storable food energy or to improve the reliability of the subsistence system, i.e., to increase resource availability beyond the level normally required so that in poor years there would still be enough sources of food to meet the need.

Bronson (1977) contributes an insightful discussion of the concept of scarcity and the roles it may play in the origins of cultivation and agricultural intensification. Scarcity may be produced by an unfavorable change in the population:resource ratio or by some form of locational constraint. Demographic pressure on resources may increase because of a regional growth in population density, because of the formation of larger, more aggregated settlements without overall population growth, or because of environmental deterioration. Bronson's "two-staple" model for the origins of agriculture provides an illustration of locational scarcity: e.g., a hunter-gatherer group that settles down in a rich fishing location may experience a shortage of nuts in the vicinity of the settlement even though they are abundantly available at some distance in the uplands. Growing *Iva* could bring two staples—fish protein and seed fats—into locational proximity.

With the development of economic integration based on redistribution, a new form of scarcity may come into existence: the scarcity perceived by individuals or kin groups who by amassing and distributing material goods (including food) in a calculated fashion gain prestige and create bonds of social indebtedness.

At this level of explanation, reasons for the development of agriculture transcend the specific economic attributes of *Iva annua*, and a more comprehensive approach is required involving examination of economic, social, and political organization, as well as evidence of population growth and regulation.

## SUMMARY

Properly processed, *Iva annua* makes a palatable food. It stores easily because of its low moisture content and protective shell. The kernels are a concentrated form of food energy and provide a more than adequate level of protein. They are also an excellent source of calcium, iron, and B-complex vitamins.

*Iva* grows on the open, wet, disturbed ground of floodplains, often appearing in small, dense stands. Within such stands the yield of achenes per unit area and per unit time is very roughly comparable to several wild seed plants from other parts of the world which were developed into major crops. The short, unpredictable season of availability and the small total area of natural stands would have been the chief factors limiting its potential as a wild source of storable food energy.

Archaeologists have characterized the lower Illinois Valley as an area teeming with wild plant and animal resources in the premodern era. In particular, estimates of nut and acorn masts were so large that prehistoric wild exploitation and cultivation of *Iva* did not seem to make economic sense. This paper makes a major downward revision in estimates of the annual mast. Also, the availability of the nut and acorn mast for human use is restricted more by its dispersion and concentration in time and space, and by competition with other animals, than by the average level of production per hectare of forest. In general, harvest potential may be much less than measures of biological yield would seem to indicate, and actual collecting experience is an essential part of resource evaluation. Estimates of food density can give a misleading impression of resource potential when the food item has a patchy distribution and a short season of availability.

To extend the area of *Iva* stands in the floodplain, it would be necessary to limit the growth of competing grasses or weeds—a step probably much more important than preparing the ground and planting seeds. In floodplain cultivation, the process of domestication would be impeded by the cultivated population's lack of isolation from the wild gene pool. *Iva* can grow in dry soil but is not an aggressive dry-soil pioneer. It is unlikely that it rose to the status of a cultigen by first becoming a weedy campfollower. Upland cultivation seems feasible, but only if *Iva* is deliberately planted and grown in dense stands. Genetic isolation from wild populations would be achieved by upland plantings.

The prehistoric wild range of *Iva* may be much larger than generally described. Thus, the value of range information as evidence of its cultivation is questionable. However, domestication remains the most reasonable explanation for the general occurrence of large prehistoric achenes. A search for wild populations with large achenes or harvest methods yielding large achenes proved unsuccessful.

A quantitative archaeological record of *Iva annua* in the lower Illinois Valley shows that fairly intensive use of the plant began before 3000 B.C. It is the first annual seed plant to appear in significant numbers. By Woodland times, *Iva* achenes are larger than modern wild populations. In fact, there is no evidence of wild *Iva* harvests during that period. Utilization of *Iva* remained at more or less the level of its Archaic exploitation and did not increase when the native starchy seeds of the Eastern Agricultural Complex became prominent. Use of *Iva* continued for some time after maize became the dominant crop plant. *Iva*'s economic demise probably relates to development of improved sunflower varieties and also perhaps because it was poorly suited for the developing polycultural field systems.

**Acknowledgments**

Part of the botanical work performed for this study was supported by NEH Grant # RO-21489-75-700. Thomas R. Styles and Mrs. Mary Nichols provided valuable assistance in our harvests and field observations. Gloria Caddell assisted in processing *Iva* harvests and drafted the illustrations. This paper would not have been possible without the efforts of the many archaeologists who excavated and flotated the plant remains.

## REFERENCES


Asch, David L., and Nancy B. Asch
1977 Chenopod as Cultigen: A Re-evaluation of Some Prehistoric Collections from Eastern North America. Mid-Continental Journal of Archaeology 2:3-45.

Asch, Nancy B., Richard I. Ford, and David L. Asch
1972 Paleoethnobotany of the Koster Site: The Archaic Horizons. Illinois State Museum, Reports of Investigations 24.

Benn, David W.
1974 Seed Analysis and Its Implications for an Initial Middle Missouri Site in South Dakota. Plains Anthropologist 19:55-72.

Birks, Harry J. B.
1976 Late-Wisconsinan Vegetational History at Wolf Creek, Central Minnesota. Ecological Monographs 46:395-429.

Black, Meredith
1963 The Distribution and Archaeological Significance of the Marsh Elder *Iva annua* L. Michigan Academy of Science, Arts and Letters, Papers 48:541-47.

Blake, S. F.
1939 A New Variety of *Iva ciliata* from Indian Rock Shelters in the South-Central United States. Rhodora 41:81-86.

Braun, E. Lucy
1943 An Annotated Catalog of Spermatophytes of Kentucky. Cincinnati: John S. Swift.

Bronson, Bennet
1977 The Earliest Farming: Demography as Cause and Consequence. In: Origins of Agriculture, edited by Charles A. Reed, pp. 23-48. The Hague: Mouton.

Christesen, Donald M., and Leroy J. Korschgen
1955 Acorn Yields and Wildlife Usage in Missouri. North American Wildlife Conference Transactions 20:337-57.

Claassen, P. W.
1919 A Possible New Source of Food Supply. Scientific Monthly 9:179-85.

Deam, Charles C.
1940 Flora of Indiana. Indianapolis: Indiana Department of Conservation, Division of Forestry.

Dendy, D. A. V., Bernice Emmett, and O. L. Oke
1975 Minor Food Seeds. In: Food Protein Sources, edited by N. W. Pirie, pp. 19-26. Cambridge: Cambridge University Press.

Downs, Albert A.
1949 Trees and Food from Acorns. In: Yearbook of Agriculture, pp. 571-73. Washington: U.S. Department of Agriculture.

Downs, Albert A., and William E. McQuilkin
1944 Seed Production of Southern Appalachian Oaks. Journal of Forestry 42: 913-20.

FAO (Food and Agriculture Organization)
1973 Energy and Protein Requirements. Report of a Joint FAO/WHO Ad Hoc Expert Committee. U.N., FAO Nutrition Meetings Report Series 52.

Fernald, Merritt Lyndon
1950 Gray's Manual of Botany. 8th ed. New York: American Book Co.

Flannery, Kent V., and Richard I. Ford
1972 A Productivity Study of Teosinte (*Zea mexicana*). University of Michigan, Museum of Anthropology (Mimeo.).

FNB (Food and Nutrition Board, National Research Council)
1974 Recommended Dietary Allowances. 8th ed. Washington: National Academy of Sciences.

Gates, Frank C.
1940 Annotated List of the Plants of Kansas: Ferns and Flowering Plants. Kansas State University, Department of Botany, Contribution 391.

Gilmore, Melvin R.
1931 Vegetal Remains of the Ozark Bluff-Dweller Culture. Michigan Academy of Science, Arts and Letters, Papers 14:83-102.

Gleason, Henry A.
1952 The New Britton and Brown Illustrated Flora of the Northeastern United States and Adjacent Canada. 3 vols. New York: New York Botanical Garden.

Harlan, Jack R.
1967 A Wild Wheat Harvest in Turkey. Archaeology 20:197-201.

Harlan, Jack R., J. M. J. de Wet, and E. Glen Price
1973 Comparative Evolution of Cereals. Evolution 27:311-25.

Harper, Roland M.
1944 Preliminary Report on the Weeds of Alabama. Geological Survey of Alabama Bulletin 53.

Hus, Henri
1908 An Ecological Cross Section of the Mississippi River in the Region of St. Louis, Missouri. Missouri Botanical Garden Annual Report 19:127-258.

Jackson, R. C.
1960 A Revision of the Genus *Iva* L. University of Kansas Science Bulletin 41: 793-876.

Jones, Volney H.
1936 The Vegetal Remains of the Newt Kash Shelter. In: Rockshelters in Menifee County, Kentucky, by W. S. Webb and W. D. Funkhouser, pp. 147-65. University of Kentucky Reports in Archaeology and Anthropology 3:101-67.

Korstian, Clarence F.
1927 Factors Controlling Germination and Early Survival in Oaks. Yale University School of Forestry Bulletin 19.

Ladizinsky, G.
1975 Collection of Wild Cereals in the Upper Jordan Valley. Economic Botany 29:264-67.

Landers, J. Larry, and A. Sydney Johnson
1976 Bobwhite Quail Food Habits in the Southeastern United States with a Seed Key to Important Foods. Tall Timbers Research Station Miscellaneous Publication 4.

Le Page du Pratz, Antoine S.
1758 Histoire de la Louisiane. 3 vols. Paris: De Bure.

McAndrews, John H.
1966 Postglacial History of Prairie, Savanna, and Forest in Northwestern Minnesota. Torrey Botanical Club Memoirs 22(2):1-72.

McFarland, Frank T.
1942 A Catalogue of the Vascular Plants of Kentucky. Castanea 7:77-108.

Mackenzie, Kenneth K.
1902 Manual of the Flora of Jackson County, Missouri. Kansas City, Mo.

Martin, John H., and Warren H. Leonard
1967 Principles of Field Crop Production. 2nd ed. London: Macmillan.

Morris, Robert T.
1912 The Hickories. Northern Nut Growers Association Proceedings 2:16-20.

Nickel, Robert K.
1977 The Study of Archaeologically Derived Plant Materials from the Middle Missouri Subarea. In: Trends in Middle Missouri Prehistory: A Festschrift Honoring the Contributions of Donald J. Lehmer, edited by W. Raymond Wood. Plains Anthropologist Memoir 13: 53-58.

Nixon, Charles M., Milford W. McClain, and Robert W. Donohoe
1975 Effects of Hunting and Mast Crops on a Squirrel Population. Journal of Wildlife Management 39:1-25.

Nixon, Charles M., D. Michael Worley, and Milford W. McClain
1968 Food Habits of Squirrels in Southeast Ohio. Journal of Wildlife Management 32:294-305.

Orr, Martha L., and Bernice K. Watt
1957 Amino Acid Content of Foods. U.S. Department of Agriculture, Agricultural Research Service, Home Economics Research Report 4.
Parmalee, Paul W., and Walter E. Klippel
1974 Freshwater Mussels as a Prehistoric Food Resource. American Antiquity 39:421-34.
Payne, Willard W., and Volney H. Jones
1962 The Taxonomic Status and Archaeological Significance of a Giant Ragweed from Prehistoric Bluff Shelters in the Ozark Plateau. Michigan Academy of Science, Arts and Letters, Papers 47:147-63.
Radford, Albert E., Harry E. Ahles, and C. Ritchee Bell
1968 Manual of the Vascular Flora of the Carolinas. Chapel Hill: University of North Carolina Press.
Robbins, Louise M.
1971 A Woodland "Mummy" from Salts Cave, Kentucky. American Antiquity 36: 200-06.
Robson, John R. K., Richard I. Ford, Kent V. Flannery, and J. E. Konlande
1976 The Nutritional Significance of Maize and Teosinte. Ecology of Food and Nutrition 4:243-49.
Sampson, Homer C.
1921 An Ecological Survey of the Prairie Vegetation of Illinois. Illinois Natural History Survey Bulletin 13:523-77.
Schaffner, John H.
1926 Observations on the Grasslands of the Central United States. Columbus: Ohio State University Press.
Segelquist, Charles A., and Walter E. Green
1968 Deer Food Yields in Four Ozark Forest Types. Journal of Wildlife Management 32:330-37.
Sharp, Ward M.
1958 Evaluating Mast Yields in the Oaks. Pennsylvania State University, College of Agriculture, Agricultural Experiment Station Bulletin 635.
Sharp, Ward M., and Vance G. Sprague
1967 Flowering and Fruiting in the White Oaks. Pistillate Flowering. Acorn Development, Weather, and Yields. Ecology 48:243-51.
Smith, Christopher C., and David Follmer
1972 Food Preferences of Squirrels. Ecology 53:82-91.
Spiller, Gene A., and Elizabeth A. Shipley
1977 Perspectives in Dietary Fiber in Human Nutrition. World Review of Nutrition and Dietetics 27:105-31.
Spinner, George P., and James S. Bishop
1950 Chemical Analysis of Some Wildlife Foods in Connecticut. Journal of Wildlife Management 14:175-80.
Stevens, Orin Alva
1950 Handbook of North Dakota Plants. Fargo: Knight Printing Co.
Steyermark, Julian A.
1963 Flora of Missouri. Ames: Iowa State University Press.
Strausbaugh, P. D., and Earl L. Core
1964 Flora of West Virginia. Part IV. West Virginia University Bulletin, Series 65, No. 3-2.
Struever, Stuart
1964 The Hopewell Interaction Sphere in Riverine-Western Great Lakes Culture History. In: Hopewellian Studies, edited by Joseph R. Caldwell and Robert L. Hall, Illinois State Museum Scientific Papers 12:85-106.
1968 Woodland Subsistence-Settlement Systems in the Lower Illinois Valley. In: New Perspectives in Archeology, edited by Sally R. Binford and Lewis R. Binford, pp. 285-312. Chicago: Aldine.

Telford, Clarence J.
1926 Third Report on a Forest Survey of Illinois. Illinois Natural History Survey Bulletin 16:1-101.
Thompson, Donald R.
1958 Miscellaneous Field Surveys. Wisconsin Wildlife Research 16(4):37-45.
Turner, Lewis M.
1934 Grassland in the Floodplain of Illinois Rivers. American Midland Naturalist 15:770-80.
1936 Ecological Studies in the Lower Illinois River Valley. Botanical Gazette 97:689-727.
Ungar, Irwin A., and William C. Hogan
1970 Seed Germination in *Iva annua* L. Ecology 51:151-54.
U.S. Forest Service
1948 Woody-Plant Seed Manual. U.S. Department of Agriculture Miscellaneous Publication 654.
Verme, Louis Joseph
1953 Production and Utilization of Acorns in Clinton County, Michigan. Unpublished Master's thesis. Michigan State University, Department of Fisheries and Wildlife.
Wadlow, Walter
1953 Barter Objects. Illinois State Archaeological Society Journal 3:88-90.
Wainio, Walter W., and Ernest B. Forbes
1941 The Chemical Composition of Forest Fruits and Nuts from Pennsylvania. Journal of Agricultural Research 62:627-35.
Watson, Patty Jo, editor
1974 Archeology of the Mammoth Cave Area. New York: Academic Press.
Watson, Patty Jo, et al.
1969 The Prehistory of Salts Cave, Kentucky. Illinois State Museum Reports of Investigations 16.
Watt, Bernice K., and Annabel L. Merrill
1963 Composition of Foods. U.S. Department of Agriculture, Agriculture Handbook 8.
Watts, William A., and Robert C. Bright
1968 Pollen, Seed, and Mollusk Analysis of a Sediment Core from Pickerel Lake, Northeastern South Dakota. Geological Society of America Bulletin 79: 855-76.
Watts, William A., and Thomas C. Winter
1966 Plant Macrofossils from Kirchner Marsh, Minnesota—A Paleoecological Study. Geological Society of America Bulletin 77:1339-60.
Weaver, John E.
1954 North American Prairie. Lincoln: Johnson Publishing Co.
White, Philip L., et al.
1955 Nutrient Content and Protein Quality of Quinua and Canihua, Edible Seed Products of the Andes Mountains. Agricultural and Food Chemistry 3:531-34.
Will, George F., and George E. Hyde
1917 Corn Among the Indians of the Upper Missouri. Cedar Rapids: Torch Press.
Wilson, Gilbert L.
1917 Agriculture of the Hidatsa Indians: An Indian Interpretation. University of Minnesota Studies in the Social Sciences 9.
Winton, Andrew L.
1932 The Structure and Composition of Foods. Vol. 2: Vegetables, Legumes, Fruits. New York: John Wiley.
Yarnell, Richard A.
1965 Early Woodland Plant Remains and the Question of Cultivation. Florida Anthropologist 18:77-82.
1972 *Iva annua* var. *macrocarpa*: Extinct American Cultigen? American Anthropologist 74:335-41.

1977 Native Plant Husbandry North of Mexico. In: Origins of Agriculture, edited by Charles A. Reed, pp. 861-75. The Hague: Mouton.

Zawacki, April Allison, and Glenn Hausfater

1969 Early Vegetation of the Lower Illinois Valley. Illinois State Museum Reports of Investigations 17.

Zohary, Daniel

1969 The Progenitors of Wheat and Barley in Relation to Domestication and Agricultural Dispersal in the Old World. In: The Domestication and Exploitation of Plants and Animals, edited by Peter J. Ucko and G. W. Dimbleby, pp. 47-66. London: Duckworth.

# PART V
# PREHISTORIC ECONOMICS AND PALEOETHNOBOTANY

## INTRODUCTION

The study of archaeological vegetable parts remains inseparable from ethnobotany. Indeed, the material which inspired John Harshberger to define the field of ethnobotany in the first place consisted of prehistoric floral items from Mancos Canyon in southwestern Colorado. The field was more quaint than scientific until the 1930s when Volney Jones, in an inspired series of seminal papers, brought standards of botanical exactness and new methods of identification to the material. In addition, his knowledge of ethnography and plant ecology enabled him to interpret archaeological plants in an anthropologically significant manner. He professionalized paleoethnobotany and laid the foundation for the hundreds of studies produced in the past two decades.

Plant evidence for past economic activities is quite heterogeneous. Harshberger was fortunate because his was a well sorted assemblage of desiccated corn husks and cobs, beans, fiber sandals, and other artifacts. Today few rock shelters and cliff dwellings are available for careful excavation; most sites are open-air locations that primarily yield highly fragmented carbonized seeds and charcoal. A botanical knowledge of gross morphology is now almost incidental to routine, detailed examination of microscopic anatomical features. The indispensible laboratory research microscope is necessary for identification of seeds, charcoal, phytoliths, pollen, and spores. Together all these categories of plant remains are the basis for reconstructing past environments, human economic pursuits, and the consequences of these activities for cultural development.

Until flotation (water separation) became widespread in American archaeology, the ethnobotanist dealt mainly with carbonized plant parts that were caught in screens or that were visually obvious in the fill. This has changed, as tons of soil are annually floated, and laboratories are inundated with bags of minute seeds, corn cob cupules, and charred wooden splinters. These macro-remains are excellent evidence of past economic exploits. Minnis demonstrates their value with a case study from the Rio Mimbres in New Mexico. Here the identification of seeds collected for food and the wood for fuel yield a dramatic picture of shifting preferences for resources and extensive clearing of the alluvial bottomlands for agriculture during the population peak in the Classic Mimbres period. Too often archaeologists willingly accept climatic changes but ignore the human impact on the natural resource base.

Knowledge of plant geography and physiology is indispensible to the paleoethnobotanist. The distribution of plants in space, their preference for particular habitats, and their growth and reproductive response to environmental parameters aid the interpretation of archaeological artifact assemblages and provide an explanation for site locations. Since prehistoric processing tools were often made to cope with specific physical properties of plants, their presence at a site, even in the absence of botanical evidence, can assist an understanding of past economic behavior. Conversely, if the vegetation of a site catchment can be demonstrated to have changed little through the centuries, then even without specific kinds of tools, the association between site location and plant community may facilitate interpretation. Phytogeography provided the inspiration that helped Fitting to interpret otherwise anomalous sites in the Safford area of southeastern Arizona.

Understandably, when a constellation of archaeobotanical evidence from a particular site or general region can be systematized, then a range of economic patterns may be discerned. Other than studies of several regions in Europe and Yarnell's *Aboriginal Relations between Culture and Plant Life in the Upper Great Lakes Region* in the United States, detailed regional summaries of archaeobotanical evidence are nonexistent. Pearsall has rectified the situation for western South America. Her compilation demonstrates the close relation between subsistence and cultural changes in the evolution of complex economic agricultural systems in South America, and despite the lack of extensive assemblages of plant remains from numerous sites, she has established a baseline for future archaeological research. Overviews are possible for other culture areas of the world and should be undertaken for similar planning and research purposes.

Paleoethnobotany has progressed measurably and has become standard practice throughout the world since Volney Jones' pioneering activities. Nevertheless, despite its promise, most botanical reports are relegated to an appendix or a separate chapter in an unintegrated site report. Plant remains are as much artifacts of cultural behavior as are the more conspicuous handmade objects. When appreciated from this perspective, the contribution of paleoethnobotany for understanding prehistoric life will be more significant than it is today.

# PALEOETHNOBOTANICAL INDICATORS OF PREHISTORIC ENVIRONMENTAL DISTURBANCE: A CASE STUDY[1]

*Paul E. Minnis*
University of Michigan
and Mimbres Foundation

> Because of popular conceptions of "primitives" as mystically more "natural" than "civilized" peoples, anthropological literature is often cited as evidence that human society is capable of a finely tuned, balanced harmony with its environment, comparable to that achieved in a successional climax. Unfortunately, it is all too easy to discover examples of native populations living in obvious disequilibrium with their environment. [George A. Collier, *Fields of the Tzotzil*, 1975]

In his definitions of ethnobotany, Volney H. Jones focused attention away from the traditional plant use list to a wider perspective, which stressed a whole constellation of plant/people interactions: "plant lore, properties and value of economic plants, the origin of cultivated plants, plants in archaeological sites, and plant names and plant knowledge of primitive peoples" (1941; see also 1954). The effects of human populations on the pattern and diversity of vegetation is one such class of interrelationships explored by Professor Jones.

The alteration, and often degradation, of flora is an important and pertinent field of study. Archaeology is particularly well suited for the analysis of man-induced vegetational change because of the time depth available through it and the excellent data base it provides for understanding the size and distribution of human populations.

---

[1] Contribution #11 of the Mimbres Foundation.

Unfortunately, the number of archaeological tools available for the study of vegetational change is severely limited. Palynology is the most commonly used measure of vegetation change. Other measures, such as changes in settlement and land use patterns (Bryan 1954), changes in artifact assemblages (Smith and Young 1972), and ethnohistorical survey (Day 1953) are occasionally used to infer change less directly. Generally, these indirect measures are used to offer hypotheses that are then tested with pollen analysis. The need for additional measures of vegetational change for the archaeologist is clear.

This paper explores the potential for using macroplant remains (plant remains larger than pollen grains and phytoliths) recovered from archaeological contexts as a relative measure of vegetational disturbance resulting from human activity. In particular, the effects of agricultural practices will be stressed, as agriculture has a profound impact on plant diversity and patterning. Agriculture disturbs "pristine" vegetation by clearing the native vegetation, maintaining artificial communities, and modifying successional relationships.

Data from archaeological sites in an arid to semi-arid region in southwestern New Mexico will be used to demonstrate the use of macroplant remains as direct measures of vegetational change. These changes will be related to a general discussion of agricultural disturbance of local vegetation.

## AGRICULTURAL DISTURBANCE OF VEGETATION PATTERNING

All humans utilize plants. Numerous needs are met by exploiting naturally occurring botanical resources. The magnitude of the effects upon the local vegetation of the procurement of these resources depends on many factors such as the size of the resource base and conservation practices. Particularly important is the human need for and the "maximum sustainable yield" of the resources.

For this study, the human need for sustenance is considered most critical because all of it ultimately must be derived from botanical resources. Agriculture generally has a profound effect on vegetation. With the large-scale use of agriculture, the local ecosystem is not only altered but is managed to increase the yield of energy available to the human group. In a very real sense, the ecosystem becomes dominated by humans practicing agriculture. More than the cultigens and cultivators are affected by domestication; the ecological relationships of most, if not all, living organisms are affected. The effects of agricultural practices on the local ecosystem will be considered generally.

Clearly, the specific effects on vegetation will be different for different agricultural systems; the effects of a complex swidden system in New Guinea will be different from those of hybrid maize monoculture in Iowa. Despite the specific differences, agricultural systems share a similar general model of effects on local vegetation. Basically three stages are recognized in the cycle of disturbance. First, the pristine vegetation must be cleared. Secondly, the artificial, less diverse, and consequently less stable ecosystem of cultivated fields must be maintained. Thirdly, after the termination of maintenance (fallow or abandonment), a successional cycle is set in motion. This three-staged model is visually summarized in Figure 1.

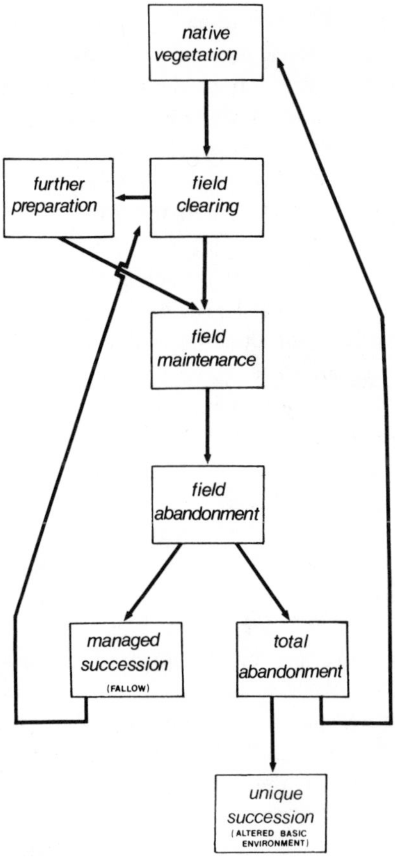


Fig. 1. Flow diagram of agricultural disturbance of native vegetation.

Preparation of land for fields includes removal of the native vegetation and soil preparation. Normally, most of the large trees and understory are removed. Firing of the dried debris is a common practice for its removal and also for enrichment of the soil. Further preparation may be necessary. Mounding for yam cultivation, construction of terraces, and deep chisel plowing are but three specific examples. Both processes, clearing the native vegetation and soil preparation, are recognizable archaeologically. Large-scale forest clearance is documented, particularly for Neolithic Europe (Iversen 1949). Prehistoric terrace systems have been located (e.g., Woodbury 1961), and their intact condition documents their efficacy in preventing erosion. Remote sensing has become a particularly effective technology for detecting subtle remains of soil alteration for agriculture (Berlin et al. 1977).

Cultivation involves the maintenance of a simplified ecosystem. Because agricultural ecosystems are inherently less stable than most natural ecosystems, energy must be expended to perpetuate them. Increased organization and information exchange are commonly associated with the maintenance of agriculture systems. Some of the numerous problems of perpetuating fields have been summarized (Nye and Greenland 1973).

The composition of fields is much different than pristine vegetation. In addition to cultigens, plants adapted to disturbed habitats thrive. Their presence as weeds can be discouraged, but in many cases some are encouraged or tolerated because of their economic potential. The presence of these plants in the archaeological record provides evidence for agriculture, as do the remains of cultivated plants themselves. The change in the ratio between arboreal pollen and non-arboreal pollen or indicator pollen types (such as *Plantago lanceolata* in northwestern Europe) has been used most commonly as a sign of field maintenance.

With the cessation of maintenance activities, there is a successional change in the composition and patterning of vegetation. If basic environmental variables have not been radically affected, the successional cycle may be similar to a natural pattern. Fallowing is simply a technique of managing succession in order to increase fertility of fields. With total abandonment, a theoretically complete successional cycle to a climax is allowed. However, often the ecosystem is over-exploited, and drastic changes in basic environmental conditions occur. Here the pattern of succession will take a different course from the natural pattern. Archaeological detection of succession following field abandonment is less common than the other two stages, but again analysis of pollen is the most commonly used technique. Extreme cases of the alteration of the environment have been recognized (Jacobsen and Adams 1958; Lisitsina 1976).

This cycle of clearing and preparation, field maintenance, and succession following abandonment is cumulative in that greater intensification or "extensification" will effect a greater disturbance on the local vegetation. This cumulative effect is of particular interest in this paper.

## AN ARCHAEOLOGICAL CASE STUDY OF AGRICULTURAL DISTURBANCE OF VEGETATION

Two paleoethnobotanical data bases will be used to examine the model of vegetational disturbance due to farming practices. All data were recovered by the Mimbres Foundation between 1974 and 1977, and excavation in the Mimbres Valley of New Mexico continues. The research design focuses on spatial and temporal variation. Sites were excavated in spatial clusters along the valley in different environmental conditions. Within each cluster, a site was excavated from each time period with a few gaps remaining. Thus, it is possible that several sites from each period are represented, and spatial variation within each temporal period is represented.

The first data set consists of identified specimens of charcoal, which are the remains of woods used for fuel and for construction. Identifications were made by the author and the Laboratory of Tree-Ring Research, University of Arizona. Charcoal and wood from archaeological sites are rarely analyzed except for their use as a source for radiocarbon and dendrochronological dating. However, some studies have attempted to use wood and charcoal as indicators of cultural and biological processes (cf. Salisbury and Jane 1940; Godwin and Tansley 1941; Conrad and Koeppen 1972; Minnis and Ford 1977; Asch, Ford, and Asch 1972; Minnis 1977; Schweingruber 1976).

The second data base includes seeds and reproductive structures recovered from flotation samples. While archaeological seeds have not been neglected to the extent of charcoal and wood, usually they have simply been interpreted as the remains of food resources used. Again, their deposition and preservation reflects cultural and biological factors, a fact which is just beginning to be dealt with systematically in archaeology (cf. Bohrer and Adams 1977; Spector 1970).

### Vegetation of the Study Area

This study area centers around the Rio Mimbres in southwestern New Mexico, which drains parts of the Gila Mountains and the Black Range (see Fig. 2). The Rio Mimbres flows through a relatively narrow valley for 60 km above ground before emptying into the desert plains around Deming and eventually recharging the great underground reservoirs of northern Chihuahua, Mexico. From 20 km north of Deming to its southernmost extent, the Rio Mimbres channel flows above ground only after torrential storms.

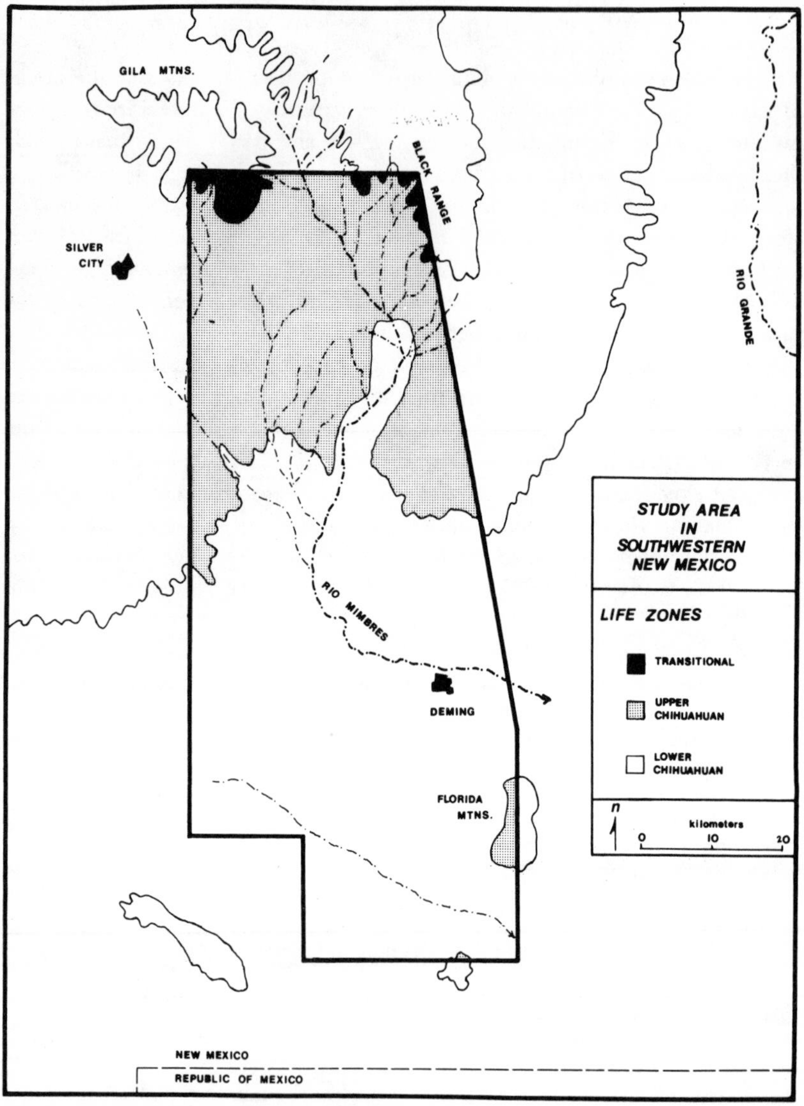


Fig. 2. Southwestern New Mexico. The enclosed area is the Mimbres Foundation study area.

General plant distributions in the Southwest are largely determined by moisture received and elevation. Specific distributions are determined by more particular variables such as slope orientation and angle, soil substrate and mineral composition, solar radiation, and local hydrological regime. The diverse intersection of these factors within the study area is seen in the diversity of vegetation associations. Three broad life zones are delimited for the Rio Mimbres area: Lower Chihuahuan, Upper Chihuahuan, and Transitional. A brief summary of the study area will be presented, but more detailed descriptions are available (Castetter 1956; Bailey 1913; Lowe 1964; Dick-Peddie 1975).

It must be remembered that each life zone is an aggregate of numerous plant associations and that by careful manipulation of environmental variables, prehistoric agriculturalists were able to successfully grow crops in zones considered generally marginal for modern cultivation. For example, in a zone with a short growing season, utilization of south-facing slopes for field location maximizes solar radiation and minimizes a cold air drainage effect. Consequently, the frost-free period is longer on these slopes than in surrounding areas.

The Lower Chihuahuan zone (1220 m to 1680 m in elevation) encompasses desert basin and range topography, which is characterized by isolated mountains rising above the flat desert plains. Unlike the desert valleys to the west, the Deming bolson is very broad and lacks distinctive playas. Except for the Rio Mimbres and the smaller Seventy-six Draw, drainage patterns in this area are ephemeral with occasional swales which hold surface water for up to a few days after rain. While modern economic activity has drastically lowered the water table, it is safe to assume that water availability was still a critical problem for prehistoric occupants of the area.

The vegetation pattern consists of stands of desert shrubs: mesquite (*Prosopis juliflora*), creosote bush (*Larrea tridentata*), Mormon tea (*Ephedra trifurca*), and yucca (*Yucca elata*) interspersed with grasslands (primarily *Bouteloua* spp. and *Hilaria mutica*). A xeric riparian assemblage is present: hackberry (*Celtis reticulata*), desert willow (*Chilopsis linearis*), and rabbitbrush (*Chrysothamnus nauseosus*).

The Lower Chihuahuan zone is marginal for agriculture. While the frost-free period is long (220 days), the annual precipitation is low (25 cm). Strong spring winds, high solar radiation and temperature, and low humidity further increase the deficiency of available moisture in relation to the high potential evapotranspiration rates of cultigens. All modern agriculture is irrigated from deep wells. The presence of prehistoric agriculturalists in this area attests to their expertise and to the likelihood of a higher water table in the past.

Most archaeological work has focused in the Upper Chihuahuan zone (1515 m to 2290 m) where the broad desert plains give way to a topography dominated by mountain ranges and narrow river valleys; water flows year around in most of the major channels. Snow pack melt and heavy summer rains permit seasonal flow in the smaller channels.

The dominant vegetation in the Upper Chihuahuan zone is woodland composed of juniper (*Juniperus* spp.), oak (*Quercus* spp.), and piñon (*Pinus edulis*). Junipers and oaks are located in the more xeric habitats, lower elevations and on south-facing slopes. Piñon is confined to the more mesic north-facing slopes and to higher elevations in this zone. Numerous shrubs, grasses, succulents, and forbs are present. Along water courses with deep alluvium, a well-developed floodplain community is established. Cottonwoods (*Populus* spp.) dominate this assemblage, but numerous other woody plants are common: willow (*Salix* spp.), walnut (*Juglans major*), box elder (*Acer negundo*), alder (*Alnus oblongifolia*), and ash (*Fraxinus velutina*).

The Upper Chihuahuan zone is the best region for agriculture within the study area. The frost-free period is adequate, the major river valleys have a deep, rich alluvium and high water table, and the precipitation is adequate (41 cm). The precipitation is bimodally distributed between a winter period of snow and an early summer period of rain. May, a critical month for crop planting and germination, is the driest month of the year, a pattern characteristic of southern New Mexico. Modern agriculture is most abundant in this zone, and crops include maize, fruit trees, and irrigated forage plants.

The Transitional zone (2135 m to 2750 m) is the northernmost zone within the study area. It is mountainous, with deep and narrow valleys. Ponderosa pine (*Pinus ponderosa*) is the characteristic plant, with other common woody plants present: junipers and Gambel's oak (*Quercus gambellii*).

This zone is again marginal for agriculture, and none is practiced at present except for small home gardens. Although precipitation is adequate (50 cm), the frost-free period is short (around 100 days), and the soil mantle is generally shallow. As 120 days are needed for maturation of maize, successful harvest is limited by the short frost-free period; the low site density of prehistoric agricultural populations supports the interpretation that plant husbandry was difficult in this zone. However, maize recovered from sites in this zone and the presence of terraced field systems demonstrates that some prehistoric cultivation was practiced.

Changes in vegetation patterns in the study area since the prehistoric period are not well understood. The effects of historic human occupation have been documented for the Lower Chihuahuan and Upper Chihuahuan zones (cf. York and Dick-Peddie 1969; Hastings and Turner 1965; Humphrey

Fig. 3. The Rio Mimbres valley in the Upper Chihuahuan life zone looking south. Note the fields and distinctive floodplain vegetation.

1958), but any effects on the Transitional zone are unknown. Probably logging and grazing have had an effect.

## Culture History of the Study Area

Human occupation of southwestern New Mexico dates from the Paleo-Indian period. Documented prehistoric occupation within the immediate study area spans 1250 years, A.D. 200-1450. For purposes of this analysis, this time span is divided into five periods: Early Pithouse, Late Pithouse, Classic Mimbres, Animas, and Salado. The correlation of these periods to the more traditional typology is available elsewhere (LeBlanc 1975, 1976, and 1977). There is a recognized cultural continuity from Early Pithouse through Classic Mimbres. The Animas and Salado each represent distinct cultural, adaptive, and probably population discontinuities.

Early Pithouse sites (A.D. 200-600) range from 1 to 50 pithouses and are generally located on high, isolated knoll tops up to 270 m above the Rio Mimbres floodplain and its major tributaries. Few pithouse villages are found above 1890 m, nor is there much of an Early Pithouse period occupation in the mountainous areas which surround the Mimbres Valley and in the desert areas of the southern part of the study area. The presence of maize is documented for the Early Pithouse period, though the dietary importance of cultigens is not clear.

After A.D. 600, there is a change in settlement pattern. During the Late Pithouse period (A.D. 600-1000), large aggregates of pithouses formed villages (up to 100 pithouses) on the first bench directly above the Rio Mimbres floodplain. A greater, but still sparse, occupation is found in the mountains and desert areas. The bulk of the population was still located along the well-watered mountain valleys. The settlement pattern change and site size increase probably reflect internal population growth rather than immigration into the study area.

The Classic Mimbres period (A.D. 1000-1150) exhibits the greatest series of changes. Architecture changes from pithouses to surface, cobble-walled pueblos. There is an increase in the number of sites, which reflects a continued population increase. The large sites (up to 200 rooms) are built in the same location as the large Late Pithouse villages. Many smaller pueblos and fieldhouses are scattered between the large pueblos. The greatest settlement pattern change occurs in the mountains where Classic pueblos are found wherever there is arable land. In addition, there is the appearance of water control features such as terracing and checkdams. Within a short period of time, the Classic Mimbres system decomposes. The cause of the collapse is not well understood. I suggest that the increased difficulty in nutritionally provisioning an expanding agricultural population in an arid to semi-arid environment with wide fluctuations in basic environmental conditions is critical to understanding this time period. Other considerations for understanding the Classic Mimbres collapse have been suggested (LeBlanc 1976).

The Animas period is less well understood than the preceding periods. It seems to have occurred directly after the Classic Mimbres, if indeed it did not briefly overlap with the extreme end of the Classic Mimbres period. Animas ceramics, burial practices, and architectural patterns are much different from the Classic Mimbres. The Animas is a desert adaptation with the highest density in the area around Deming and a decreasing site density further up the major mountain river valleys (such as the Mimbres Valley). No Animas sites are known from the mountains themselves; the relationships between the Animas occupation and the complex archaeological systems in northern Mexico (e.g., Casas Grandes) are unclear, but some cultural commonalities are present. The Animas seem to be a local variant on a desert-dwelling tradition which occupied the area from El Paso to eastern Arizona and into northern Mexico. Like the Pithouse and Classic Mimbres populations, the Animas people were agriculturalists. More detailed culture history of the Animas awaits further analysis by the Mimbres Foundation.

The last prehistoric occupation is the Salado (ca. A.D. 1425-1475) that has been reported in detail (LeBlanc and Nelson 1976). This puebloan

Fig. 4. Firewood collecting, as depicted on a Classic Mimbres black-on-white bowl. (Photograph used by permission of The Maxwell Museum, University of New Mexico and by permission of the School of American Research.)

tradition is known from only four sites within the study area. All are located directly next to the Rio Mimbres floodplain, and most are located in the Upper Chihuahuan zone. This period seems culturally unrelated to the Animas and probably represents a limited migration from the heavily populated Gila River drainage to the west of the Rio Mimbres. Again, the Salado were agriculturalists.

Relative population size for each period in the Mimbres Valley, excluding the desert region, was calculated from survey data (Hastorf 1977). These data are subject to revision based on further surveys. Floor areas were estimated for each site and corrected for the length of occupation. The results were normalized so that the base figure is 100 for the Early Pithouse.

| PERIOD | RELATIVE POPULATION |
|---|---|
| Early Pithouse | 100 |
| Late Pithouse | 509 |
| Classic Mimbres | 1466 |
| Animas | 503 |
| Salado | 243 |

Though the Late Pithouse period population may be somewhat underestimated, the size of the Classic Mimbres population is contrasted with the much smaller populations during the other periods.

Agriculture is present in the area from A.D. 300, as evidenced by maize from Early Pithouse flotation samples (Hogg 1977). The importance of cultigens is, however, not clear. A rough index of the role of maize, the chief cultigen, is the ubiquity of maize fragments. This ubiquity is measured by the number of flotation samples (containing prehistoric material) that included maize, mostly cupule fragments. It is simply assumed that a relative increase in use of a resource will be reflected in a relative increase in its presence in the archaeological record.

Figure 5 shows that maize is the most ubiquitous taxon recovered (after "weed" seeds) in all periods and that it seems to increase in importance through time. The very high percentage of maize in Animas samples is skewed, as a result of burning of the pueblos that resulted in better preservation than from sites of other periods. It is clear that maize was an important resource in all periods.

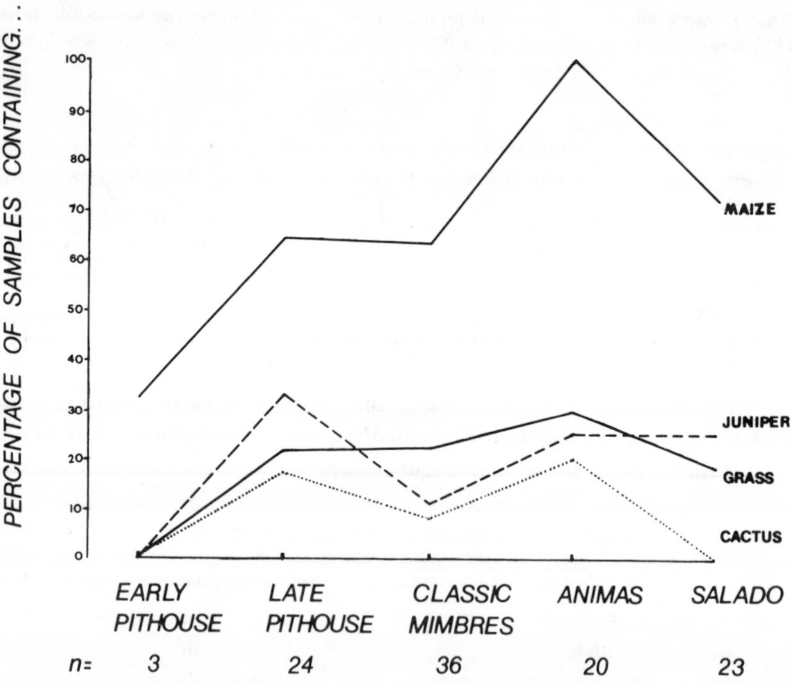


Fig. 5. Percentage of selected taxa of seeds recovered from flotation samples.

## Example One—Charcoal

The first effect of agriculture upon the local vegetation is clearing of the pristine vegetation for fields. Woody as well as herbaceous vegetation would be affected. With an increase in farming activity, one would expect an increase in field clearance. In the study area, archaeological evidence suggests that an increase in agriculture included intensification as well as more extensive agriculture.

The Rio Mimbres floodplain is an excellent zone for farming. Here the alluvium is rich and deep, and the water table is high. Limited experiments with maize suggest that the floodplain is two to four times more productive than the next best zone, the terraces directly above the floodplain. Clearly then, the greatest effect of agriculture would be on the vegetation of the floodplain.

To test this expectation, charcoal from flotation samples and pieces recovered directly by excavators were identified. The identified charcoal was then grouped into two categories: woods from taxa growing on the floodplain and woods from taxa growing on terraces and mountains around the sites. The first group includes cottonwood/willow, ash, box elder, sycamore (*Platanus wrightii*), and walnut. The most common types of the second category include juniper, oak, piñon, and mountain mahogany (*Cercocarpus brevifolia*). In all, 1940 identifications were made.

Most of these specimens, which reflect the remains of fuelwoods, were collected from hearths or from ash lenses in trash deposit proveniences. It is believed that changes in the ratio between floodplain woods and nonfloodplain woods recovered represent the changes in their *relative* usage. Figure 6 summarizes the percentages of floodplain charcoal recovered (computed by counts of individual specimen identifications) for the five archaeological periods. This is compared with a relative index of the human population for these periods. As can be seen, the percentage of floodplain wood is low for the Early Pithouse and Classic Mimbres periods and is high for the Late Pithouse, Animas, and Salado periods.

The low percentage for the Early Pithouse is probably a result of the fact that these sites are located on high knolls further away from the floodplain than sites of other periods, and hillside fuelwood was more easily secured.

The low occurrence of floodplain woods during the Classic Mimbres is not due to the distance of these sites from the floodplain as the Late Pithouse, Animas, and Salado sites are located in the same area, on the first bench directly above the floodplain. Figure 6 suggests that with an increased population during the Classic Mimbres, more fields were cleared with the resultant decimation of the natural woody vegetation and a decrease in the availability of floodplain fuelwood.

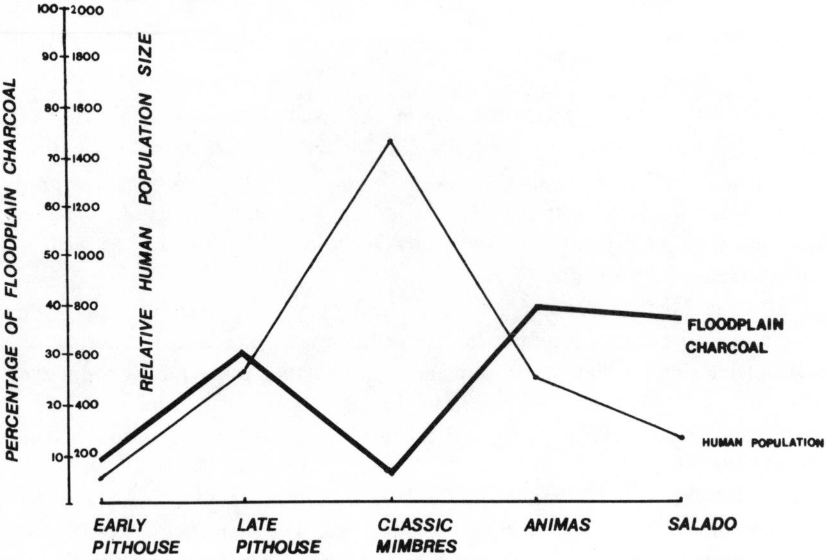


Fig. 6. Percentage of floodplain charcoal to all species. Dendrochronological samples are not included. Note the relative size of the human population.

Analysis of 600 dendrochronological samples submitted to the Laboratory of Tree-Ring Research, University of Arizona, reveals a similar pattern for the five sites with the largest samples:

| SITE | PERIOD | FLOODPLAIN TAXA |
|---|---|---|
| LA 12110 | Early Pithouse | 9% |
| LA 635 | Late Pithouse | 22% |
| LA 676 | Classic Mimbres | 0% |
| LA 12076 | Classic Mimbres | 4% |
| LA 12077 | Salado | 21% |

The specimens identified at the University of Arizona were usually taken from large pieces of charcoal that were used primarily for construction rather than for firewood. Thus, two relatively independent procurement strategies (fuel and construction wood) show a similar pattern. This finding further reinforces the conclusion that the basic floodplain pattern had been radically altered during the Classic Mimbres period.

The same genera of floodplain plants were recovered from Classic sites as from sites of other periods. It is the relative frequencies with which these are found that changes. This argues against a drastic qualitative change in the woody riparian vegetation. The high occurrence of floodplain charcoal during the Animas and Salado periods would thus reflect a succession back to a more natural association. The lower human populations during these periods in the Upper Chihuahuan zone probably would not have produced as great a pressure on the riparian vegetation.

Of further interest are the frequencies of floodplain wood types recovered from late Pithouse and Salado sites. The Animas samples are not useful here because most of the samples are from areas to the south of the other samples and are therefore not adequately comparable. If our interpretation is correct, the floodplain wood recovered from the Late Pithouse proveniences would represent the floodplain vegetation before extreme denuding. The Salado samples would represent a woody assemblage after the period of greatest disturbance and would presumably be in a successional cycle.

The following table shows the frequency of floodplain wood from Late Pithouse and Salado sites, by percentage. Two points are of interest. First, cottonwood/willow is the dominant taxon during the Late Pithouse period. No single type is dominant from Salado proveniences. Second, there is a greater diversity of taxa present in the Salado samples. If these samples reflect the floodplain vegetation during these periods, then the Late Pithouse sample resembles a mature cottonwood-dominated floodplain assemblage with a relatively open understory. The Salado reflects a more diverse successional stage where cottonwood had not yet become dominant (Campbell and Dick-Peddie 1964).

| TAXON | LATE PITHOUSE | SALADO |
|---|---|---|
| Cottonwood/Willow | 88.5% | 27.6% |
| Sycamore | – | 2.3% |
| Ash | 4.2% | 2.3% |
| Alder | – | 2.3% |
| Walnut | 1.4% | 18.4% |
| Box Elder | 4.2% | 47.1% |
| Sample Size | 71 | 87 |

This last interpretation is tenuous. A detailed study of the successional patterns in the floodplain is not available. In addition, as there is a 100-year hiatus between the Animas and the Salado periods and a 225- to 275-year time span between the Classic Mimbres and the Salado periods, the effects of the Animas occupation upon the floodplain vegetation in the Upper Chihuahuan zone are unclear.

### Example Two–Seeds

With increased agriculture as a response to the increased needs of an expanding population, one would expect an increase in species adapted to disturbed soils–"weeds." There is a positive correlation between the number of weeds growing in an area and the number of weed seeds produced (Jensen 1969). All things being equal, an increase in disturbance should be mirrored in an increase in weed seeds produced in relation to other seeds. The charred seeds recovered from flotation samples taken from archaeological sites within the study area will be used to test this expectation.

Seeds were recovered in two conditions: charred and uncharred. Uncharred seeds are assumed to be from modern "seed rain" for three reasons. First, numerous seeds are naturally present in soil. Quick (1961) estimates from over 1.5 million to 7 million viable seeds per hectare, but figures as high as 3.5 billion have been reported. Five surface (0-10 cm) samples taken from soil away from archaeological sites in the study area yielded 100-2100 seeds per liter. Subsurface samples contained numerous seeds but in decreasing numbers at greater depths from the surface. Second, none of the 6000 seeds identified from these samples were charred, which suggests that there is an insignificant background of naturally charred seeds. Third, it is unlikely that uncharred prehistoric seeds would be preserved until the present time in this location.

Charred seeds are assumed to be prehistoric. For this study, charred seeds of goosefoot (*Chenopodium* spp.), pigweed (*Amaranthus* spp.) and purslane (*Portulaca* sp.) are considered the prehistoric weed seeds. Charred seeds from other weeds are present, but their occurrence is numerically insignificant. These three types of seeds are assumed to be prehistoric seed rain which was accidentally charred in archaeological contexts, such as blowing into hearths. This is not to say that these species were not being utilized but rather suggests that their presence in this archaeological record better reflects their natural occurrence. Several points support this interpretation. These seeds are small (around 1 mm diameter), produced in enormous quantities, and are easily dispersed by natural processes such as wind. These three seeds are the most ubiquitous, being found in most samples in low numbers. This distribution resembles a background contamination rather than being the result of processing accidents where large numbers of seeds in a few samples would be expected.

From the five periods, 2550 charred seeds were recovered. The proportion of charred weed seeds to all seeds is plotted in Figure 7 for the five archaeological periods. This is compared with the relative index of human population size for these periods. The pattern is clear. An increase in population is positively correlated with an increase in weed seeds. This suggests that an increase in an agricultural population results in a greater degree of disturbance as evidenced by an increase of plants adapted to disturbed soils. This index further suggests that these annuals were an increasing potential food source, but not necessarily that they were being exploited to a greater degree.

The percentage of Cheno-Am pollen for the five periods in the study area is remarkably similar to the weed seed index in Figure 7 (Carl D. Halbirt 1976: personal communication), supporting the above interpretation. It is recognized that the family Chenopodiaceae is used in this pollen analysis and the seeds represent the genus *Chenopodium*. However, in the study area *Chenopodium* is the most common Chenopodiaceous plant, and thus the pollen and seed data are in fact quite comparable.

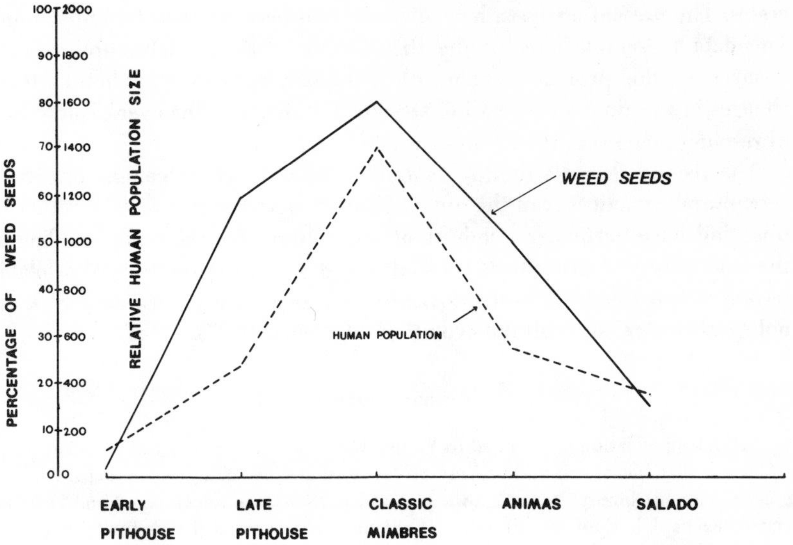


Fig. 7. Percentage of "weed" seeds to all seeds from flotation samples compared with the relative size of the human population.

## DISCUSSION

This study has two objectives. The first is to analyze charcoal and archaeological seeds as artifacts—the products of biological and cultural processes. Analysis of seeds recovered from archaeological sites needs to go beyond simple listing of recovered types and their recorded ethnographic use. The analysis of wood and wood charcoal has even further to go in North America. Archaeologists need to recognize that wood is more than a source of datable material; it constitutes a valuable source of information similar to ceramics and lithics. It is interesting to note that European archaeologists have utilized botanical artifacts extensively (e.g., Godwin 1975).

The addition of another archaeological tool for understanding vegetational change is important. Too often, the results of individual analyses have less than convincing interpretations. A strength of archaeological research is the comparison of numerous independent tests. For the case study presented here, analysis of faunal remains (Powell and Langenwalter 1977), pollen (Halbirt 1976: personal communication), lithics (Rugge 1977), and catchment analysis (Hastorf 1977) reach similar conclusions to this study. From this perspective, the analysis of macroplant remains should be a welcome addition to an always limited data base for understanding prehistoric processes. The present analysis, has, of course, not been taken as far as it should. The data presented here for the Rio Mimbres drainage only show relative changes in the probable extent of floodplain agriculture. Whether these changes in the riparian vegetation are actual evidence of the over-exploitation of resources requires other analysis which is in progress.

The second purpose of this paper is to show what the general effects of agricultural practices can be on vegetation patterning. Considerations of this kind have particularly important implications for the understanding of the interaction of prehistoric populations, the over-exploitation of available resources, problems of food provisioning, agricultural intensification, technological change, and cultural evolution (cf. Cohen 1977).

## Acknowledgments

Special appreciation is extended to Volney Jones, whose humility will probably never let him realize his contribution to my professional and personal development. His generosity with his library, time, knowledge, and puns will be remembered and is appreciated. Richard I. Ford and Steven A. LeBlanc deserve special thanks for their general counsel. The Laboratory of Tree-Ring Research, University of Arizona is thanked for providing data. Appreciation is extended to Pat Gilman for commenting on this paper and contributing toward the maintenance of sanity. As I am unable to find anyone willing to accept responsibility for errors, I must.

## REFERENCES

Asch, Nancy B., Richard I. Ford, and David L. Asch
1972 Paleo-ethnobotany of the Koster Site: The Archaic Horizons. Illinois State Museum Reports of Investigation 24.

Bailey, Vernon
1913 Life Zones and Crop Zones of New Mexico. North American Fauna No. 35. United States Department of Agriculture, Bureau of Biological Survey.

Berlin, G. Lennis et al.
1977 Identification of a Sinagua Agricultural Field by Aerial Thermography, Soil Chemistry, Pollen/Plant Analysis, and Archaeology. American Antiquity 42:588-600.

Bohrer, Vorsila L., and Karen R. Adams
1977 Ethnobotanical Techniques and Approaches to Salmon Ruin, New Mexico. Eastern New Mexico University Contributions in Anthropology 8(1).

Bryan, Kirk
1954 The Geology of Chaco Canyon, New Mexico, in Relation to the Life and Remains of Prehistoric Peoples of Pueblo Bonito. Smithsonian Miscellaneous Collections 122(7).

Campbell, C. J., and William A. Dick-Peddie
1964 Comparison of Phreatophyte Communities on the Rio Grande in New Mexico. Ecology 45:497-502.

Castetter, Edward F.
1956 The Vegetation of New Mexico. New Mexico Quarterly 26:257-88.

Cohen, Mark Nathan
1977 The Food Crisis in Prehistory: Overpopulation and the Origin of Agriculture. New Haven: Yale University Press.

Collier, George A.
1975 Fields of the Tzotzil. The Ecological Bases of Tradition in Highland Chiapas. Austin: University of Texas Press.

Conrad, Lawrence A., and Robert C. Koeppen
1972 An Analysis of Charcoal from the Brewster Site (13-Ck-15), Iowa. Plains Anthropologist 17:52-54.

Day, Gordon M.
1953 The Indian as an Ecological Factor in the Northeastern Forest. Ecology 34:329-46.

Dick-Peddie, William A.
1975 Vegetation of Southern New Mexico. 26th Field Conference Guidebook. New Mexico Geological Society.

Godwin, Harry A.
1975 History of the British Flora: A Factual Basis for Phytogeography. 2nd ed. London: Cambridge University Press.

Godwin, H. A., and A. G. Tansley
1941 Prehistoric Charcoal as Evidence of Former Vegetation, Soil, and Climate. Journal of Ecology 29:117-26.

Hastorf, Christine A.
1977 A Predictive Model for Changing Food Resources in the Prehistoric Mimbres Valley, New Mexico. M.A. thesis, Anthropology Department, University of California, Los Angeles.

Hastings, James R., and Raymond M. Turner
1965 The Changing Mile. Tucson: University of Arizona Press.

Hogg, Don Jack
1977 Report on the Excavation of Three Mogollon Pit Houses on the Upper Mimbres River, New Mexico. M.A. thesis, Eastern New Mexico University.

Humphrey, Robert A.
1958 The Desert Grassland. Tucson: University of Arizona Press.

Iversen, Johannes
1941 The Influence of Prehistoric Man on Vegetation. Danmarks Geologiske Undersoegelse R. 2:20-26.

Jacobsen, Thorkild, and Robert M. Adams
1958 Salt and Silt in Ancient Mesopotamian Agriculture. Science 128:1251-58.

Jensen, Hans A.
1969 Content of Buried Seeds in Arable Soil in Denmark and its Relation to the Weed Population. Dansk Botanisk Arkiv 27(2):1-56.

Jones, Volney H.
1941 The Nature and Status of Ethnobotany. Chronica Botanica 6:219-21.
1954 The Development and Present Status of Ethnobotany in the United States. Congrès International de Botanique, Huitième 13:52-53.

LeBlanc, Steven A.

1975 Mimbres Archaeological Center: Preliminary Report of the First Season of Excavation, 1974. Los Angeles: Institute of Archaeology, University of California, Los Angeles.

1976 Mimbres Archaeological Center: Preliminary Report of the Second Season of Excavation 1975. Journal of New World Archaeology 1(6):1-23.

1977 The 1976 Field Season of the Mimbres Foundation in Southwestern New Mexico. Journal of New World Archaeology 2(2):1-24.

LeBlanc, Steven A., and Ben Nelson

1976 The Salado in Southwestern New Mexico. The Kiva 42:71-80.

Lisitsina, G. N.

1976 Arid Soils—The Source of Archaeological Information. Journal of Archaeological Science 3:55-66.

Lowe, Charles H.

1964 Arizona's Natural Environment. Tucson: University of Arizona Press.

Minnis, Paul E.

1977 The Analysis of Wood Charcoal and Population Dynamics in the Rio Mimbres Drainage, Southwestern New Mexico. Museum of Anthropology, University of Michigan, Ethnobotanical Laboratory Report No. 495.

Minnis, Paul E., and Richard I. Ford

1977 Analysis of Plant Remains from Chimney Rock Mesa. In: Archaeological Investigations at Chimney Rock: 1970-1972, edited by Frank W. Eddy, pp. 81-91. Colorado Archaeological Society Memoir 1.

Nye, P. H., and D. J. Greenland

1960 The Soil Under Shifting Cultivation. Technical Communication 51. England: Commonwealth Bureau of Soils.

Powell, Susan, and Paul Langenwalter

1977 Changes in Prehistoric Hunting Practices in the Mimbres River Valley. Unpublished paper presented at the Society for American Archaeology annual meeting, New Orleans.

Quick, Clarence R.

1961 How Long Can a Seed Remain Alive. In: Seeds, The Yearbook of Agriculture, edited by Alfred Stefferud, pp. 94-98. Washington, D.C.: U.S. Dept. of Agriculture.

Rugge, Margaret C.

1977 Temporal Variation in Chipped Stone Artifacts from Mimbres Valley, New Mexico. Paper presented at the Society for American Archaeology annual meeting, New Orleans.

Salisbury, E. J., and F. W. Jane

1940 Charcoal from Maiden Castle and Their Significance in Relationship to the Vegetation and Climate Conditions in Prehistoric Times. Journal of Ecology 28:310-25.

Schweingruber, F. H.

1976 Prähistorisches Holz. Academica Helvetica 2.

Smith, Philip E. L., and T. Cuyler Young, Jr.

1972 The Evolution of Early Agriculture and Culture in Greater Mesopotamia: A Trial Model. In: Populational Growth: Anthropological Implications, edited by Brian Spooner, pp. 1-59. Cambridge: M.I.T. Press.

Spector, Janet D.

1970 Seed Analysis in Archaeology. The Wisconsin Archaeologist 51:163-90.

Woodbury, Richard B.

1961 Prehistoric Agriculture at Point of Pines, Arizona. Society for American Archaeology Memoir 17.

York, John C., and William A. Dick-Peddie

1969 Vegetational Changes in Southern New Mexico During the Past Hundred Years. In: Arid Lands in Perspective, edited by W. McGinnies and B. J. Goldman, pp. 157-66. Tucson: University of Arizona Press.

# ARCHAEOLOGICAL INTERPRETATION BASED ON ETHNOBOTANICAL INFERENCES IN THE UPPER GILA REGION

*James E. Fitting*
Gilbert/Commonwealth

## BACKGROUND

Early in 1976, Commonwealth Associates, a member of the Gilbert/Commonwealth Companies, was awarded a contract for the Mitigation of Adverse Effects to Archaeological Resources on the Foote Wash Conservation and Development Project, Graham County, Arizona, by the Interagency Archaeological Service–San Francisco. An earlier survey report and a mitigation design prepared by the Arizona State Museum at the University of Arizona served as a basis for the work described in this article. The field work for this project was carried out in March of 1976, and a final report was submitted to the National Park Service in January of 1977 (Commonwealth Associates Inc. 1977).

This project involved the intensive study of a series of small features, cobble clusters, and cobble circles, as well as one large scattering of cultural material which contained no features. Neither the features, nor the associated artifacts, were of particular assistance in interpreting the significance of these sites. In the course of the study, it became apparent that the archaeological interpretations for this project would need to be based on inferences drawn as to the reasons for these occupations. During the analysis, Volney Jones was consulted in Ann Arbor for his knowledge of plant utilization in the lower Sonoran life zone.

Location of Foote Wash and No Name Wash Study Area.

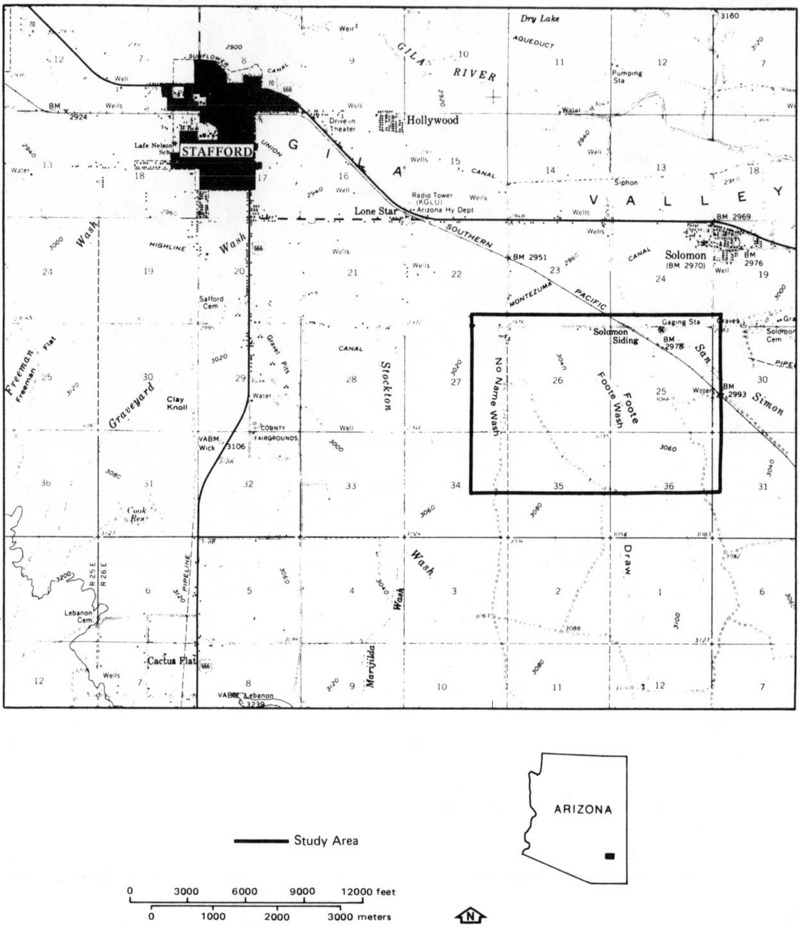


Fig. 1. Location of Foote Wash and No Name Wash study area.

## THE STUDY AREA

The archaeological studies at Foote Wash described herein were carried out along both Foote Wash and No Name Wash, small tributary valleys of the Gila located approximately 5 mi southeast of Safford in Graham County, Arizona (Fig. 1). These intermittently flowing tributaries run northward to the wide floodplain of the "inner valley," which is surrounded by mountain ranges to the north and south (Wilson and Moore 1958).

The valley floor near Safford is only 2900 ft above sea level. To the south of Safford are the Pinaleno Mountains, a Precambrian outcropping of granite and gneiss (Wilson 1965), which rise to over 10,000 ft, with Mt. Graham marking the highest point. The mountains of the north side of the valley are both lower (4000 to 6000 ft) and younger. They are composed of andesites and rhyolites (Wilson 1965) of Cenezoic age and could be considered as an outlying portion of the Mogollon Rim itself. The coarser fills of the valley, exposed in the terrace cuts of small washes like Foote and No Name, contain large cobbles of granite, gneiss, basalt, andesite, and rhyolite as well as quartzite, and pebbles ranging from weathered nodules of volcanic obsidian to a variety of siliceous pebbles. They attest to a long period of intensive weathering of the nearby surrounding mountain ranges.

Down-cutting in the area has alternated with periods of stabilization, and a sequence of five major terraces can be recognized, although they have not been dated. The oldest is represented by remnants along the base of the Pinaleno Mountains. The most recent terraces, quite possibly of Althithermal age (ca. 8000-3000 B.C.), are located along the north side of the river (the pre-agricultural cut bank) and southwest of Safford along Graveyard Wash, Freeman Wash, and Cottonwood Wash. Foote Wash, to some extent No Name Wash, and the still larger Stockton Wash complex to the west of No Name Wash, cut through the $T_3$ system which is probably of late Pleistocene or early Holocene age.

The microenvironmental resources of the Foote and No Name Wash region are dependent on two factors, climate and soils. These factors determine the plant cover and animal communities of the area. The soils, in turn, were formed under specific climatic regimes. Climate in this area is partially a function of elevation. The variables of rainfall and temperature are the significant factors for human occupation.

From the formation of soils and the establishment of the general parameters of contemporary plant and animal communities, it is possible to make the assumption that, on a larger scale, there has been relatively little climatic change in the area over the past 10,000 years (Martin 1963). This is a true statement when these conditions are contrasted with the full glacial conditions of the Southwest, when, with lower temperatures and higher rainfall, the "inner valley" was probably a spruce-fir-pine forest surrounded by spruce parklands on the terraces (Martin and Mehringer 1965:439).

In some parts of Arizona, the contemporary vegetation cover represents essentially the same cover that might have been encountered during the prehistoric period (Goodyear 1975). The Foote and No Name Wash areas present a very different picture. They have been utilized for nearly a century for grazing cattle. In addition to periods of heavy and continued grazing, there have been major alterations in drainage, with the three major check dams already in existence in the upper portions of Foote Wash and with the construction of cement-lined canal systems through the fields of the "inner valley." The plant cover has been managed for grazing by bulldozing the mesquite over large areas at regular intervals. There is evidence in several places of either controlled or uncontrolled massive burning, particularly along the eastern edge of Foote Wash near its mouth and along the western fork of Foote Wash below the first check dam.

Therefore, it can be assumed that the modern vegetation cover has little resemblance to that which might have been exploited by prehistoric peoples. As an example, mesquite cover is sparse and low in the wash bottoms today, except behind check dams where it is luxuriant. Before construction of these dams and before bulldozing, it was probably low and sparse in the areas where it is dense today and luxuriant in the areas where it is sparse today. Contemporary range management policies tend to discourage such species as cholla, prickly pear, and barrel cactus that might have been more numerous in prehistoric times.

The species of plants that would have been prevalent under aboriginal conditions are similar to those which are present today, but with considerably altered densities and a somewhat altered distribution. The terrace tops would have been marked by perhaps more cholla and prickly pear and by larger creosote and saltbush than they are today. The valley slopes might have had more barrel cactus and prickly pear. The stands of mesquite in the bottoms would have been larger and denser in the areas near the mouths of the washes and possibly for several miles up the washes.

## SITES AND COLLECTION METHODS

The archaeological resources along Foote and No Name Washes were diffuse. As one of the preliminary survey reports suggested, the entire area could be considered as a large, low density site. At least several flakes could be collected from every quarter section and several clusterings of cultural material were designated as sites in the survey report. The Commonwealth contract called for an examination of ten of these areas.

Site CC:2:17 (Locus 1) was the most distinctive. It was a large, low density, artifact scatter along a Cenezoic land surface that had been exposed by erosion in the bottoms of Foote Wash. Within the generally densely vegetated wash bottom, this Cenezoic remnant was bare, and a separate geological study indicated that it probably had little or no plant cover since its late Pleistocene exposure.

All of the other sites were either cobble clusters, cobble circles, or other features, located on terraces above the wash bottoms. One reported cobble circle on the wash bottom was found to be a recent natural feature, and several other previously reported cobble features had been destroyed between the time of the survey and the mitigations. One of the cobble clusters was found to be a fairly recent historic feature.

A variety of collection and recording techniques was used at these sites. All were mapped and general surface collections were taken at all sites. Controlled surface collections, with the plotting of individual artifacts around surface features and recording of other artifact clusters, were taken at all but one site. Test pits were opened at more than half of the sites. These test pits suggested that there were no subsurface features and virtually no subsurface cultural material. The location of these sites within the study is presented in Figure 2, and a summary of the recording techniques and features is presented in Table 1.

## ARTIFACTS

The largest category of cultural material recovered from the Foote Wash and No Name Wash mitigations was that of chipped stone, represented by a sample of 1739 items with a total weight of over 63,000 g, or approximately 140 lb. It would be charitable to characterize this industry as crude. While clearly the product of human workmanship, there was little variety in the tools that were recovered, and few of the items show any signs of use after knapping. There were no finished bifacial tools in the sample and no flakes of bifacial retouch. Nine items show signs of bifacial chipping, but these have been formally classified as bifacial cores rather than preforms.

It was observed during the preliminary survey that basalt was the most frequent type of raw material encountered among the chipped stone assemblages in the study area. This was supported by the controlled collections where basalt accounted for more than 30% of the entire sample by both count and weight. Actually three types of igneous material, rhyolite, andesite and basalt, dominated the collections accounting for 75% of all chipped stone by count and nearly 90% by weight. These materials are all common in cobble form in the Pinaleno deposits of the area. These cobbles are derived from outcroppings of this material located to the north and east of the study area at distances of ten to fifteen miles (Wilson and Moore 1958).

Formal classification of chipped stone included the categories of cobble cores, block cores, plano-convex cores, bifacial cores, small cores, primary

Site Locations Within Foote Wash and No Name Wash Study Area.

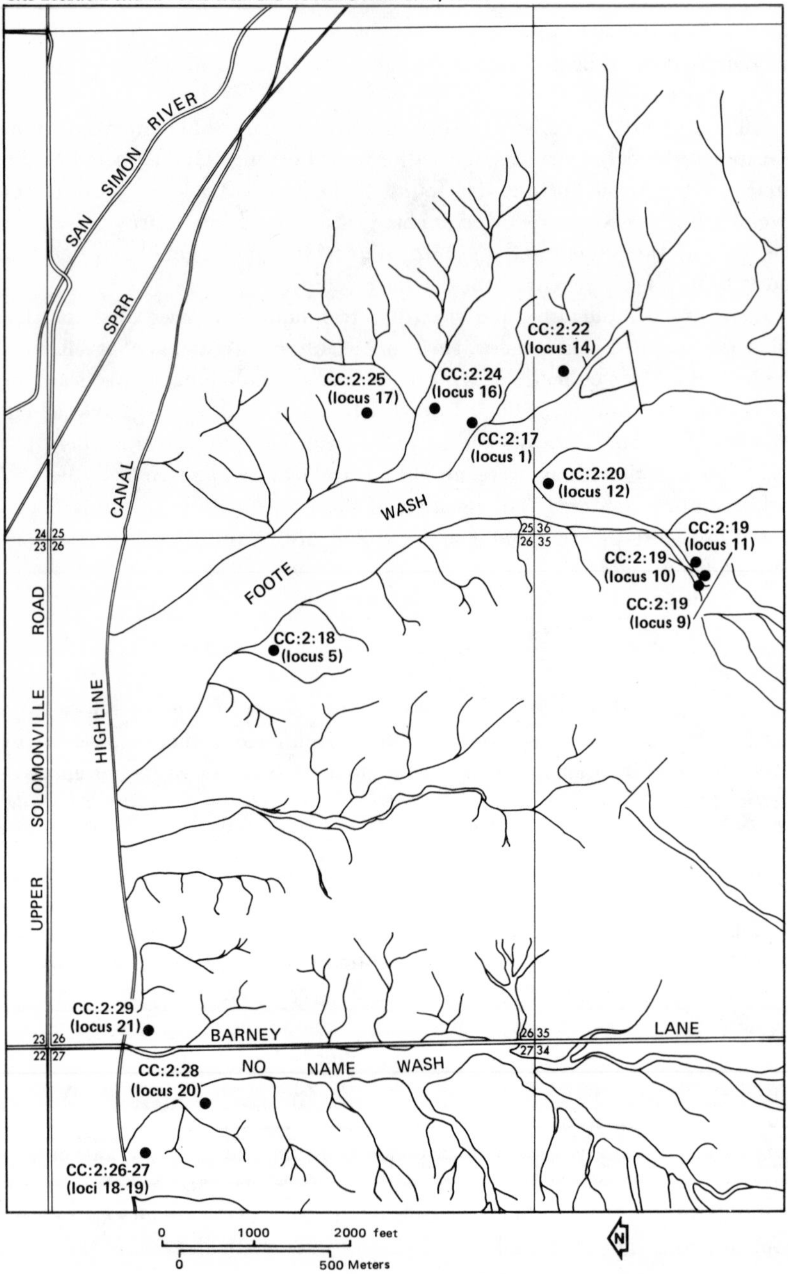


Fig. 2. Site locations within Foote Wash and No Name Wash study area.

TABLE 1
SITES AND COLLECTION METHODS

| Site | Mapping | General Surface Collection | Controlled Surface Collection | Test Pit(s) | Number of Cobble Clusters | Number of Cobble Circles | Number of Artifact Concentrations | Number of Historic Features |
|---|---|---|---|---|---|---|---|---|
| CC:2:17 (Locus 1) | X | X | X | X | 0 | 0 | 31 | 0 |
| CC:2:18 (Locus 5) | X | X | X | X | 1 | 0 | 0 | 0 |
| CC:2:19 (Loci 9-10) | X | X | X | X | 2 | 1* | 3 | 0 |
| CC:2:20 (Locus 12) | X | X | X | – | 1 | 0 | 1 | 0 |
| CC:2:22 (Locus 14) | X | X | X | – | 1 | 0 | 0 | 0 |
| CC:2:24 (Locus 16) | X | X | X | X | 1 | 2 | 0 | 0 |
| CC:2:25 (Locus 17) | X | X | X | – | 1 | 1 | 1 | 0 |
| CC:2:26-27 (Loci 18-19) | X | X | X | X | 4 | 2 | 0 | 0 |
| CC:2:28 (Locus 20) | X | X | X | X | 1 | 0 | 0 | 0 |
| CC:2:29 (Locus 21) | X | X | – | – | 0 | 0 | 0 | 1 |

*Recent Na'ural Feature.

and secondary decortication flakes and flat flakes, retouched knives manufactured on these flake types, endscrapers, spokeshave scrapers and miscellaneous retouched flakes (Table 2).

Overall, the chipped stone sample from the Foote Wash-No Name Wash area presents a uniform picture. Cobble cores and all flake types were recovered from all of the sites. All but the smallest of the lithic assemblages contained flaked knives, the most common tool type. Based on microscopic study, the industry was clearly not directed toward woodworking, hunting and butchering, or chopping and cutting of hard substances. The assemblage does not reflect an industry devoted to the preparation of blanks or preforms, or bifacial tools of any type.

Instead, it seems to represent an immediate, convenient or expedient tool industry. Local cobbles were used for the production of flake tools. Decortication flakes were used for tools as readily as flat flakes, and these tools, most commonly knives, were used for an immediate task and then discarded. There seems to have been little curation of tools, and given the ready availability of raw materials in almost all parts of the project area, with the notable exception of CC:2:17, tools seem to have been manufactured, used, and discarded at a single site.

Ceramics were found at six of the ten sites, although the one wash bottom site, CC:2:17 (Locus 1) produced 178 of the 273 sherds recovered from the area. The majority of all sherds were either plain polished red wares or brown wares (Table 3). The most common decorated type was Encinas Red on Brown. The design pattern on the sherds that has been classified as red on buff was similar to that of the Encinas Red on Brown sherds and may be a variant of this type. The five white-slipped sherds have no decoration, but the faint slip is similar to that of Mimbres Bold Face Black on White, Three Circles Red on White, or Cerros Red on White. The corrugated sherds appear to be from an Alma Neck Banded jar. A total of nine individual vessels were identified including five bowls and four jars.

Several conclusions can be drawn from the ceramics recovered in 1976. First, they can be interpreted as representing a single time period of roughly A.D. 700-1100, or possibly A.D. 800-1000. This is significant because they predate the period of most intensive occupation of this portion of the Gila Valley (A.D. 1100-1400). Second, bowl forms are more common than jars at all sites where vessel form could be determined. This might be expected if these vessels were used in food transportation or collection, rather than cooking or storage. Finally, at three of the five ridge top sites, the majority of the ceramics were found along the wash side. At one, only a single sherd was found while at the other, a single vessel was found away from the wash. On the whole, the orientation seems to have been toward activities on the wash slopes or in the wash itself.

TABLE 2
CORES, FLAKES AND TOOLS BY SITE (COUNT)

| | Cores | | | | | Flakes | | | | Tools | | | | | | |
|---|---|---|---|---|---|---|---|---|---|---|---|---|---|---|---|---|
| | Cobble Cores | Blocky Cores | Plano-Convex | Bifacial Cores | Small Cores | Prim. Dec. Flks. | Sec. Dec. Flks. | Flat Flakes | Blocky Flks. | Knives on PDF | Knives on SDF | Knives on FF | Endscrapers | Spokeshaves | Ret. Flks. | Total |
| CC:2:17 (Locus 1) | 1 | 1 | – | – | – | 1 | 1 | 11 | 1 | – | – | – | – | – | 3 | 19 |
| CC:2:18 (Locus 5) | 24 | 14 | 2 | 3 | 7 | 67 | 81 | 166 | 116 | 14 | 13 | 31 | 2 | 6 | 18 | 564 |
| CC:2:19 (Locus 10) | 7 | – | – | 1 | 1 | 34 | 20 | 31 | 46 | 5 | 1 | 9 | – | – | – | 155 |
| CC:2:20 (Locus 12) | 10 | 11 | 7 | 1 | – | 40 | 29 | 41 | 54 | 19 | 8 | 12 | – | – | 2 | 234 |
| CC:2:22 (Locus 14) | 9 | 1 | 1 | – | – | 14 | 13 | 11 | 17 | 4 | 7 | 3 | – | – | – | 80 |
| CC:2:24 (Locus 16) | 10 | 1 | 3 | – | 2 | 13 | 3 | 10 | 10 | 9 | 6 | 6 | 1 | 1 | 1 | 76 |
| CC:2:25 (Locus 17) | 8 | 1 | 3 | 1 | – | 52 | 30 | 37 | 36 | 10 | 1 | 10 | – | – | – | 189 |
| CC:2:26-27 (Loci 18-19) | 10 | 1 | 1 | 3 | 2 | 25 | 21 | 17 | 36 | 4 | 3 | 7 | – | – | 1 | 131 |
| CC:2:28 (Locus 20) | 5 | – | – | – | – | 51 | 51 | 57 | 79 | 1 | 6 | 3 | – | 1 | 4 | 258 |
| CC:2:29 (Locus 21) | 3 | – | – | – | – | 4 | 7 | 7 | 7 | 3 | 2 | – | – | – | – | 33 |
| Total | 87 | 30 | 17 | 9 | 12 | 301 | 256 | 388 | 402 | 69 | 47 | 81 | 3 | 8 | 29 | 1739 |

TABLE 3
SHERD COUNT SUMMARY

| | Plain Brown Ware | Corrugated Brown Ware | Polished Red Ware | Red on Brown | Red on Buff | White Slip | Total |
|---|---|---|---|---|---|---|---|
| CC:2:17 (Locus 1) | 43 | 2 | 84 | 49 | – | – | 178 |
| CC:2:19 (Locus 10) | 7 | – | – | – | – | – | 7 |
| CC:2:22 (Locus 14) | 19 | – | 8 | – | 20 | 5 | 52 |
| CC:2:25 (Locus 17) | – | – | 8 | – | – | – | 8 |
| CC:2:26-27 (Loci 18-19) | 1 | – | – | – | – | – | 1 |
| CC:2:28 (Locus 20) | 27 | – | – | – | – | – | 27 |
| Total | 97 | 2 | 100 | 49 | 20 | 5 | 273 |
| % of Total | 35.53 | 0.73 | 36.63 | 17.95 | 7.33 | 1.83 | |

Only one ground stone mano was found in 1976, and this was collected at CC:2:17 (Locus 1). No animal bones or plant remains were recovered from any of the test excavations.

The pattern of industrial activity and the type of tools produced seem to be similar at all sites, with variation attributable to chance natural distribution of raw materials and variation in skill levels of the knappers. The most significant result of this comparison is the demonstration of the extraordinarily close similarity of the chipped stone type distribution at CC:2:17 (Locus 1) to the cumulative distribution for all sites, indicating that the users were participating in the same industrial and economic system as the users of the mesa top sites.

The differences between ceramic and non-ceramic localities may be related to a least-effort model for site distribution. Ceramic containers were heavy and would have been carried the shortest possible distance, particularly when full. It is possible that they were utilized in the economic activities carried out at these sites. Goodyear (1975:95) has observed that ". . . there is no doubt that prehistoric work groups would have required drinking water in order to perform gathering and processing tasks adequately. . . . Jars, on purely morphological and functional grounds, are automatically implied even in the absence of known ethnographic patterns. . . ." Jars form a minority vessel type at these sites, but bowls could be related to food transport as well as plant processing.

It is clear that base camps, or semi-permanent villages, for the people who used these sites are not in the immediate study area. It is probable that people entered the study area on collection trips from near the wash mouth.

The sites closest to the mouth of No Name Wash are CC:2:26-17 (Loci 18-19) and CC:2:28 (Locus 20). CC:2:18 (Locus 5) is the site which we investigated closest to the mouth of Foote Wash. Only one site, CC:2:26-27 (Loci 18-19) was ceramic, and this produced a single sherd. Among the sites studied, CC:2:28 (Locus 20) is the furthest site away from the mouth of No Name Wash, and sites CC:2:19 (Locus 10) and CC:2:22 (Locus 14) are the furthest from the mouth of Foote Wash along the two main branches. All three of these sites were ceramic sites.

The largest ceramic site was CC:2:17 (Locus 1), located in a specialized area in the bottom of Foote Wash (see Fig. 3). The two mesa top sites on either side of Foote Wash, CC:2:20 (Locus 12) and CC:2:24 (Locus 16), are non-ceramic. The next furthest site from CC:2:17 (Locus 1) on the east side of Foote Wash, CC:2:25 (Locus 17), also contained ceramics.

It has been suggested that the functional purpose and chronological position of all of these sites were similar, without regard to the presence or absence of ceramics. The presence or absence of ceramics appears to be a

Wash Bottom Site CC:2:17 (locus 1)

Fig. 3. Wash bottom site CC:2:17 (Locus 1).

least-effort locational variable. If one were working near the mouth of the wash, it was not necessary to carry in ceramic vessels, but carrying in ceramic vessels would be necessary if one were working further back in either wash. For instance, CC:2:17 (Locus 1) seems to have served as a central area for several mesa top sites. If one were working out from a mesa top site immediately adjacent to CC:2:17 (Locus 1), one could return to this area for food or water. If one were working at a greater distance from this base, carrying food or water would require less effort than returning to CC:2:17 (Locus 1).

## ECONOMIC ACTIVITY

The absence of plant and animal remains among the archaeological materials recovered from sites on Foote and No Name Washes means that interpretation of the economic activities that took place at these sites must be largely inferential. We must start with the basic assumption that these are utilitarian sites rather than ceremonial areas, which seems to be a fairly safe assumption given the nature of the cultural materials recovered. The inferences about site economy need to be drawn from: (1) the economic potential of the areas where cultural materials were recovered and (2) the type of response to available resources that might have been exploited, as represented in the cultural assemblages.

The drastic changes in vegetation that seem to have taken place over the past century further complicate these interpretations. However, as a starting point, we can make the inference that plant collection was more significant than either agriculture or hunting in the activities represented at these sites. The Gila Valley around Safford was intensively utilized by agricultural groups in a period postdating the Foote Wash-No Name Wash sites. These agricultural peoples occasionally utilized the bottoms of large washes, as demonstrated in nearby Stockton Wash (Brown 1974), but they do not seem to have utilized Foote or No Name Wash. Although it is not possible today, mesa top dry farming may have been practiced under situations of population stress in the Gila Valley bottoms or during periods of high rainfall. However, the time period represented by the collections from the study area was not one of population stress in the valley. If the users of these sites were an agricultural group, agricultural activities would have been pursued elsewhere.

The lithic assemblage is totally lacking in projectile points and cutting and butchering implements. If hunting took place, kills were taken elsewhere for butchering. It is possible that snaring of small animals might have taken place, but there is no bone refuse to indicate that butchering or processing took place at these sites.

Therefore, we will assume that these sites were used for plant collecting activities. The major potential food plants in the area would have been mesquite and cactus, with both cactus buds and fruit representing other potential food sources. It is probable that mesquite would have covered the wash bottoms, with prickly pear and cholla growing on the slopes, and cholla scattered on the mesa tops.

Fortunately, Goodyear (1975) has carried out an extremely detailed study of the procurement systems used for these food sources and the archaeological manifestations of these procurement systems. Three of Goodyear's subsystems are potentially applicable to the study area: the bud-flower subsystem, the prickly pear subsystem and the leguminous seed subsystem.

## The Bud-Flower Subsystem

To summarize Goodyear's (1975:58-76) presentation, the buds and flowers of prickly pear and cholla are available in the early spring, roughly from April through early June. They provide a food source with critical vitamins and minerals at a time when the groups using them have been dependent on stored foods. Ethnographically, they were collected in baskets, utilizing wooden tongs, transported to a central area, and prepared in roasting pits with heated rocks.

Goodyear has suggested that the material culture associated with this procurement system, involving baskets and wooden tongs, would not be preserved in the archaeological record. The archaeological manifestations which could be expected would be: (1) subsurface facilities with indications of carbonization (roasting pits with charcoal), (2) rocks in association with these subsurface features, and (3) pits with diameters of three to four feet. He further suggested that cholla pollen might be expected with such pits but did not find this association in the features which he identified as roasting pits.

The original survey in the study area offered the hypothesis that many of the cobble features were surface manifestations of subsurface roasting pits. The test excavations indicated that they were surface features not associated with pits, or with hearths of any type. They were not dominated by fire-cracked rock either. Therefore, it would not appear that the sites within the study area were directed toward the exploitation of cactus buds and flowers.

## The Prickly Pear Subsystem

The fruit of the prickly pear is available from June through August, with the major period of potential harvest in July (Goodyear 1975:120-60). Ethnographically, these fruits were collected with wooden tongs, carried to a central area, and brushed with creosote branches to remove the spines. The fruit was eaten fresh and was dried or made into syrup for storage.

Goodyear developed several test implications for archaeological expressions of this system, three of which are applicable to the present study area. The archaeological manifestations of this system would include: (1) statistical dominance of jars over bowls since they would be needed for boiling and storage of syrup, (2) the presence of acute-angle cutting edges on tools utilized in slicing leaves and preparing tools, and (3) the absence of grinding tools.

Bowls were more common than jars at the Foote and No Name Wash sites. This was true for both the wash bottom site and the mesa top sites. There were acute-angle cutting tools at all sites which might have been utilized for this activity. However, they might also have been utilized for cutting and processing other types of plants. The final criteria, the absence of grinding implements, is met by most of the sites. Only CC:2:17 (Locus 1) produced a grinding implement. This, however, seems far from definitive evidence for demonstrating such a subsystem.

Therefore, cactus fruit procurement could have been an activity carried out in the study area, but the evidence for this is ambiguous. The vessel type ratio suggests that it was not the major activity represented at these sites, and the other evidence could be attributed to other activities as well as to cactus fruit procurement.

## The Leguminous Seed Subsystem

The most prevalent potential food species postulated for Foote and No Name Wash is mesquite, the dominant element in Goodyear's (1975:161-97) Leguminous Seed Subsystem. This is a food source which would be available for most of the summer months. It has been identified as one of the most important wild plant food sources available in the Sonoran desert and some historic groups relied heavily on it. It could be picked while still moist, in which case it would have a high sugar and carbohydrate content, or while dry, in which case the seeds would be important for the protein content. There are ethnographic accounts indicating that the seeds were collected and removed from the collection area for processing. They were either pounded or ground, and sometimes parched using open bowls.

The three archaeological manifestations which Goodyear suggests for this subsystem are: (1) the occurrence of manos and metates for the grinding of parched seeds, (2) the prevalence of wide-mouthed bowls for the parching of seeds, and (3) a spatial correlation of sites with this technology in zones of availability. We have already noted the lack of metates and near absence of manos in the study area. Goodyear, however, has suggested that this might be expected if the harvesting zones were within easy walking distance of the villages.

There is a dominance of bowls over jars in the study area. Quite possibly, seeds were parched at these sites, to lighten the load, and carried to nearby villages for grinding. The most convincing point is the distribution of these sites in, and along the edges of, areas of high availability.

Actually, the activities carried out at these sites probably involved both cactus fruit procurement and mesquite utilization. Both resources would have been available in close proximity to each other and during the same season. It would be unrealistic to assume that, in such a local situation, aboriginal collectors would have maintained distinct sites for these activities. The distinctive distribution found by Goodyear was the result of an entirely different type of environmental situation.

The data presented by Goodyear also provide some suggestions for the function of the cobble circles and cobble clusters. Goodyear (1975), citing Raab (1972) as well, has suggested that cobble circles served as stands for baskets used in exploitation of plant foods. The cobble circles in the Foote and No Name Wash areas are, on the average, larger than those reported by Goodyear. The possibility that they functioned as basket rests during collecting activities should, however, be considered. Another possibility is that they served as collection and processing areas for the cactus fruit; that is, the spines of the fruit were removed by brushing within such a confined area (See Fig. 4, 5).

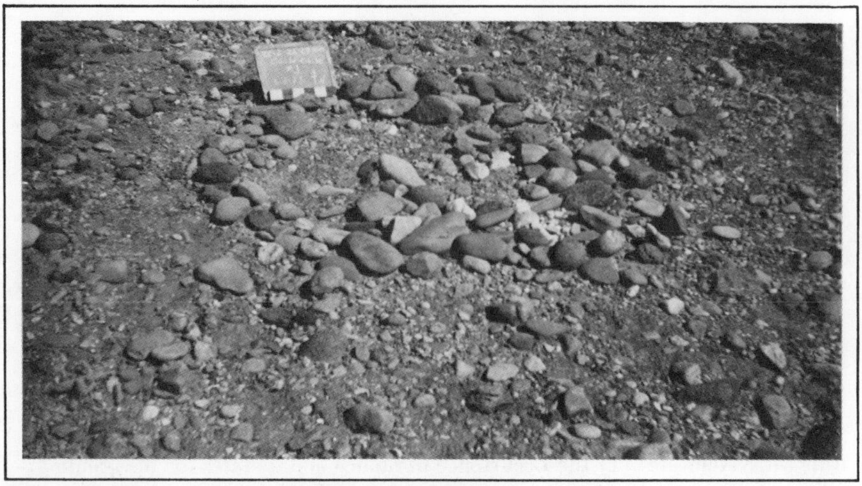

A. CC:2:26-27 (loci 18-19) Cobble Circle No. 1

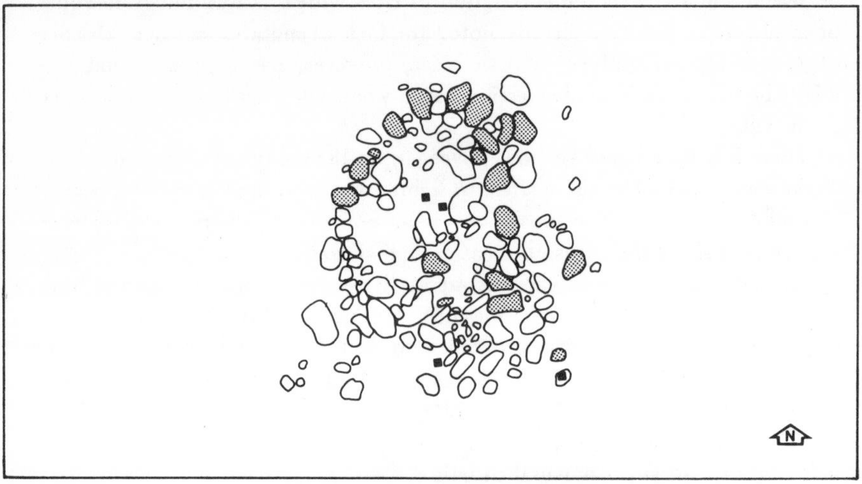


B. CC:2:26-27 (loci 18-19) Cobble Circle No. 1

■ Debitage
▲ Pottery
Basalt Cobbles

0 10 20 30 40 in.
0 50 100 cm.

Fig. 4. Cobble Circle No. 1 at CC:2:26-27 (Loci 18-19).

A. CC:2:28 (locus 20) Cobble Cluster

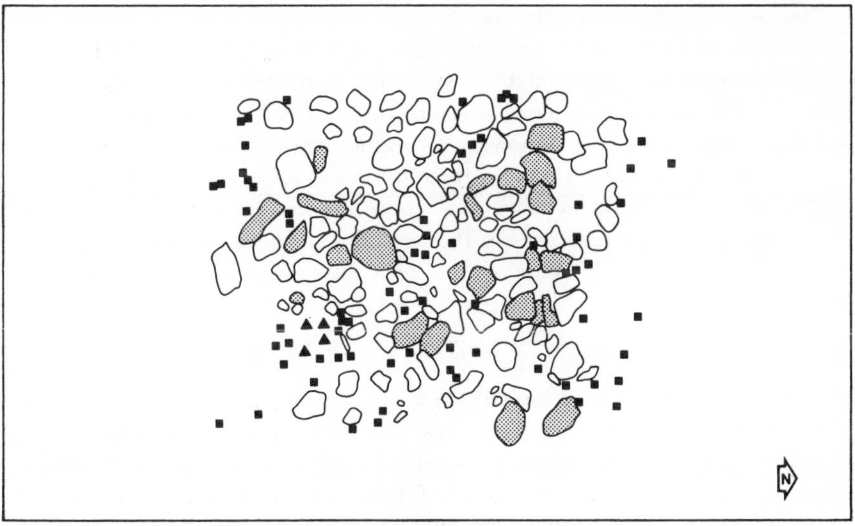


B. CC:2:28 (locus 20) Cobble Cluster

■ Debitage
▲ Pottery
Basalt Cobbles

0 10 20 30 40 in.
0 50 100 cm.

Fig. 5. Cobble cluster at CC:2:28 (Locus 20).

The information on mesquite exploitation may also contain a suggestion for the use of cobble clusters. It has been suggested that, under aboriginal conditions, the bottoms of Foote and No Name Washes would have been covered by a dense growth of mesquite similar to that existing behind check dams today. The exception, of course, would be CC:2:17 (Locus 1) where a clear area probably existed at the time these sites were utilized since soil conditions did not allow a thick plant growth. We found that walking through thick mesquite was a slow, and often painful, process. Travel between two points in the draw bottoms was often circuitous, but the sparse vegetation of the mesa tops made foot travel relatively easy. The fastest way to walk from the mouth of Foote Wash to CC:2:19 (Locus 10), for example, would have been along the mesa between Foote and No Name Wash, rather than along the wash bottom. If the prehistoric peoples who utilized this area took advantage of this natural characteristic, collecting areas in the wash bottoms would have been entered from the sides of the wash.

These were also the areas where the naturally occurring cobbles, used to manufacture flake tools, were found. Rather than being carried from the base village to the work area, these tools would have been most effectively manufactured on these mesa top sites just before people ventured down the slopes into the wash bottoms.

Goodyear (1975:81) notes that saguaro collecting took place within well defined territories and that these collecting territories were recognized by those utilizing this collecting system. This observation was not reported for mesquite collection. However, if specific mesquite collecting territories were utilized by distinct groups, these territories might possibly be marked. It would be difficult to mark these territories in the wash bottoms with a dense plant cover. However, they could have been marked by cobble clusters in the staging areas on the mesa tops above the collection areas.

## SEASONALITY AND SETTLEMENT PATTERN

It is clear that a very limited range of economic activities is represented at the Foote and No Name Wash sites. The study of the archaeological contexts suggests that the people who were using these sites had settled villages in other areas and that collecting of wild food plants was carried out in addition to both agricultural and hunting activities, which were conducted elsewhere. Furthermore, it has been inferred that these collecting activities took place during a limited time period in the summer.

Martin and Plog (1973:155-79) have dealt with the regional settlement systems which might be expected in areas such as those surrounding Foote and No Name Washes. They found that within the Southwest, ethnographic accounts indicated a dependence of between 0 and 60% on agriculture as a food source. Even groups with a primary dependence on agriculture utilized other food sources extensively, particularly wild food plants.

From their initial observations on subsistence, they utilized ethnographic sources to identify five settlement types. These included: (1) nucleated villages, (2) small permanent villages, (3) homesteads, (4) semi-permanent villages, and (5) camps. Only the latter settlement type is represented in the Foote and No Name Wash study area.

These five settlement types were found to articulate in six different patterns. The Walapai, Yavapai, Papago settlement pattern involved valley camps, semi-permanent villages in the uplands and mesa tops, and camps in the mountains. The Havasupai pattern involved camps and semi-permanent villages in the valleys, small permanent villages and camps in the uplands, and camps in the mountains. The Western Apache pattern included valley, upland, and mountain camps, and semi-permanent villages. The settlement pattern of the Chiricahua Apache included only camps for all areas, while the Pima, Yuma, Maricopa settlement pattern was dominated by nucleated villages in the valleys, with camps in the uplands, and no utilization of the mountains. The Mohave, Cocopa pattern included homesteads in the valleys, camps in the valleys and uplands, and no settlement in the mountains.

Although they closely border the Gila Valley, the Foote and No Name Wash sites could generally be characterized as upland sites, with the exception of CC:2:17 (Locus 1). As such, they could fit within any of the settlement systems outlined by Martin and Plog. Clearly, it will be necessary to use other archaeological data to even deal inferentially with the settlement system of which they are a part.

If we accept the probable dating of these sites as A.D. 800-1000, it is possible to use the San Simon Village (Sayles 1945), a semi-permanent upland village, as an additional site in the hypothetical settlement system of this time period. It is possible that such villages existed in the Safford valley and that one such village may be covered by a later occupation at V:16:10 in the Bylas area (Johnson and Wasley 1966).

Even with upland semi-permanent villages and camps, we do not have a complete example of any of Martin and Plog's settlement systems. The closest approximation is to the Walapai, Yavapai, Papago settlement system. In order to complete this pattern, we would need to postulate both valley bottom camps and mountain sites. Tuohy's (1960) CC:2:9A and CC:2:16 sites may be examples of valley bottom camps with an orientation similar to the Foote and No Name Wash upland collecting stations. Mountain sites are known for this time period in the Reserve area; Fewkes (1904:187) reported small cave occupations in the mountains on the north side of the valley. He related these to the later valley bottom villages, but it is possible that they contain earlier occupations as well.

The chronological placement and economic activities of the Foote and No Name Wash sites are far from certain on the basis of the information obtained in 1976. This information has to be compared with even less substantial inferences to speculate on the settlement system for a poorly known period in the Gila Valley. Until more information is available, the settlement system identification must remain tentative, and we feel that no further attempt can be made to define the range or seasonality of the postulated sites within the system beyond the limits of our study area.

## CONCLUSIONS

Neither the mesa top nor wash bottom sites in Foote and No Name Washes represent intensive or long-term occupations. All represent short periods of site use on what was probably a seasonally limited basis. The cobble clusters and cobble circles were not used as either hearths or storage features. They appear to have been purely surface features, constructed after initial site use. Both ceramics and chipped stone were encountered in essentially natural depositional conditions on the unaltered surface under several of these features. No plant or animal remains were associated with any of these features.

The lithic industries at all of these sites are similar. The range of knapping activities carried out is limited and essentially the same at all sites. Locally available raw material sources were used to derive flakes, which were used for cutting soft materials. An expedient, or immediate, industry is represented, and there does not appear to have been any curation of tools. Materials suitable for the manufacture of manos and metates were readily available but grinding implements are not well represented in the collections.

One extensive, but low density, wash bottom site, CC:2:17 (Locus 1), produced the majority of the ceramics but only a small amount of lithic material. The lithic material, however, was essentially similar to that on the mesa top sites and reflected the same economic pattern. Bowls were more common at the mesa top sites than had been originally thought. Bowls were more common than jars, and the functional position of ceramics was probably the same at all sites. The distribution of ceramic and non-ceramic sites in the study area is related more probably to a least-effort model than to chronological differences in site use.

The most probable use of these sites was as base localities for the collection of wild food plants. Mesquite beans and cactus fruit were undoubtedly collected in the area. The lithic industry was an expedient industry directed toward these activities, although final processing may have taken place elsewhere. These sites were probably utilized between June and September, with the most intensive utilization in July.

The function of the cobble clusters and cobble circles is obscure. They do not appear to have served any specific economic purpose and were constructed after initial site utilization. The cobble circles may have served as collection areas, but this cannot be proved. It has been suggested that the cobble clusters on the mesa points served as base localities for the collection of mesquite beans from defined collecting territories in the wash bottoms. The postulated dense mesquite cover in the wash bottoms, in all areas other than CC:2:17 (Locus 1), would have made movement through the bottom difficult. Travel to and from collection areas along the mesa tops would have been easier than along the wash bottoms, with cobble clusters representing "staging points" for collection in the bottoms.

The ceramics represent a fairly well circumscribed time horizon, probably between A.D. 700 and 1100, and possibly between A.D. 800 and 1000. Given the similarity of the lithic industries, all sites were assigned to this time horizon.

This time horizon is one that has been previously identified as only a minor component in the Safford region. It predates the most intensive period of exploitation of this portion of the Gila Valley by several centuries. Although several major sites of this latter time period have been reported near the Foote Wash and No Name Wash study area, none was located in the study area.

The sites within the project area are a part of a Cerros or Encinas phase settlement system. This system includes semi-permanent terrace top pit-house villages, such as that represented by the San Simon Village site; terrace top special purpose sites, such as those found along Foote and No Name Washes; similar bottomland sites, such as CC:2:17 and CC:2:9A, are included in this system; and cave sites may also be a part of it. On the basis of present data, this conforms most closely to the Walapai, Yavapai, Papago settlement system.

**Acknowledgments**

The research on which this study is based was collected as a part of Contract No. CS 8880-6-0020 between the United States Department of the Interior, National Park Service, Interagency Archeological Services-San Francisco, and Commonwealth Associates Inc. It has been reported in full in Commonwealth Associates Inc. Report N. 1786, submitted in final form to the National Park Service in January of 1977.

While this study contains no citations to Volney Jones' published works, his suggestions provided many of the insights used in interpretation, and he made his personal library available for use on the project. This type of aid, advice, and helpful suggestion, is something that is fondly remembered and appreciated by the students whom Volney has instructed and guided during his years at The University of Michigan.

**REFERENCES**


Brown, Jeffrey L.
1974 Pueblo Viejo Salado Sites and Their Relationship to Western Pueblo Culture. The Artifact 12(2):1-53.

Commonwealth Associates Inc.
1977 Mitigation of Adverse Effects to Archaeological Resources on the Foote Wash Conservation and Development Project. Report No. 1786, submitted to the National Park Service, Interagency Archeological Services-San Francisco.

Fewkes, Jesse W.
1904 Two Summer's Work in Pueblo Ruins. Bureau of American Ethnology, 22nd Annual Report (1900-01):3-196.

Goodyear, Albert C. III
1975 Hecla II and III: An Interpretative Study of Archaeological Remains from the Lakeshore Project, Papago Reservation, South Central Arizona. Arizona State University, Anthropological Research Paper 9.

Johnson, Alfred E. and William W. Wasley
1966 Archaeological Excavations near Bylas, Arizona. The Kiva 31(4):205-53.

Martin, Paul Schultz
1963 The Last 10,000 Years: A Fossil Pollen Record from the American Southwest. Tucson: University of Arizona Press.

Martin, Paul Schultz, and P. J. Mehringer, Jr.
1965 Pleistocene Pollen Analysis and the Biogeography of the Southwest. In: The Quaternary of the United States, edited by H. E. Wright and David G. Frey, pp. 433-51. Princeton: Princeton University Press.

Martin, Paul Sidney, and Fred Plog
1973 The Archaeology of Arizona: A Study of the Southwest Region. Garden City, New York: Natural History Press.

Raab, Mark L.
1972 The Powerline Site: A Prehistoric Fruit Processing Station in Papagueria. Unpublished manuscript in possession of the author.

Sayles, Edwin B.
1945 The San Simon Branch: Excavations in Cave Creek and in the San Simon Valley. Gila Pueblo, Medallion Papers 34.

Tuohy, Donald R.
1960 Archaeological Survey and Excavation in the Gila River Channel Between Earven Dam Site and Buttes Reservoir Site, Arizona. Tucson: Arizona State Museum.

Wilson, Eldred D.
1965 Guidebook 1–Highways of Arizona: U.S. Highway 666. Arizona Bureau of Mines, Bulletin 174.

Wilson, Eldred D., and Richard T. Moore
1958 Geology of Graham and Greenlee Counties, Arizona (Map). Arizona Bureau of Mines.

# PALEOETHNOBOTANY IN WESTERN SOUTH AMERICA: PROGRESS AND PROBLEMS

*Deborah M. Pearsall*
University of Missouri-Columbia

The study of the interrelationship of people and the plant world in the past is a challenging part of the field of ethnobotany. In spite of the problems caused by the imperfect nature of the archaeological record and available recovery techniques, sufficient clues to patterns of people/plant interrelationships often emerge. In his studies of plant remains from numerous American sites, Volney Jones provides an example of expertise and cautious interpretation that can be profitably applied in any area of the world. In this paper, I will review paleoethnobotanical work which has been done in western South America. The focus will be on research completed since Margaret Towle's 1961 work, *The Ethnobotany of Pre-Columbian Peru*, and will emphasize the hunting and gathering and early agricultural stages.

In the past 15 years, there has been increasing interest among archaeologists in investigating patterns of subsistence and the origin of agriculture in South America. After an early overemphasis on coastal Peru, recent work in the Peruvian sierra, the Meridional Andes, and the Intermediate area has given a more balanced picture of early pre-Columbian western South America. Patterns of people/plant interrelationships emerging from studies in Peru can be seen in more sketchy form in the less well-known regions of the Andes and can provide hypotheses for testing in these areas. Even the sketchy evidence we now have suggests that the Meridional and Intermediate areas were more important in early domestication and agricultural development than previously suspected.

Fig. 1. Archaeological sites and regions discussed in the text.

The following discussion is organized into three geographic sections: Central Andes, Meridional Andes, and Intermediate area. Cultural periods as presented by Willey (1971) are regrouped, primarily for convenience in discussing patterns, which often last several periods. Tables 1-4, the paleoethnobotanical data for the Central and Meridional Andes, were compiled from a variety of sources, differing in precision of identification (common or scientific name) and in details of description and cultural associations of remains. The accuracy of the identifications, particularly with regard to wild/cultivated status, is evaluated in the text whenever possible. Because this paper is a discussion of actual paleoethnobotanical remains, the tropical lowlands are sadly underrepresented. An attempt will be made to fit this important region into the better known Andean areas in the concluding section.

## CENTRAL ANDES

The Central Andes extend from the northern Peruvian border to the Lake Titicaca basin, including the Bolivian shore (Fig. 1). There are three major geographic divisions within this area: the desert coastal strip, the high Andean Cordillera, and the eastern tropical lowlands. The coastal strip is crossed by about 40 small, intermittent rivers. Within the three ranges of the Peruvian Cordillera, the complex topography of high mountain peaks, deeply dissected river valleys, and high rolling plains, or puna, creates many diverse ecological zones. To the east of the Cordillera Oriental, the slopes drop off rapidly to the tropical lowlands, an area of meandering rivers and dense forest cover.

### Early Preceramic (Periods I-IV, 11,000-4200 B.C.)

This long time period encompasses the emergence in the sierra of post-Pleistocene hunting, gathering, and early agricultural patterns prior to the beginning of camelid domestication, and an initial pre-agricultural utilization of the coastal zone.

The earliest botanical remains from the Central Andes are rind fragments of bottle gourd (*Lagenaria*) from Pikimachay cave (Table 1). Pickersgill suggests that these were wild rather than domesticated gourds and cautions that the remains may be intrusive (quoted in Flannery 1973). Guitarrero Cave lacks plant remains in the earliest stratum. Pachamachay Cave has plant remains in all strata (Rick 1978).

The 32 occupation levels of Pachamachay contain a consistent assemblage of botanical remains, suggesting a remarkably stable relationship of people and plants over the millennia (Pearsall, in press). Seeds, wood, and tuberous root remains were all from puna dwelling species, suggesting utilization of the high altitude lakeshore, grassland, rocky strata, and stream edge zones of the Junin puna for plant gathering. There is no evidence for the use of cultivated plants. The lack of evidence of plants brought in from outside the puna suggests that lower altitude zones were not used. Other evidence from Pachamachay supports a model of a year-round puna centered subsistence system. Rick has argued that hunting populations exploiting primarily vicuña, a camelid which occupies restricted territories year round, could become sedentary long before actual domestication of camelids occurred (Rick 1977). He suggests that Pachamachay was a hunting base camp, occupied by such a sedentary population (Rick 1978).

Adaptation in the lower altitude, inter-Andean valley zone, as demonstrated in Stratum II of Guitarrero Cave and the Jaywa and Piki complexes in the Ayachucho area, shows people/plant interrelationships to be different from those of the puna-centered gathering system. Plant remains from Guitarrero Cave include cultivated common bean (*Phaseolus vulgaris*), lima bean (*P. lunatus*) and aji (*Capsicum chinensis*) (Table 1). The identification of these remains as cultivars is secure, although the disturbance of deposits above Stratum II suggests some caution in dating. The presence of fully domesticated beans and aji by about 6000 B.C. in association with wild plants from zones both above and below the cave itself shows the early development of a pattern of multizonal gathering, combined with cultivation of exotics, by a semi-sedentary population. The two species of beans probably had multiple domestications in both South America and Mesoamerica (Heiser 1965: 937; Kaplan 1965:363, 1971:416-27; Gentry 1969:55-59). Northwestern Argentina is one suggested area of origin for both species, although the wild related forms extend north to Peru and Bolivia (Burkart and Brucher 1953; Burkart 1952). Aji (*Capsicum chinensis*) was probably domesticated from *C. frutescens*, which is widespread throughout lowland South America (Pickersgill 1969b; Smith 1977:91). The presence of these cultivars outside their probable areas of origin suggests interregional contact of some type. The presence of achiote seeds and *Crescentia* rinds at Ayacucho also supports contact between the inter-Andean area and the lower tropics (MacNeish et al. 1975:20-24). The cultivated status of the quinoa and amaranth seeds from these levels has not yet been demonstrated.

The earliest botanical remains from the Peruvian coast date to about 8000 B.C., from caves in the Chilca Valley of the central coast (Table 2). These finds have not been described in detail, but are said to be wild (Douglas Yen, as quoted in Engel, 1973). The report of lima beans from the central coast at 5800 B.C. must be viewed cautiously, since references to this find are all personal communication or brief newspaper reports. The occurrence of bottle gourd in the Siches complex (El Estero site) of the far north coast is much more secure and contemporary with gourd remains from the sierra. Patterns of people/plant interrelationships on the coast of Peru are not clear for this time period, but there are indications that patterns better understood in Preceramic V times had their beginning several millennia earlier. Lanning (1963, 1967) has suggested that the occupations found in the central coast lomas area during this time period were seasonal occupations, utilized during the South American winter by populations which lived in the sierra in the summer.

## Preceramic V (4200-2500 B.C.)

On the coast a variety of new plants appear in the subsistence system (Table 2). Carbon 14 dates place the Cabeza Larga, Encanto Lomas sites, and the Cilca Valley site in this time range (Moseley 1975; Lanning 1967; Engel 1963). The dating of the Yacht Club and Pampa sites is less secure and can be placed either at the end of this period or the beginning of the next (Lanning 1967; Moseley 1975:124-35). As mentioned above, occupation in the lomas zone has been linked by Lanning to sierran-coastal seasonal transhumance patterns. The appearance of sites at around 3600 B.C. outside the lomas area, on the shore or in the lower river valleys, is interpreted by Lanning as a shift away from transhumance to utilization of marine and riverine resources during the summer months, with winter utilization of the lomas gradually falling off due to climate change. Patterson and Moseley (1968:123) suggest, however, that the summer occupation area for populations using the lomas in the winter may have been the coastal river valleys all along. Cardich also argues against transhumance, suggesting a pattern of Andean regional nomadism for this time period (Cardich 1976). There seems to be no convincing evidence at present for seasonal sierran-coastal migrations during this or previous time periods.

If the pattern during the early Preceramic periods and early Preceramic V on the coast is one of terrestrial hunting and gathering in the coastal zones, there is a transition from this pattern to the beginning of agriculture on the one hand, and increasing emphasis on marine fishing and shell fishing on the other, during Preceramic V and VI. A shift in settlement to seaside locations supports the contention that marine resources were of increasing importance throughout the final Preceramic periods (Moseley 1975:40-48). Of the cultivars present in late Preceramic V times, none can be considered a subsistence staple (Pickersgill 1969a:57). The presence of aji, beans, and squash all argue for contact with sierran cultivators, or perhaps for direct contact to the eastern Andes or northwest Argentina (see above, also Whitaker and Cutler 1965; Cutler and Blake 1971, Pickersgill 1969a:55; Heiser 1965:937).

In the sierra (Table 1), while previous patterns of plant utilization continue at Pachamachay, occupation in the Ayacucho area dating to the Chihua complex shows the addition of cotton, lucuma, bean(?), *Sapindus*, and corn (Confite Morocho) to the subsistence system. The corn, however, is said to be from disturbed contexts (MacNeish 1969:42). If the suggested date of the corn remains can be accepted, it is rather interesting that corn does not show up on the coast until late in the next period, especially since beans, squash, and aji were introduced into the coast through the sierra, and cotton from the coast to the sierra. If sierran-coastal contacts were sporadic during this time period, such irregular distribution of cultivars could be expected, however. MacNeish et al. (1975:28-32) suggest an expansion of agriculture in the Ayacucho area at this time period, with both dry season agriculture in the low puna and wet season agricultural camps now classified as macroband settlements. The evidence for dry season agriculture on the puna is not entirely convincing, however.

The recent literature dealing with the domestication of the llama and alpaca cannot be fully discussed here, although protein procurement is relevant to people/plant interrelationships. Studies using age/sex death ratios and size to determine wild or domesticated status of camelid bones have given contradictory results (Wing 1977; Pires-Ferreira et al. 1977; Wheeler Pires-Ferreira et al. 1977). By the end of this period, and perhaps as early as its beginning, camelid domestication was underway in the Junin puna. The exact time is irrelevant; animal domestication is more important in its later ramifications rather than in its initial effects. Domestication of camelids took place during millennia of gradually changing use of these animals, beginning with hunting, shifting to mixed hunting and control of semi-domesticates, and finally to pasturing of fully domesticated forms, with maintenance of the earlier patterns throughout (Wheeler Pires-Ferreira et al. 1977:155-59). For a group of sedentary hunters, the domestication of camelids would involve little change in plant utilization patterns, as shown by the Pachamachay remains. More nomadic hunters, the pattern suggested by Cardich for Lauricocha, would become more sedentary as utilization of domesticates increased (Cardich 1976).

## Preceramic VI (2500-1800 B.C.)

On the coast, at least 30 sites date from this time period, and unlike those discussed for Preceramic V, these settlements are considered to represent fully sedentary life (Lanning 1967:59). Although settlements are located typically on the shoreline, the orientation vis-à-vis shore, lagoon, and riverine zones is not the same at all sites. Sites such as Aspero, Piedra Grande, Culebras, and Chuquitanta are located in the river valley mouth areas, while sites such as Punta Grande and the Tank site are located in desert regions away from the valley areas (Moseley 1975:47). This difference in settlement orientation, if the sites are all year-round settlements, may therefore reflect a pattern of specialization in subsistence strategies, with the sites located near the valley mouths relying more heavily on agriculture. Older patterns of plant gathering in the valleys and the lomas have not disappeared, however (Table 2).

The appearance of sedentism and site specialization is associated with the appearance of new cultivars on the coast and the first ceremonial or public architecture (e.g.: Aspero, Chuquitanta). The Aspero site, with several other northcentral coast sites, has the earliest coastal occurrence of maize. Maize occurs as early or earlier in the sierra and earlier to the north in coastal Ecuador. The presence of *Spondylus*, native to Ecuadorian waters, at Aspero (Feldman 1977) and striking similarities in style between carved gourds from Huaca Prieta and Valdivia ceramics (Bird 1963; Lathrap 1974) suggest north-south contact at this time period. Of the large number of new cultivars which appear on the coast, lima bean, lucuma, maize, squash, and *Canavalia* occur at the same time or earlier in the sierra, while others have no secure sierran occurrence. The sierran grain/tuber complex does not occur on the coast, although there is evidence of guinea pig raising (hutches at Culebras I site, Lanning 1967:67). Cucurbits, beans, aji, guava, achira, and sweet potato all have suggested origins east of the Andes, in northwestern Argentina, or in northern South America (Pickersgill 1969a). The crops appear to come in sporadically, suggesting trade as a mechanism, but this is difficult to assess because of dating problems. The nature of sierran-coastal contacts in the Late Preceramic is one of the most intriguing questions for this period, but also one which requires additional research, particularly in the sierran and eastern zones.

In the sierra, cotton and *Canavalia* are reported from the Cachi complex at Ayacucho, at about the same time they occur on the coast. Data on the origin of these cultivars is incomplete, but they may have been domesticated on the Peruvian coast, from introduced ancestral forms native to coastal Ecuador. Of all the cultivars present on the coast, only cotton shows a developmental sequence in the area, with remains at earlier levels (Preceramic VI) more related to wild *Gossypium barbadense* L., and size of seed and fiber diameter increase over time, showing development towards modern cultivated types in the area by the Initial period (Stephens and Moseley 1974:109-22). *Canavalia plagiosperma* may also have been domesticated west of the Andes. Two possible ancestral forms, *C. brasiliensis* and *C. maritima* occur wild in Ecuador and the Tumbes region of Peru (Sauer and Kaplan 1969). Fully domesticated *C. plagiosperma* may be earliest on the north coast, lending some support to a northern origin. *C. piperi* and

*C. dictyota*, two other possible ancestors, occur in eastern South America and the eastern slopes of the Andes, however, suggesting this region as another possible source. During this same time period in the Ayacucho area, MacNeish et al. (1975:32-37) suggest that permanent agricultural settlements were now present, with llama herding and high altitude agriculture being seasonally practiced in the puna zones by a separate population. Again, the cultivated status of the quinoa, amaranth, and tubers from this time period is questionable.

From the Pachamachay deposits, size increases in *Chenopodium* seeds in the Late Preceramic period and the early Ceramic deposits may reflect a pattern of intensification of plant utilization (Pearsall, in press). While there is no evidence of fully domesticated quinoa, the increasing size throughout the later levels suggests selection for larger seeds, which may reflect tending or actual cultivation of forms retaining wild characteristics.

## Ceramic Periods (Initial Period and Early Horizon)

Very few plant remains have been recovered from the early Ceramic periods in the sierra. From the Junin area, occupation continues at Pachamachay until approximately A.D. 200. At about 1800 B.C., there is a structure associated with a house in the cave mouth, which may be a corral (Rick 1978). This identification is not definite, but suggests that herding may be taking precedence over hunting around the Preceramic/Ceramic transition. By 800-400 B.C., open-air settlements begin to appear on the shores of Lake Junin (Rick 1978; Matos M. 1975). A house excavated in the town of Ondores was observed to be built into a corral-like structure (Rick 1978). A shift in year-round settlement to the lake shore would give access to good grazing land and potential agricultural land.

Just off the Junin puna, in the Tarma river watershed, a recently completed intensive survey by joint University of Michigan/Universidad de San Marcos teams failed to uncover any Formative sites (Initial or Early Horizon) below the puna (Hastings 1977). Although this absence of sites may be due in part to destruction of sites or the difficulty of the terrain, it is very similar to patterns in the Ayacucho area, where no initial period sites were found (MacNeish 1969; MacNeish et al. 1970). The reason for this apparent lack of intensive use of the upper valley zones after the Preceramic remains obscure.

In the Lake Titicaca basin, botanical remains from the Chiripa site (Table 1), suggest patterns strikingly similar to those in the lake Junin area. There is no evidence of plant exploitation in micro-environmental zones outside the altiplano (Erickson 1977, Horn and Erickson n.d.). The site represents sedentary occupation from its earliest levels. *Chenopodium* seems to be at the stage of early genetic manipulation, and the presence of chuño, freeze-dried *Solanum* tubers, may represent cultivation (Erickson 1977). As in the Junin area, the importance of camelid utilization and the richness of the wild lacustrine flora may have caused retention of patterns of puna-centered intensive gathering and cultivation, delaying the emergence of full dependence on agriculture.

On the coast of Peru, many Preceramic VI sites are abandoned when pottery appears (Patterson and Moseley 1968:124-26). Public or ceremonial building becomes more common. The pattern of emphasis in subsistence toward agriculture and use of marine resources continues, with increasing emphasis on agriculture over time. Site location eventually is shifted inland in response to the development of canal irrigation, which opened up increased cultivation area (Moseley 1975:48-50). Crops introduced in the Late Preceramic spread throughout the coast, and a variety of new cultivars showing contacts with diverse areas outside the coast are added. As Collier and later authors have pointed out, however, the initial introduction of pottery and the early spread of maize did not have a "revolutionary" effect on the coast (Collier 1961:104-07). The interrelationships of people and plants in the coastal zone had strong continuity from the Late Preceramic through the early Ceramic periods. The pattern of localized seasonal gathering and cultivation of industrial crops such as cotton and gourd, with a variety of introduced fruits and vegetables, only gradually gave way to full emphasis on starchy staples such as maize and manioc.

## MERIDIONAL ANDES

The Meridional, or Southern Andean area, includes Bolivia, central and northern Chile, and northwestern Argentina (Fig. 1). Willey (1971:194-97) divides this diverse area into twelve natural environmental and cultural zones. Four of these zones are important for the occurrence of significant botanical remains: Atacama, Valliserrana, Quebrada de Humahuaca, and Mendoza-Neuquén.

### Early Preceramic (Periods IV-V: 6000-2500 B.C.)

Although several important lithic traditions have been defined for this time period in the sierra, there are no reported plant remains. González and Pérez (1966) note the presence of grinding stones at Intihuasi Cave and suggest a hunting and gathering subsistence system, with seasonal occupation of sites. By analogy to sierran sites of these time periods in Peru, interrelationships of people and plants were probably already complex. Patterns of multizonal gathering in the inter-Andean valley areas can be hypothesized. The lack of concrete data on patterns of plant use in northwestern Argentina and Bolivia during the Early Preceramic is particularly frustrating, since this area has been proposed as a center of domestication of early cultivars such as beans, achira, aji, peanut, and possibly *Canavalia* (Pickersgill 1969a; Krapovickas 1969). Subsistence-oriented research is urgently needed.

On the coast, sites are located close to the sea and show utilization of marine resources (Willey 1971:202-08; Nuñez A. 1974:114-59). The earliest plant remains are algarrobo seeds from the Tarapacá region (Table 4). Gathering of these wild tree pods is a common pattern in the coast and sierra throughout later time periods as well. Quiani II-Chinchorro (burial component) deposits, dated just prior to the end of this period, include remains of quinoa and cotton. The quinoa remains, since they are not described or illustrated, are best considered questionable cultivars. Cotton was probably introduced from the coast of Peru, where it occurs at the same time period. Other cultigens occurring at this time period on the Peruvian coast do not appear, however, If these coastal sites represent seasonal occupations of hunters and gatherers who also occupied sites up the coastal valleys for terrestrial hunting (Nuñez A. 1974:134-35), a pattern of coastal/regional nomadism is suggested. Cultivation in the coastal oases was apparently not as important a component of the seasonal round as it was further to the north in Peru.

## Preceramic VI (2500 B.C.–first pottery)

The first remains of maize and gourd on the coast date to this time period (Table 4). The finds come from on top of a Quiani II midden, dating possibly to 2000 B.C. (Willey 1971:202-15), but perhaps much later (Nuñez A. 1974: 114-59). The presence of the bottle gourd is not surprising, since remains are reported from the Quebrada de Humahuaca from this time period. It is curious that this useful container does not occur earlier. Preceramic maize occurs in the central sierra and the northcentral coast of Peru during the Late Preceramic (Tables 1, 2). Maize is securely known in the Meridional Andes in association with pottery, as early as 1760 B.C. at Tulan Cave (Popper 1977). Just as in Peru, where maize occurs in the Preceramic periods as a minor crop prior to its development as an important staple, the Quiani II maize, if early, does no more than imply sierran-to-coastal exchange of a crop that will not be important in the region until much later. Subsistence on the coast remains tied to exploitation of sea resources and collecting in the coastal valleys, after agriculture has taken on increased importance elsewhere.

From the sierra, remains of algarrobo seeds and chañar (*Geoffrea decorticans*) show the first botanical evidence of a pattern of localized gathering hypothesized for the preceding Preceramic period as well (Table 3). The presence of a seed identified as squash at the La Gruta I site at about

2500 B.C. is difficult to interpret, since no scientific name is given, and it is a single occurrence with no other plant remains listed. In a brief research note, remains of maize, *Capsicum*, bottle gourd, *Phaseolus*, and *Solanum tuberosum* are reported from Late Preceramic contexts from the Quebrada de Humahuaca. Whether any of these besides maize were actually cultivated, cannot be evaluated until the remains are more fully reported. It is in regions such as this in northwestern Argentina where early cultivation of indigenous wild peppers and beans may have occurred, but at even earlier time levels, since cultivated types occur to the north by 6000 B.C. The occurrence of maize, if confirmed, lends further support to its Preceramic spread throughout the Andean area. Sites of this time period in the sierran zone seem to represent the temporary camps of hunting and gathering populations (Nuñez A. 1974:174-81). So few sites are known from this period, however, that little can really be said about settlement patterns (González and Pérez 1966). These authors feel that the pattern of seasonal sedentism proposed for the hunting and gathering population at Intihausi cave and other earlier sierran sites extended into this period and formed a basis on which incipient agriculture and herding could begin in the sierra.

## Initial Ceramic Period (First pottery-A.D. 600-800)

The earliest occurrence of pottery from the Meridional Andes is from Tulan Cave in the San Pedro de Atacama Oasis, dated by Carbon 14 to 1760 B.C. (Nuñez A., in Popper 1977). From the level where the date was taken (IX), and just below it, come a variety of plant remains, identified by Virginia Popper of the University of Michigan (Table 4). Besides cultivated maize and the possibly cultivated bottle gourd, wild plant remains show patterns of local gathering of plants for fuel, industrial construction, rope, thatch, basketry, food, and medicine. Later levels show fewer remains, perhaps reflecting sporadic use of the cave. A series of grave lots from the Oasis document the occurrence of *Cucurbita maxima* by the end of this period. Although over 5000 graves and 40 sites with structures have been discovered in the Puna de Atacama, problems with dating, quality of investigation, and lack of emphasis on questions of subsistence result in only sketchy evidence for cultivation from this important zone (Popper 1977).

On the coast, the sites associated with the extensive use of marine resources and seasonal use of inland valley resources formed a stable exploitation pattern throughout the Preceramic VI period. Some were then abandoned and show no pottery remains, while others, such as the Conanoxa site, show later occupation with pottery and cultivated plants. Even when pottery and gourd, maize, and cotton appear at the Pichalo III site, marine resources remain important (Willey 1971:314). Many remains from the coast are from grave lots, which makes it difficult to answer questions of the degree of sedentism and settlement pattern. Nuñez A. describes the period from 500 B.C. to A.D. 800 as the early pottery horizon for this region and lists the economic plants which occur (Table 4). Most sites seem to be either on the shore or in the lower reaches of the small river valleys draining to the sea. The excellent preservation of the coastal zone shows that most cultivars had arrived in this zone by A.D. 800, some by introduction from the Bolivian/Peruvian sierra (maize, quinoa, potatoes) and others from lower elevations in inter-Andean valleys or the eastern lowlands (sweet potato, bean, peanut, pacae).

The Rio Loa subarea of Chile (Table 4) has recently been studied by Pollard (1971) and Pollard and Drew (1975). Ceramic complexes stretching from 800 B.C. to A.D. 1535 show a progression from small hunting and gathering campsites to fortified villages practicing irrigation agriculture and llama herding. Vega Alta I is characterized by small, open hunting and gathering campsites. By the end of the Vega II occupation (about 200 B.C.), a village of ten circular pithouses near the Rio Loa suggests semi-permanent settlement. The following Loa I period shows an increase in the number of sites that are riverine in orientation. By A.D. 100, and the beginning of Loa II, Pollard and Drew suggest that full sedentism was achieved. Evidence includes cultivation terraces and irrigation ditches, a llama corral, larger site size, and the expansion of the subsistence base to include cultivars. These patterns are intensified and elaborated in the following Lasana phase. Analysis of *Lama* sp. bone from Vega Alta II and Loa II deposits revealed that both wild and domesticates were used at each, with Vega Alta II domesticates being in the initial stages of domestication, and Loa II domesticates more fully domesticated (Pollard and Drew 1975). This data is strikingly similar to patterns from the central highlands of Peru: no clear hunting-herding shift, rather a mixed strategy that changes in emphasis over time. The occurrence of maize in Loa II deposits is interpreted by Pollard and Drew as a shift in subsistence in response to population pressure. The lack of plant remains from Loa I leaves open the possibility that agriculture initially began at this period and became gradually more elaborated. The races of maize represented—Capio Chico Chileño, Polulu, and Chutucuno Chico—are related, respectively, to Confite Puñeno, Confite Morocho, and Confite Puntiagudo of Peru (Mangelsdorf and Pollard 1975).

In the Valliserrana and Mendoza-Nenquén regions of the sierra (Table 3), Initial Ceramic sites with plant remains are more or less contemporary with the Loa and early Lasana phases of the Loa subarea, with the exception of Tafí, which may be either earlier or later, depending on the interpretation of the pottery. From site descriptions of Alamito, Cienega, and Tafí (González 1963; González and Pérez 1966, 1968; Nuñez R. 1971), settlement patterns appear similar to those discussed for Loa I and II: small villages located along river terraces, sometimes with artificial mounds, faced platforms, plazas, and associated agricultural terraces or old field outlines. González (1963) notes the mixture of altiplano, Chilean, and Amazonian traits evident both in the settlement pattern and material culture of this time period. The presence of *Cucurbita maxima*, *Bixa orellana*, and peanut from Cienega II suggests contact with the eastern foothills of the Andes.

### Middle and Late Ceramic Periods (A.D. 600/800-contact)

The pattern of sedentary agricultural settlement along major river valleys and oases in the Rio Loa, Valliserrana, San Pedro de Atacama, and coastal Chile areas, which began to emerge by the end of the Initial Ceramic period, continues in the later ceramic periods. In the Rio Loa area, site size increases, and irrigation canals and well defined field complexes are established. Although hunting and gathering remain part of the subsistence system, agriculture and herding become of predominant importance. Sites of the Aguada culture have a settlement pattern like that of the Initial Ceramic cultures of the sierra (González 1963; González and Pérez 1966). In the Late Ceramic period, population growth continues, and settlement size increases with the development of urban centers. The full, complex agricultural complex is present (Tables 3, 4).

## INTERMEDIATE AREA

The Intermediate area includes Ecuador, Colombia, and the far western portion of Venezuela (Fig. 1). This area includes very diverse environments, from the savannahs of southwestern Ecuador and the tropical forest of the western Colombian coast, to paramo grassland and fertile intermontane valleys and basins of the Andes. The northern floodplain of the Cauca and Magdelena rivers and the drier littoral of western Venezuela provide two other coastal environments. Because of the paucity of botanical remains from this region, references to remains will be made in the text.

### Preceramic IV (7000-5000 B.C.)

Strata II and III of El Inga Cave, Las Casitas, and the middle strata of the El Abra Rockshelters are among occupations dated to this time period (Willey 1971:57-60; Rouse and Cruxent 1963:30-34; Hurt et al. 1976). Plant remains are entirely lacking. Faunal remains from El Abra (Rockshelter 4, Zone 4) show that deer and guinea pig, in equal proportions, were the most common types (Hurt et al. 1976:19). The location of El Abra and El Inga in mid-altitude sierran basins (2600 m) suggests that the rich plant resources accessible in the zone were also utilized. Drawing an analogy to gathering and early cultivation patterns at Guitarrero Cave is not unreasonable. None of the sites discussed for this period appear to be base camps or long-term occupation sites. Willey suggests El Inga may be a highland hunting station in a sierran-coast transhumance system but admits there is no real evidence for this (Willey 1971:57-60).

On the coast of Ecuador, sites of the Vegas culture date to between 6650 and 5300 B.C. (Stothert 1976:88-89). These sites occur in the western area of the Santa Elena Peninsula, 1-5 km from the sea, along fossil drainage systems, near beaches, and close to dry lagoons, with access to swamp, riverine, littoral, and forest areas (Stothert 1976; 1977). Analysis of faunal remains from OGSE-80, a large Vegas site 3.5 km from the sea, shows a predominance of terrestrial protein, with a variety of fish also present (Byrd 1976). Stothert considers OGSE-80 a year-round settlement. Rather than specialized shell collecting stations, Vegas sites seem to represent generalized hunting, gathering, and perhaps agricultural settlements (Stothert 1976; 1977). The location of the Vegas sites and the varied faunal assemblage from OGSE-80 seem to support the view that at least some of these sites were in long-term, multiple use, with perhaps seasonal, more specialized camps making up another component of the settlement system. The lack of plant remains makes it difficult to evaluate Stothert's (1977) suggestion of root crop agriculture. As was seen for the coast of Peru, the first cultivars in a setting of rich marine and hunting resources tend to be industrial plants (bottle gourd, cotton), with other crops being a very minor component and not really a staple in the diet.

In recent excavations of several small cave sites on the southern Guayas coast of Ecuador, Carl Spath discovered the sporadic use of the shelters as camps from before 7000 B.C. until Late Ceramic times. Located between two extensive mangrove swamps, the caves seem to have served as a resting place for populations traveling between these areas. Plant remains from the caves are from taxa that occur today in the area of the shelters and do not include food plants (C. Spath, 1978: personal communication).

## Preceramic V (5000-3000 B.C.)

This is the first half of the Meso-Indian period of Rouse and Cruxent (1963:20), represented by the Cubagua complex on the eastern coast of Venezuela and El Heneal on the western coast. A strong marine orientation is suggested for the Cubagua and later complexes in this area by site location (coast and islands) and shell remains. Information from the coast of Colombia is lacking for this time period, as it was for the Preceramic IV.

Reichel-Dolmatoff proposes a gradual adaptation to sea resources in response to the disappearance of the big game animals, so that by 3000 B.C., a well-defined pattern of shell mound-dwelling emerges (Reichel-Dolmatoff 1965; 51). In the sierra, seasonal occupation continues at the El Abra rockshelters. In zone 3 (7000-2500 B.C.) of Rockshelter 4, bones of guinea pig are five times as numerous as those of deer (Hurt et al. 1976:19). This gradual shift in utilization, which continues into the next zone, is reminiscent of the shift from a half camelid, and half deer exploitation pattern to primarily camelid utilization at the 5500-4200 B.C. time period in Peru (Wheeler Pires-Ferreira et al. 1977:156). Elizabeth Wing suggests guinea pig domestication occurred during the 4200-2500 B.C. time range, beginning perhaps even earlier. The wild form of the domesticated guinea pig (*Cavia porcellus*), occurs in northern South America (Wing 1977).

On the coast of Ecuador, the Vegas complex lasts until about 5300 B.C., by Carbon 14 dating (Stothert 1976). From then until the appearance of an Achallan complex site OGSE-63 just before 3000 B.C., there are no sites on the Santa Elena peninsula dated to this time period. Whether this lack of sites represents abandonment of the peninsula requires further survey and dating of now undated sites.

## Early Ceramic Period (3000-1500 B.C.)

Ceramic using cultures appear in the Intermediate area long before pottery occurs in Peru or the Southern Andes. Among the earliest pottery complexes are those appearing on the Santa Elena peninsula in southwestern Ecuador (Achallan, San Pablo, Valdivia), at the Puerto Hormiga site in Colombia, and at the Rancho Peludo site in Venezuela (Stothert 1976; Bischof and Viteri Gamboa 1972; Meggers, Evans, and Estrada 1965; Reichel-Dolmatoff 1965; Rouse and Cruxent 1963).

In southwestern Ecuador, the Achallan complex (Site OGSE-63) and the San Pablo complex (at the Valdivia type site), occur before the more widespread Valdivia ceramic complex. Analysis of faunal remains from OGSE-63 shows a continuation of Vegas exploitation patterns of terrestrial hunting with some fishing (Byrd 1976). Data on patterns of plant use are lacking for both complexes. With the introduction of Valdivia ceramics, probably from the east (Lathrap 1977; Zevallos et al. 1977), patterns begin to change. Faunal analysis indicates a switch to primarily aquatic protein sources at the peninsular, coastal sites, with terrestrial hunting being predominant at the inland sites (Loma Alta, Real Alto) and at the type site, located north of the peninsula (Byrd 1976). These data suggest a pattern of year-round subsistence specialization similar to that suggested for the Peruvian coast in the Late Preceramic and Initial periods. The evidence for the presence of maize at the Real Alto site (Zevallos et al. 1977; Pearsall 1978) suggests further that the terrestrial orientation may correlate with agriculture. Carbonized maize from the site of Cerro Narrio in the Ecuadorian sierra, dated to 2000-1800 B.C., adds confirming evidence for the cultivation of maize during this time period (Collier and Murra 1943; D. Collier, 1978: personal communication confirming identification).

Even less information of people/plant interrelationships is available from this time period in Colombia and western Venezuela. In Colombia, sites dated to this period are scattered over a wide area of the northern coast and the lagoons of the lower river courses (Reichel-Dolmatoff 1965:51-60). No site has been reported from the interior, suggesting to Reichel-Dolmatoff that most cultural development was occurring on the coast. Puerto Hormiga is described as a seasonal fishing and shell fishing village (Reichel-Dolmatoff 1965:51-54). Settlement in Venezuela also appears to be predominantely coastal. At the Rancho Peludo site, clay griddles interpreted as manioc griddles suggest cultivation or use of wild tubers of manioc by 1860 B.C. (Rouse and Cruxent 1963:48-49). The lack of interior sites in Colombia and Venezuela may be due to a lack of systematic survey and destruction of riverine sites by alluviation or later occupation. The existence of ceramic styles related to coastal Valdivia in inland Guayas, the sierra, and the eastern lowlands of Ecuador argues against purely coastal cultural development in this time period (Braun 1971; Porras G. 1977; Lathrap 1977). The importance of a rich riverine habitat for allowing sedentary life and the development of agriculture, stressed by Reichel-Dolmatoff (1965:61-69) for the later ceramic periods, probably came into play in this time period.

## Middle Ceramic Period (1500-500 B.C.)

Little concrete plant data exist from this period, where settled, agriculturally based life is proposed throughout the Intermediate area (Willey 1971:278-86). At Momil, manioc grater chips, and later manos and metates, suggest complex patterns of agricultural development and change in the northern Colombian lowlands (Reichel-Dolmatoff 1965:61-79; Foster and Lathrap 1973). A maize/manioc "line" seems to extend through Colombia into Venezuela, suggesting an interface of primary seed and primary root cropping systems in this zone (Rouse and Cruxent 1963:51-59). Root cropping and seed cropping should not be considered mutually exclusive for a mixed subsistence base dependent on ecological factors and maintained by cultural preference probably existed. A cache of carbonized maize kernels recovered in a Chorrera period (Ecuador, 1000 B.C.) vessel by Olaf Holm and identified as a broad-kerneled race similar to Andaqui of Colombia (Pearsall 1977), gives additional "hard" evidence of this cultivar at the 1000 B.C. time period.

## CONCLUSIONS: PATTERNS OF PEOPLE/PLANT INTERRELATIONSHIPS

In this paper, I have reviewed recent paleoethnobotanical research in western South America to pull together what we know about people/plant interrelationships in this region. Three systems emerge, distinctive in ecological setting, combination of patterns, and evolution: the high altitude puna system, the mid-altitude inter-Andean valley system, and the coastal valley and littoral system. Evidence is most abundant for the Central Andes, but enough data exist for the Meridional and Intermediate areas to suggest patterns which can be tested further.

The high altitude puna system is characterized by a stable, long-term pattern of localized gathering, regional self-sufficiency, gradual transition from hunting to herding, and a delay in the development and intensification of agriculture because of rich wild plant resources. Pachamachay Cave and the Chiripa site from the Central Andean area cover the full range of this system, from earliest Pre-Ceramic hunting and gathering to full dependence on herding and the beginnings of genetic manipulation of indigenous plants by the Early Ceramic period. Other high altitude sites from western South America, for which no plant remains are reported, may be part of this system, if an abundance of wild plant and animal resources allowed early sedentism.

The mid-altitude intermontane valley system is characterized by the early emergence of a pattern of seasonal utilization of wild plant resources in different zones within the valley system by nomadic or semi-sedentary populations. Early inter-regional contact led to the spread of cultivated plants, domesticated in middle and low altitude areas, and integrated into the round of multizonal gathering. The Ayacucho caves, Guitarrero Cave, probably also the sketchily reported Preceramic sites from the Valliserrana region of the Meridional Andes, and El Inga and El Abra from the Intermediate area, are part of the early stage of this system. Following this stage, a pattern of increased utilization of cultivated plants emerges, with more permanent settlement in the river valley bottom zone, growth of larger communities, and a shift from hunting to herding. In the Meridional Andes, this stage seems to correlate with the introduction of ceramics and a variety of new cultivars of both sierran and lowland origin. Sites in the Rio Loa area, San Pedro de Atacama Oasis, and Valliserrana region show gradual size increase, riverine or oasis orientation, and finally evidence of terracing and irrigation systems. The mid-altitude zone, along with the lowland tropics (Sauer 1969; Lathrap 1970), was probably the source of many cultivated plants and an area of innovation and change.

The coastal system is characterized by an initial pattern of terrestrial gathering and hunting, oriented in Peru to the lomas and valley zones and in Ecuador to riverine and mangrove swamp, with fish and shell resources augmenting terrestrial hunting. Emphasis later shifts to exploitation of marine rather than terrestrial protein sources, with cultivated plants appearing on the Peruvian coast. The Preceramic and Early Ceramic littoral sites of Chile, Colombia, and Venezuela are also probably characteristic of this stage. In Ecuador, the situation is more complex. Here pottery appears and shortly after it, evidence of maize agriculture; specialization of sites to exploit marine or terrestrial resources, and sedentary occupation. These patterns appear later on the coast of Peru, where the utilization of cultivated plants increases over time, resulting eventually in valley-wide irrigated cropping systems.

**(References for this article appear on p. 412)**

TABLE 1
CENTRAL ANDEAN SIERRA

| Years B.C. | Guitarrero Cave | Pachamachay Cave | Ayacucho Caves | Chiripa | Years B.C. |
|---|---|---|---|---|---|
| 0 | | | | *Chenopodium*, tubers[5] | 0 |
| 1000 | Ceramic period (undated): maize: Confite Puneño, Confite Morocho[1] | | | *Amaranthus*, *Solanum* tubers, *Chenopodium*, *Scirpus*, *Juncus* *Carex*, *Opuntia*, *Stipa*, *Festuca*, *Malvastrum*, *Plantago*[6] | 1000 |
| Ceramics | | | | | Ceramics |
| 2000 | | | Cachi Complex: maize, beans, cotton, gourd, squash, lucuma, coca(?), quinoa, amaranth, *Sapindus*, *Canavalia*, tubers[4] | | 2000 |
| 3000 | | Levels 1-32: *Opuntia floccosa*, *Chenopodium*, *Amaranthus*, *Polygonum*, *Euphorbia*, *Plantago*, *Malvastrum*, Gramineae, *Scirpus*, *Sisyrinchium*, *Ranunculus*, *Luzula*, Leguminosae, Umbelliferae, Compositae, bulbs, tuberous roots, camelid dung[3] | Chihua Complex: squash, quinoa, amaranth, gourd, tubers, cotton, lucuma, maize: Confite Morocho, *Sapindus*, bean(?)[4] | | 3000 |
| 4000 | | | | | 4000 |
| 5000 | | | Piki Complex: quinoa, gourd, squash(?), amaranth, pepper(?)[4] | | 5000 |
| 6000 | Stratum II: *Fourcraea*, *Phaseolus lunatus* *P. vulgaris*, *Oxalis tuberosa*, *Capsicum chinensis*, lucuma, *Opuntia*, *Trichocereus*, bulbs[2] | | Jaywa Complex: achiote, gourd, *Crescentia*, pepper(?)[4] | | 6000 |
| 7000 | | | | | 7000 |
| 8000 | | | No plant remains in these levels[4] | | 8000 |
| 9000 | Stratum I: no remains[2] | | | | 9000 |
| 10,000 | | | Ayacucho Complex: gourd[4] | | 10,000 |

TABLE 2
CENTRAL ANDEAN COAST

| Years B.C. | Far North | North | North Central | Central | South | Years B.C. |
|---|---|---|---|---|---|---|
| 1000 | *General trends*: manioc, *Capsicum chinense* appear; *Phaseolus vulgaris*, coca, llacon, pepino, potato, *Cucurbita maxima*, cherimoya, papaya, pineapple, guanabana, granadilla added[5,7,8] | | | *General trends*: same | | 1000 |
| | *General trends*: earlier occurring plants continue and spread; maize spreads throughout coast; peanut, manioc (?) added; quantities of plant remains increase[9] | | | *General trends*: same | | Ceramics |
| Ceramics | | | | | | |
| 2000 | | *Cucurbita ficifolia, C. moschata, Phaseolus lunatus, Canavalia plagiosperma, Capsicum baccatum, Canna*, lucuma, ciruela de fraile, cotton, gourd[7,9,11] | *Cucurbita ficifolia, C. moschata, Phaseolus lunatus, Canavalia plagiosperma, Capsicum baccatum, Canna*, lucuma, ciruela de fraile, cotton, gourd, pacae, avocado, tobacco, guava, *Zea mays*[9,12,13] | cotton, *Capsicum baccatum*, gourd, guava, *Canna*, sweet potato, potato(?), *Canavalia*, legumes, *Cucurbita moschata, C. ficifolia, C. andreana*, sedges, reeds, rushes, tillandsia, beans, wild tubers, milkweed, lucuma, *Phaseolus lunatus*, pacae, jicama[7,9,14,15,16] | cotton, squash, gourd, pacae[9,16,18] | 2000 |
| | | | | Encanto: gourd, squash, grasses, sedge[7,17] | | |
| 3000 | | | | Yacht Club: cotton, gourd, aji, guava, tillandsia, reeds, roots, legumes[9,17] | Cabeza Larga: cotton, sedge, *Phaseolus lunatus, P. vulgaris*(?)[7,9] | 3000 |
| | | | | Pampa: *Cucurbita moschata, C. ficifolia, C. andreana*[9] | | |
| | | | | Chilca (3800-2650 B.C.): gourd, *Phaseolus lunatus*, wild bean, cotton(?)[9,18,19] | | |
| 4000 | | | | | | 4000 |
| 5000 | Siches Complex: Bottle gourd[10] | | | | | 5000 |
| | | | | *Phaseolus lunatus*[20] | | |
| 6000 | | | | | | 6000 |
| 7000 | | | | Chilca Caves: jicama, potato, ullucu, manioc, sweet potato, cactus fruit[19] | | 7000 |
| 8000 | | | | | | 8000 |

TABLE 3 MERIDIONAL ANDEAN SIERRA

| Years A.D. | Quebrada de Humahuaca | Valliserrana | | Mendoza-Neuquén | | Years A.D. |
|---|---|---|---|---|---|---|
| 1450 | | Late Ceramic Period: algarrobo, quinoa, peanut, gourd, *Juglans australis*, potato, *Canna*(?), maize: Morocho, Pisingallo, Perla, Capia, Chulpi, Rosero, *Phaseolus vulgaris*, *Cucurbita pepo*, *C. maxima*[22,23,24] | | | Atuel I: *Zea mays* var. *amilacea* (Capia, Guarani, Culli types), *Zea mays* var. *minima*, *Cucurbita*[28] | 1450 |
| 1000 | | | | | | 1000 |
| 900 | | | | | | 900 |
| 800 | | | Aguada Culture: potato; *Zea mays* var. *microsperma*, popcorn, Capia; *Cucurbita maxima*, algarrobo, chañar[22] | | | 800 |
| 700 | | | | | | 700 |
| 600 | | | | | | 600 |
| 500 | | | | | | 500 |
| 400 | | Alamito II: *Zea mays* var. *microsperma*, algarrobo, peanut(?), chañar, bean, squash[25] | | | | 400 |
| 300 | | Alamito I[25] | Cienega II: *Zea mays* var. *microsperma* (perla), *Cucurbita maxima*, chañar, algarrobo, *Bixa orellana*, peanut, *Capparis cynophallophora*, *Opuntia quimilo*[22] | | | 300 |
| 200 | | | | | | 200 |
| 100 | | | Cienega I: *Zizypus mistol*, *Ximenia america*, *Zea mays* var. *microsperma*, second race of maize[22] | Los Morrillos III: maize, beans, squash, algarrobo[27] | | 100 |
| B.C. 0 | | | | | | 0 B.C. |
| 100 | | | | | | 100 |
| 200 | | | | | Atuel II: guarani maize, beans, squash, quinoa, *Zea mays* var. *minima*[27] (this complex also dated 700-1000 A.D., no C 14)[28] | 200 |
| 300 | | | Tafí: maize, *Phaseolus vulgaris* quinoa, *Amaranthus caudatus*, *Chenopodium*[22] (this complex also dated 0-300 A.D.)[26] | | | 300 |
| | | | | | | Ceramics |
| 400 | | | | | Atuel III: algarrobo, chañar[27] | 400 |
| Ceramics | | | | | | |
| 500 | Huachichocana Cave: maize, *Capsicum*, *Phaseolus*, *Solanum tuberosum*[21] | Puente del Diablo: *Phaseolus*[21] | | Los Morrillos II: algarrobo[27] | | 500 |
| 2500 | Quebrada de Inca Cave: gourd[21] | | | La Gruta I: squash seed[27] | | 2500 |

TABLE 4
CHILEAN COAST AND SIERRAN OASES OF THE ATACAMA REGION

| Years A.D. | Chilean Coast | | Rio Loa Area | San Pedro de Atacama Oasis | | Years A.D. |
|---|---|---|---|---|---|---|
| 1400 | | | | | | 1400 |
| 1200 | | | | | Oasis Grave Lots: Catarpe 2: maize, *Prosopis chilensis* var. *chilensis*[31] | 1200 |
| 1000 | Arica, Pica: intensive agriculture, irrigation[27] | | Lasana Complex: beans, quinoa, aji, gourd, potato, maize: Capio Chico Chileno, Polulo, Chutucuno Chico[29,30] | Tulan Cave: | | 1000 |
| 800 | | | | (770 A.D.) | Coyo Oriental: *Cucurbita maxima*, *Cucurbita* sp., maize, *Prosopis chilensis* var. *chilensis*, gourd[31] | 800 |
| 600 | | | | Level IV: *Ephedra*, *Prosopis*, *Opuntia*, fiber rope[31] | Quitor 6, Quitor 2: *Prosopis chilensis* var. *chilensis*[31] | 600 |
| 400 | Various sites: maize, squash, gourd, cotton, quinoa, sweet potato, bean, peanut, algarrobo, pacae, chañar, potato, pallar[27] | | | Level V: *Ephedra*, *Geoffrea*, *Prosopis*[31] | | 400 |
| 200 | | | Loa II: algarrobo, gourd, maize: Capio Chico Chileño, Polulo, Chutucuno Chico[29,30] | | | 200 |
| B.C. 0 | | Pichalo III: gourd, maize, cotton[16] | Loa I: no remains[29] | Level VI: no remains[31] | | 0 B.C. Ceramics |
| 200 | | | | Level VII: no remains[31] | | 200 |
| 400 Ceramics | Conanoxa (360 B.C.): maize[27] | | Vega Alta II: algarrobo, roots, cactus buttons, gourd[29] | | | 400 |
| 600 | | Quiani II-Chinchorro: cotton, quinoa[16] | Vega Alta I: no remains[29] | Level VIII: maize kernel, Cyperaceae, *Prosopis*[31] | | 600 |
| 800 | | | | Level IX: maize kernel, Cyperaceae, *Prosopis*, *Opuntia*, fiber rope *Lagenaria*, *Ephedra*, *Cortaderia*[31] | | 800 |
| 2500 | --Conanoxa (1790 B.C.): no remains[27] | top of Quiani II midden (undated): maize, gourd, cotton[16] | | (1760 B.C.) Just above sterile: *Tessaria*, Compositae, fiber rope, *Opuntia*, *Prosopis*, *Atriplex* sp. *Atriplex microphyllum*, Cyperaceae, *Scirpus*, *Ephedra*, *Festuca*, maize kernals, Cyperaceae, *Scirpus*[31] | | 2500 |
| 4200 | Tarapacá Region: algarrobo[27] | Quiani I: no remains[16] | | | | 4200 |
| 4800 | | | | | | 4800 |

*Notes for Tables 1-4*

[1]Lynch 1971
[2]Kaplan et al. 1973; Smith 1977
[3]Pearsall in press
[4]MacNeish 1969; MacNeish et al. 1970
[5]Towle 1961
[6]Erickson 1976; 1977
[7]Pickersgill 1969*a*
[8]Whitaker and Cutler 1965
[9]Lanning 1967
[10]Richardson 1972
[11]Sauer and Kaplan 1969
[12]Kelley and Bonavia 1963
[13]Moseley and Willey 1973
[14]Patterson and Moseley 1968
[15]Stephens and Moseley 1974
[16]Willey 1971
[17]Moseley 1975
[18]Engel 1963
[19]Engel 1973
[20]Heiser 1965
[21]Research Reports, American Antiquity, 39(3) 1974
[22]González and Pérez 1968
[23]Cigliano 1968
[24]Raffino 1973
[25]Nuñez R. 1971
[26]González and Pérez 1966
[27]Nuñez A. 1974
[28]Lagiglia 1968
[29]Pollard 1971
[30]Mangelsdorf and Pollard 1975
[31]Popper 1977. The Tulan sequence is dated only at levels IV and IX. Levels V-VIII are arbitrarily spaced on the chart to fill the intervening time.

## REFERENCES


Bird, Junius B.
1963 Pre-Ceramic Art from Huaca Prieta, Chicama Valley. Nawpa Pacha 1:29-34.

Bischof, Henning, and Julio Viteri Gamboa
1972 Pre-Valdivia Occupations on the Southwest Coast of Ecuador. American Antiquity 37:548-51.

Braun, Robert
1971 Cerro Narrio Reanalyzed: The Formative as seen from the Southern Ecuadorian Highlands. Paper presented at the Primer Simposio de Correlaciones Antropologicas Andino-Mesoamericano, Salinas, Ecuador.

Burkart, Arturo
1952 Las Leguminosas Argentinas, silvestres y cultivadas. 2nd ed. Buenos Aires: Acme Agency.

Burkart, Arturo and H. Brucher
1953 Phaseolus aborigineus Burkart, die mutmassliche andine Stammform der Kulturbohne. Der Zuchter 23(3):65-72.

Byrd, Kathleen Mary
1976 Changing Animal Utilization Patterns and their Implications: Southwest Ecuador (6500 B.C.-A.D. 1400). Ph.D. Dissertation, Department of Anthropology, University of Florida.

Cardich, Augusto
1976 Vegetales y Recolecta en Lauricocha: Algunas Inferencias Sobre Asentamientos y Subsistencias Preagrícolas en los Andes Centrales. Relaciones de la Sociedad Argentina de Antropología, N.S., 10:27-41.

Cigliano, Eduardo Mario
1968 Sobre Algunos Vegetales Hallados en el Yacimiento Arqueologico de Santa Rosa de Tastil, Dept. Rosario de Lerma. La Plata Universidad Nacional, Museo, Revista Antropología 7(38):15-23.

Collier, Donald
1961 Agriculture and Civilization on the Coast of Peru. In: The Evolution of Horticultural Systems in Native South America: Causes and Consequences, a Symposium, edited by Johannes Wilbert, pp. 101-09. Antropologica Supplement 2.

Collier, Donald, and John Murra
1943 Survey and Excavation in Southern Ecuador, Anthropological Series, Field Museum of Natural History 35.

Cutler, Hugh C., and Leonard W. Blake
1971 Travels of Corn and Squash. In: Man Across the Sea: Problems of Pre-Columbian Contacts, edited by C. L. Riley, J. C. Kelley, C. W. Pennington, and R. L. Rands, pp. 366-75. Austin: University of Texas Press.

Engel, Frédéric
1963 Datations à l'aide du Radio-Carbone 14, et Problèmes de la Préhistoire du Perou. Journal de la Société des Americanistes 52:101-31.
1973 New Facts about Pre-Columbian Life in the Andean Lomas. Current Anthropology 14:271-80.

Erickson, Clark L.
1976 Chiripa Ethnobotanical Report: Flotation-Recovered Archeological Remains from an Early Settled Village on the Altiplano of Bolivia. ms.
1977 Subsistence Implications and Botanical Analysis at Chiripa. Paper presented in the Symposium: Commodity Flow and Political Development in the Andes, 42nd Annual Meetings of the Society for American Archaeology, New Orleans.

Feldman, Robert A.
1977 Life in Ancient Peru. Field Museum of Natural History Bulletin 48(6):12-17.

Flannery, Kent V.
1973 The Origins of Agriculture. Annual Review of Anthropology 2:271-310.

Foster, Donald W., and Donald W. Lathrap
1973 Further Evidence for a Well Developed Tropical Forest Culture on the North Coast of Colombia During the First and Second Millennium B.C. Journal of the Steward Anthropological Society 4(2):160-99.

Gentry, Howard S.
1969 Origin of the Common Bean *Phaseolus vulgaris*. Economic Botany 23(1): 55-69.

González, Alberto Rex
1963 Cultural Development in Northwestern Argentina. In: Aboriginal Cultural Development in Latin America: An Interpretative Review, edited by B. J. Meggers and C. Evans. Smithsonian Miscellaneous Collection 140(1):103-18.

González, Alberto Rex, and José Antonio Pérez
1966 El Area Andina Meridional. In: Actas y Memorias 36th Congreso Internacional de Americanistas, España, 1964 (1):241-65.
1968 Una Nota Sobre Etnobotanica del N.O. Argentino. Actas y Memorias 37th Congreso Internacional de Americanistas, Argentina, 1966(2):209-28.

Hastings, Charles M.
1977 Prehispanic Subsistence Strategies and Settlement Patterns in Tarma Province, Peru. Paper presented at the 5th Conference on Andean and Amazonian Archaeology, Bloomington, Indiana.

Heiser, Charles B.
1965 Cultivated Plants and Cultural Diffusion in Nuclear America. American Anthropologist 67:930-49.

Horn, Darvin and Clark Erickson
n.d. Domestication and Subsistence Implications of Plant and Animal Utilization of the Titicaca Basin. ms.

Hurt, Wesley R., Thomas van der Hammen, and Gonzalo Correal Urrego
1976 The El Abra Rockshelters, Sabana de Bogota, Colombia, South America. Indiana University Museum Occasional Papers and Monographs 2.

Kaplan, Lawrence
1965 Archeology and Domestication in American *Phaseolus* (Beans). Economic Botany 19:358-68.
1971 *Phaseolus*: Diffusion and Centers of Origin. In: Man Across the Sea: Problems of Pre-Columbian Contacts, edited by C. L. Riley, J. C. Kelley, C. W. Pennington, and R. L. Rands, pp. 416-27. Austin: University of Texas Press.

Kaplan, Lawrence, Thomas Lynch, and C. Earle Smith
1973 Early Cultivated Beans (*Phaseolus vulgaris*) from an Intermontane Peruvian Valley. Science 179:76-77.

Kelley, David H., and Duccio Bonavia
1963 New Evidence for Pre-Ceramic Maize on the Coast of Peru. Nawpa Pacha 1:39-42.

Krapovickas, Antonio
1969 The Origin, Variability, and Spread of the Groundnut (*Arachis hypogaea*). In: The Domestication and Exploitation of Plants and Animals, edited by P. J. Ucko and G. W. Dimbleby, pp. 427-41. Chicago: Aldine Publishing Co.

Lagiglia, Humberto
1968 Plantas Cultivadas en el Área Centro-Andina y su Vinculación Cultural Contextual. In: Actas y Memorias, 37th Congreso Internacional de Americanistas, Argentina 1966, 2:229-33.

Lanning, Edward P.
1963 A Pre-Agricultural Occupation on the Central Coast of Peru. American Antiquity 28:360-71.
1967 Peru Before the Incas. New Jersey: Prentice-Hall, Inc.

Lathrap, Donald W.
1970 The Upper Amazon. New York: Praeger Publishers.
1974 The Moist Tropics, the Arid Lands, and the Appearance of Great Art Styles in the New World. The Museum of Texas Tech University Special Publication 7:115-58.
1977 Olmec-Chavin Connections. Paper presented at the Symposium: Evidence of Population Movements out of the Moist Tropics of South America, 42nd Annual Meetings of the Society of American Archaeology, New Orleans.
Lynch, Thomas
1971 Preceramic Transhumance in the Callejón de Huaylas, Peru. American Antiquity 36:139-48.
MacNeish, Richard S.
1969 First Annual Report of the Ayacucho Archaeological-Botanical Project. Andover, Mass.: Peabody Foundation for Archaeology.
MacNeish, Richard S., Antoinette Nelken-Terner, and Angel Garcia Cook
1970 Second Annual Report of the Ayacucho Archaeological-Botanical Project. Andover, Mass.: Peabody Foundation for Archaeology.
MacNeish, Richard S., T. C. Patterson, and D. L. Browman
1975 The Central Peruvian Prehistoric Interaction Sphere. Andover, Mass.: Peabody Foundation for Archaeology.
Mangelsdorf, Paul C., and G. C. Pollard
1975 Archaeological Maize from Northern Chile. Harvard University Botanical Museum Leaflets, 24(3):49-64.
Matos M., Ramiro
1975 Prehistoria y Ecología Humana en las Punas de Junin. Revista del Museo Nacional, Lima, Peru, 41:35-80.
Meggers, Betty J., Clifford Evans, and Emilio Estrada
1965 The Early Formative Period on Coastal Ecuador: The Valdivia and Machalilla Phases. Smithsonian Contributions to Anthropology 1.
Moseley, Michael E.
1975 The Maritime Foundations of Andean Civilization. Menlo Park, California: Cummings Publishing Co.
Moseley, Michael E., and Gordon R. Willey
1973 Aspero, Peru: A Reexamination of the Site and Its Implications. American Antiquity 38:452-68.
Nuñez A., Lautaro
1974 La Agricultura Prehistorica en los Andes Meridionales. Santiago, Chile: Editorial Obre.
Nuñez R., Victor A.
1971 La Cultura Alamito de la Subarea Valliserrana Del Noroeste Argentina. Journal de la Société des Américanistes 60:7-64.
Patterson, Thomas C., and Michael E. Moseley
1968 Late Pre-Ceramic and Early Ceramic Cultures of the Central Coast of Peru. Nawpa Pacha 6:115-33.
Pearsall, Deborah M.
in press Pachamachay Ethnobotanical Report: Plant Utilization at a Hunting Base Camp. In: The Preceramic Cultural Ecology of the Central Peruvian Puna: High Altitude Hunters, edited by John W. Rick. New York: Academic Press.
1977 Maize and Beans in the Formative Period of Ecuador: Preliminary Report of New Evidence. Paper presented in the symposium: Intensification of Agricultural Systems in the New World. Was it a Unitary Phenomenon? 42nd Annual Meetings of the Society for American Archaeology, New Orleans.
1978 Phytolith Analysis of Archeological Soils: Evidence for Maize Cultivation in Formative Ecuador. Science 199:177-78.

Pickersgill, Barbara
1969a The Archeological Record of Chili Peppers (*Capsicum* sp.) and the Sequence of Plant Domestication in Peru. American Antiquity 34:54-61.
1969b The Domestication of Chili Peppers. In: The Domestication and Exploitation of Plants and Animals, edited by P. J. Ucko and G. W. Dimbleby, pp. 443-50. Chicago: Aldine Publishing Co.

Pickersgill, Barbara, and Charles B. Heiser, Jr.
1977 Origins and Distribution of Plants Domesticated in the New World Tropics. In: Origins of Agriculture, edited by Charles A. Reed, pp. 803-35. The Hague: Mouton.

Pires-Ferreira, Edgardo, Jane Wheeler Pires-Ferreira, and Peter Kaulicke
1977 Utilización de Animales Durante el Período Precerámico en la Cueva de Uchcumachay y Otros Sitios de Los Andes Centrales del Perú. Journal de la Société des Américanistes 64:149-54.

Pollard, Gordon C.
1971 Cultural Change and Adaptation in the Central Atacama Desert of Northern Chile. Nawpa Pacha 9:41-64.

Pollard, Gordon C., and Isabella M. Drew
1975 Llama Herding and Settlement in Prehispanic Northern Chile: Application of an Analysis for Determining Domestication. American Antiquity 40: 296-305.

Popper, Virginia S.
1977 Prehistoric Cultivation in the Puna de Atacama, Chile. Undergraduate Honor Thesis, Harvard University.

Porras G., Pedro I.
1977 Fase Pastaza: Nueva Evidencia del Formativo Temprano en el Sudoriente del Ecuador. Paper presented at the symposium: The First Inhabitants of the Tropical Alluvium, 42nd Annual Meetings of the Society for American Archaeology, New Orleans.

Raffino, Rodolfa A.
1973 Agricultura Hidraulica y Simbiosis Economica Demografica en la Quebrada del Toro, Salta, Argentina. Revista del Museo de la Plata (Nueva Serie), Sección Antropología 7(49):297-332.

Reichel-Dolmatoff, Gerardo
1965 Colombia. New York: Praeger Publishers.

Richardson, James B., III
1972 The Pre-Columbian Distribution of the Bottle Gourd (*Lagenaria siceraria*): A Re-evaluation. Economic Botany 26:265-73.

Rick, John
1977 Identifying Prehistoric Sedentism in Hunter-Gatherers: An Example from Highland Peru. Paper presented at the symposium: The Andean Pre-Ceramic, 76th Annual Meetings of the American Anthropological Association, Houston.
1978 The Preceramic Cultural Ecology of the Central Peruvian Puna: High Altitude Hunters. Ph.D. Dissertation, Department of Anthropology, The University of Michigan.

Rouse, Irving, and José M. Cruxent
1963 Venezuelan Archaeology. New Haven: Yale University Press.

Sauer, Carl O.
1969 Seeds, Spades, Hearths, and Herds. 2nd (1952) ed. Cambridge, Mass.: M.I.T. Press.

Sauer, Jonathan, and Lawrence Kaplan
1969 Canavalia Beans in American Prehistory. American Antiquity 34:417-24.

Smith, C. Earle, Jr.
1977 Recent Evidence in Support of the Tropical Origin of New World Crops. In: Crop Resources, edited by D. S. Seigler, pp. 79-95. New York: Academic Press.

Stephens, S. G., and Michael E. Moseley
1974 Early Domesticated Cottons From Archaeological Sites in Central Coastal Peru. American Antiquity 39:109-22.

Stothert, Karen E.
1976 The Early Prehistory of the Sta. Elena Península, Ecuador: Continuities Between the Preceramic and Ceramic Cultures. Actas del 41st Congreso Internacional de Americanistas, México, 1974 (2):88-98.
1977 Preceramic Adaptation and Trade in the Intermediate Area. Paper presented at the symposium: The Andean Preceramic, 76th Annual Meeting of the American Anthropological Association, Houston.

Towle, Margaret
1961 The Ethnobotany of Pre-Columbian Peru. Chicago: Aldine Publishing Co.

Wheeler Pires-Ferreira, Jane, Edgardo Pires-Ferreira, and Peter Kaulicke
1977 Domesticación de los Camélidos en los Andes Centrales Durante el Período Precerámico: Un Modelo. Journal de la Société des Américanistes 64: 155-65.

Whitaker, Thomas W., and Hugh C. Cutler
1965 Cucurbits and Cultures in the Americas. Economic Botany 19:344-49.

Willey, Gordon R.
1971 An Introduction to American Archaeology, Vol. 2: South America. Englewood Cliffs, New Jersey: Prentice-Hall, Inc.

Wing, Elizabeth S.
1977 Animal Domestication in the Andes. In: Origins of Agriculture, edited by Charles A. Reed, pp. 837-59. The Hague: Mouton.

Zevallos M., Carlos W. C. Galinat, D. W. Lathrap, E. R. Leng, J. G. Marcos, and K. M. Klumpp
1977 The San Pablo Corn Kernel and Its Friends. Science 196:385-89.

# PART VI
# PUBLICATIONS OF VOLNEY H. JONES

# PUBLISHED WORKS OF VOLNEY H. JONES

*Compiled by Richard I. Ford*


1935

A Chippewa Method of Manufacturing Wooden Brooms. Papers of the Michigan Academy of Science, Arts, and Letters 20(1934):23-30.

Vegetal Remains (of Jemez Cave, N.M.). In: Report on the Excavation of Jemez Cave, by Herbert G. Alexander and Paul Reiter, pp. 60-65. Albuquerque: The University of New Mexico and The School of American Research.

1936

Some Chippewa and Ottawa Uses of Sweet Grass. Papers of the Michigan Academy of Science, Arts, and Letters 21(1935):21-31.

A Summary of Data on Aboriginal Cotton of the Southwest. In: Symposium on Prehistoric Agriculture, edited by D. D. Brand. University of New Mexico Bulletin, Whole Number 296, Anthropological Series 1(5):51-64.

The Vegetal Remains of Newt Kash Hollow Shelter. In: Rock Shelters in Menifee County, Kentucky, edited by W. S. Webb and W. D. Funkhouser. University of Kentucky Reports in Archaeology and Anthropology 3(4): 147-65.

1937

Notes on the Preparation and the Uses of Basswood Fiber by Indians of the Great Lakes Region. Papers of the Michigan Academy of Science, Arts, and Letters 22(1936):1-14.

## 1938

An Ancient Indian Food Plant of the Southwest and Plateau Regions. El Palacio 44(5-6):41-53.

## 1940

Some Notes on Uses of Plants by the Comanche Indians (co-author with Gustav G. Carlson). Papers of the Michigan Academy of Science, Arts, and Letters 25(1939):517-42.

## 1941

The Nature and Status of Ethnobotany. Chronica Botanica 6(10):219-21.

Obituary of Melvin R. Gilmore. Chronica Botanica 6(16):381-82.

The Plant Materials from Winona and Ridge Ruin. In: Winona and Ridge Ruin, by John C. McGregor. Museum of Northern Arizona Bulletin 18 (Pt. 1):295-300.

Prehistoric Lima Beans in the Southwest (co-author with Charlie R. Steen). El Palacio 48(9):197-203.

Review of Paul A. Vestal and Richard Evans Schultes, The Economic Botany of the Kiowa Indians as It Relates to the History of the Tribe. American Antiquity 6(3):289-90.

## 1942

Fossil Bones as Medicine. American Anthropologist 44(1):162-64.

The Location and Delimitation of Archaeological Sites by Means of Divergent Vegetation. Society for American Archaeology Notebook 2(4):64-65.

A Native Southwestern Tea Plant. El Palacio 49(12):272-80.

Notes on the Manufacture of Rush Mats Among the Chippewa (co-author with Vernon Kinietz). Michigan Academy of Science, Arts, and Letters 27(1941):525-37.

Review of Willis H. Bell and Edward F. Castetter, The Utilization of Yucca, Sotol, and Beargrass by the Aborigines in the American Southwest. American Antiquity 8(2):190-91.

## 1943

Review of Paul C. Mangelsdorf and James W. Cameron, Western Guatemala, a Secondary Center of Origin of Cultivated Maize Varieties. American Anthropologist 45(2):283-84.

1944

Vegetal Material. Appendix D in: Early Stockaded Settlements in the Governador New Mexico, by Edward Twitchell Hall, Jr. Columbia Studies in Archaeology and Ethnology 2(Pt. 1):79-81.

Was Tobacco Smoked in the Pueblo Region in Pre-Spanish Times? American Antiquity 9(4):451-56.

1945

Plant Materials. Appendix II in: Archaeological Studies in Northeast Arizona, a Report on the Archaeological Work of the Rainbow Bridge-Monument Valley Expedition, by Ralph L. Beals, George W. Brainerd, and Watson Smith. University of California Publications in American Archaeology and Ethnology 44(1):159-63.

The Use of Honey-Dew as Food by Indians. The Masterkey 19(5):145-49.

1946

Plant Materials from Alkali Ridge Sites. Appendix C in: Archaeology of Alkali Ridge, Southeastern Utah, by John Otis Brew. Papers of the Peabody Museum of American Archaeology and Ethnology, Harvard University, 21:330-33.

Review of George F. Carter, Plant Geography and Culture History in the American Southwest. American Antiquity 11(4):262-65.

1948

A New and Unusual Navajo Dye (*Endothia singularis*). Plateau 21(2):17-24.

Notes on Indian Maize. Pennsylvania Archaeologist, Bulletin of the Society for Pennsylvania Archaeology 18(1-2):23-24.

Notes on the Manufacture of Cedar-Bark Mats by the Chippewa Indians. Papers of the Michigan Academy of Science, Arts, and Letters 32(1946): 341-63.

Notes on Frederick S. Dellenbaugh on the Southern Paiute from Letters of 1927 and 1928. The Masterkey 22(6):177-82.

Prehistoric Plant Materials from Castle Park. Appendix III in: The Archaeology of Castle Park, Dinosaur National Monument, by Robert F. Burgh and Charles R. Scoggin. University of Colorado Studies, Series in Anthropology 2:94-99.

Review of Frank G. Speck, Eastern Algonkian Block-Stamp Decoration: A New World Original or an Acculturated Art. American Antiquity 14(2): 138-39.

1949

Maize from the Davis Site: Its Nature and Interpretation. Appendix in: The George C. Davis Site, Cherokee County, Texas, by H. Perry Newell and Alex D. Krieger. Memoirs of the Society for American Archaeology 5:239-49.

Notes on the Organic Remains from Abo Mission. Appendix 2 in: The Mission of San Gregorio de Abo, a Report on the Excavation and Repair of a Seventeenth-Century New Mexico Mission, by Joseph H. Toulouse, Jr. Monographs of the School of American Research 13:29-32.

Report on Vegetal Material (from Nalakihu, Arizona). Appendix 2 in: Nalakihu, Excavations at a Pueblo III Site on Wupatki National Monument, Arizona, by Dale S. King. Museum of Northern Arizona Bulletin 23: 152-57.

1950

The Establishment of the Hopi Reservation, and Some Later Developments Concerning Hopi Lands. Plateau 23(2):17-25.

Review of Alice Marriott, Maria, the Potter of San Ildefonso. American Antiquity 16(1):81-83.

Review of Lila M. O'Neale, Textiles of Pre-Columbian Chihuahua. American Antiquity 16(1):83-84.

1951

Garryowen. The Masterkey 25(6):186-88.

Plant Materials from the Babocomari Village Site. In: The Babocomari Village Site on the Babocomari River, Southeastern Arizona, by Charles C. DiPeso. Dragoon, Arizona: Amerind Foundation Publication 5:15-19.

"Poison Gas" Used by Arizona Indians. Ciba Symposia 11(9):1380.

Vegetable Material. In: Archeology of the Byrnum Mounds Mississippi, by John L. Cotter and John D. Corbett. U.S. Department of the Interior, National Park Service, Archeological Research Series 1:48-49.

Review of Robert E. Ritzenthaler, The Oneida Indians of Wisconsin, American Anthropologist 53(4):574.

Review of Robert E. Ritzenthaler, The Building of a Chippewa Indian Birch-Bark Canoe. American Anthropologist 53(4):574-75.

1952

Material from the Hemenway Archeological Expedition (1887-1888) as a Factor in Establishing the American Origin of the Garden Bean. In:

1952 (cont.)

Indian Tribes of Aboriginal America, Selected Papers of the XXIXth International Congress of Americanists, edited by Sol Tax, pp. 177-84. Chicago: University of Chicago Press.

1953

The American Indians as of 1492 and Later. In: Progress of Mankind: Prehistoric to Present. Ann Arbor: University of Michigan Extension Service, Telecourse Syllabus, Lesson 9.

Plant Materials from the Fuller Site. In: A Report of Excavations Made at the T. O. Fuller Site, Shelby County, Tennessee, between March 8, 1952, and April 30, 1953, Conducted by Memphis Archaeological and Geological Society, edited by Kenneth L. Beaudoin, pp. 15-21. Memphis.

Review of Jonathan Deininger Sauer, The Grain Amaranths: A Survey of their History and Classification. American Antiquity 19(1):90-92.

1954

Plant Materials from Sites in the Durango and La Plata Areas, Colorado (co-author with Robert L. Fonner). Appendix C in: Basket Maker II Sites near Durango, Colorado, by Earl H. Morris and Robert F. Burgh. Carnegie Institution of Washington Publication 604:93-115.

1955

Plant Materials from a Cave in Zion National Park (ZNP-21). Appendix 5 in: Archaeology of Zion Park, by Albert H. Schroeder. University of Utah, Department of Anthropology, Anthropological Papers 22:183-203.

1956

The Debt We Must Repay. Ceremonial Magazine 35(1):6-8.

Hermetic Sealing as a Technique of Food Preservation among Indians of the American Southwest (co-author with Robert C. Euler). American Philosophical Society, Proceedings 100(1).

Review of Paul Weatherwax, Indian Corn in Old America. American Anthropologist 58(4): 768-69.

Review of Theodore P. Bank II, Birthplace of the Winds. Michigan Alumnus Quarterly Review 62(21):367-68.

1957

Botany. In: The Identification of Non-Artifactual Archaeological Materials, edited by Walter W. Taylor. National Academy of Sciences Publication 565:35-38.

Corn from the Shepherd Site, Maryland. In: The Shepard Site Study, by Howard A. MacCord, Karl Schmitt, and Richard G. Slattery. Archaeological Society of Maryland Bulletin 1:22.

The Development and Present Status of Ethnobotany in the United States. Huitième Congrès International de Botanique, Paris 1954, Comptes Rendus des Séances et Rapports et Communications, Section 15:52-53.

1958

The Death of James H. Miller, Agent to the Navaho: Fact and Establishment. The Masterkey 32(3):88-92.

Vegetal Remains. Appendix D in: Excavations in Mancos Canyon, Colorado, by Erik K. Reed. University of Utah, Department of Anthropology, Anthropological Papers 35:201-203.

1960

A Seventh-Century Record of Tobacco Utilization in Arizona (co-author with Elizabeth Ann Morris). El Palacio 67(4):115-17.

1961

Plant Materials from the Willow Beach Site, Arizona. Appendix I in: The Archaeological Excavations at Willow Beach, Arizona, 1950, by Albert H. Schroeder. University of Utah, Department of Anthropology, Anthropological Papers 50:123-35.

Review of Lewis H. Morgan, The Indian Journals, 1859-1962 (edited by Leslie A. White). International Journal of Social Psychiatry 7(2):155-56.

1962

Seventh Century Evidence for the Use of Tobacco in Northern Arizona. Sonderdruck aus Akten des 34. Internationalen Amerikanistenkongressen, Wien:306-09.

The Taxonomic Status and Archaeological Significance of a Giant Ragweed from Prehistoric Bluff Shelters in the Ozark Plateau Region (co-author

1962 (cont.)

with Willard W. Payne). Papers of the Michigan Academy of Science, Arts and Letters 47(1961):147-63.

Review of Alexis Praus, The Sioux, 1788-1922; a Dakota Winter Count. Cranbrook Institute of Science News Letter 31(7):71-72.

Review of W. W. Newcomb, The Indians of Texas. The Michigan Quarterly Review 1(2):141-42.

1963

Review of George E. Hyde, Indians of the Woodlands: From Prehistoric Times to 1725. William and Mary Quarterly Review 21(1):64-65.

1964

The Nature and Status of Ethnobotany (abstract). Proceedings of the Ninth Pacific Science Congress (1957)4:246-47.

Textile Adhesions and Impressions on Metal Objects from Philippine Archaeological Sites. Appendix B in: The Archaeology of the Central Philippines, A Study Chiefly of the Iron Age and Its Relationships, by Wilhelm G. Solheim II. Manila: Republic of the Philippines, National Science Development Board, Monographs of the National Institute of Science and Technology, Monograph 10:219-22.

Review of Elman R. Service, Primitive Social Organization, An Evolutionary Perspective. The Michigan Quarterly Review 3(1):64-65.

1965

The Bark of the Bittersweet Vine as an Emergency Food Among the Indians of the Western Great Lakes Region. The Michigan Archaeologist 11(3-4): 170-80.

Seeds from an Early Pueblo Pit House near Albuquerque. El Palacio 72(2): 24-26.

Review of John Leighly (editor), Land and Life: A Selection from the Writings of Carl Ortwin Sauer. American Antiquity 30(3):369-70.

1966

Two Textiles from the Valley Sweets Site. The Michigan Archaeologist 12(1):22-24.

1967

Abstracts of Papers, Thirty-Second Annual Meeting, Society for American Archaeology (co-author with Richard I. Ford). Ann Arbor: The University of Michigan.

1968

Four Textile Products from the Burnt Bluff Site (B-95), Michigan. In: The Prehistory of the Burnt Bluff Area, by James E. Fitting. Museum of Anthropology, University of Michigan, Anthropological Papers 34:95-97.

1970

Review of Virgil J. Vogel, American Indian Medicine. Michigan History 54 (3):263-65.

1971

Corn from the McKees Rocks Village Site. Pennsylvania Archaeologist 38 (1-4):81-86.

Ethnobotanical Report (on Site Bc236, Chaco Canyon). Appendix B in: Site Bc236, Chaco Canyon National Monument, New Mexico, by Zorro A. Bradley, pp. 94-95. Washington, D.C.: National Park Service, Division of Archeology, Office of Archeology and Historic Preservation.

M. R. Harrington's Collecting Trip to Walpole Island. The Masterkey 45(4): 146-50.

Review of Jonathan Deininger Sauer, The Grain Amaranths: A Survey of Their History and Classification. In: Prehistoric Agriculture, edited by Stuart Struever, pp. 544-49. Garden City, N.Y.: The Natural History Press.

1972

Review of Maitland Bradfield, The Changing Pattern of Hopi Agriculture. New Mexico Historical Review 47(4):391-93.

1973

Corn from Bamert Cave, Amador County, California, Appendix VI in: The Archaeology of Bamert Cave, Amador County, California, by Robert F.

1973 (cont.)

Heizer and Thomas R. Hester, pp. 73-75. Berkeley: University of California, Department of Anthropology, Archaeological Research Facility.

Some Additional Early Collecting Trips by M. R. Harrington. The Masterkey 47(2):69-74.

1974

Job's Tears, *Coix lachryma-jobi* L., Beads from the West Ridge-Gibraltar Site, Southeastern Michigan (co-author with Richard I. Ford). The Michigan Archaeologist 20(2):105-11.

1976

Bibliography of Carl Eugen Guthe. American Antiquity 41(2):173-77.

James Bennett Griffin, Archaeologist. In: Cultural Change and Continuity. Essays in Honor of James Bennett Griffin, edited by Charles E. Cleland, pp. xxxix-lxxvii. New York: Academic Press.

1977

An Analysis of Maize from the Pulcher Site, Illinois. American Antiquity 42(3):488-90.

An Investigation of *Sophora secundiflora* Seeds (Mescalbeans) (co-author with G. M. Hatfield, L. J. J. Valdes, W. J. Keller, W. L. Merrill). Lloydia 40(4):374-83.

Published Words of James Bennett Griffin (compiled with Richard I. Ford). In: For the Director: Research Essays in Honor of James B. Griffin, edited by Charles E. Cleland. Museum of Anthropology, University of Michigan, Anthropological Papers 61:343-62.

In Press

Red Medicine: The Mescalbean (*Sophora secundiflora*) among the Indians of North America (co-author with William L. Merrill). In: Frontiers in Ethnopsychopharmacology, edited by J. L. Diaz. Centro de Estudios Economicos y Sociales del Tercer Mundo, Mexico.

Tobacco. In: Handbook of North American Indians, edited by William Sturtevant. Vol. 3.

Ethnobotanical Reports Prepared by Volney H. Jones

Report No. 86A, September 1936. In: The Excavation of Pindi Pueblo, New Mexico, by Stanley A. Stubbs and W. S. Stallings, Jr. Monographs of the School of American Research 18:140-42 (1953).

Report No. 96, January 1942. In: Pima and Papago Indian Agriculture, by Edward F. Castetter and Willis H. Bell, pp. 30-33. Albuquerque: University of New Mexico Press (1942).

Report No. 170. In: Pima and Papago Indian Agriculture, by Edward F. Castetter and Willis H. Bell, p. 35. Albuquerque: University of New Mexico Press (1942).

Report No. 339, March 1953. In: Paa-Ko, Archaeological Chronicle of an Indian Village in North Central New Mexico, by Marjorie F. Lambert. Sante Fe: School of American Research Monographs 19(1-5):162-63 (1954).

Thesis

The Ethnobotany of the Isleta Indians. Unpublished 1931 M.A. Thesis, Department of Biology, University of New Mexico.